The Right Graph

THE RIGHT GRAPH

A Manual for Technical and Scientific Authors

Harold Kirkam

Robin C. Dumas

WILEY

A John Wiley & Sons, Inc., Publication

Published by John Wiley & Sons, Inc., Hoboken, New Jersey
Published simultaneously in Canada

For general information on our other products and services or for technical support, please contact our Customer Care Department within the United States at (800) 762-2974, outside the United States at (317) 572-3993 or fax (317) 572-4002.

Wiley also publishes its books in variety of electronic formats. Some content that appears in print may not be available in electronic format. For more information about Wiley products, visit our web site at www.wiley.com.

Library of Congress Cataloging-in-Publication Data:

Kirkham, H.
   The right graph: a manual for technical and scientific authors/Harold Kirkham,
   Robin C. Dumas
       p. cm.
   Includes bibliographical references and index.
   ISBN 978-0-470-40547-5 (cloth)
1.   Technical illustration. 2.   Technical writing. 3.   Computer graphics. I. Title.
II. Dumas, Robin C.

T11.8.K57 2009
808′.0666—dc22

                                                                    2008042471

Printed in the United States of America

10   9   8   7   6   5   4   3   2   1

Contents

Preface

For a picture to be worth a thousand words, it has to be a pretty good picture. Unfortunately, in a lot of scientific and technical work these days, the percentage of pictures that are that good is getting rather low! Whether you are a beginning writer or a much published author, your graphics can benefit from this book.

We address both the "what" and the "how" of graphics. Areas covered by the book include the following:

- What the technical purpose of the graphics is, and how best to get the desired result
- What the final product should look like, for data and other graphics
- What methods lead to faster production of graphics
- How to use a variety of PC software to get the desired result
- How to do page layout with embedded artwork and what to do when the graphic won't fit
- How file format conversion problems can be worked around
- How to edit diagrams for different end-uses (one or two columns, viewgraphs)

You need this book because it will teach you the stuff that the graphics artists of old used to know, as well as the details of what your software is capable of.

Back in the old days when you wanted to write a paper or a report that included some kind of diagram, the company graphics department got involved. Maybe you had a pencil sketch of what you wanted, or maybe you had some data you had taken in an experiment. Along you went to the graphics department, where you talked to some guy who always seemed to have a cigarette going, and who had a selection of pencils, all of them pale blue.

This was your interface to the graphics artists, the only people on the planet who could make a drawing that would show what you wanted to show, be perfectly legible, and who wouldn't smudge the India ink in the process.

Mr. Blue Pencil would listen, more or less patiently, to what you said, and would write all over your drawing—in pale blue. Just what he wrote was a mystery that could only be solved by an artist. It looked somewhat like the sort of thing you might have written yourself, years before, when you were exploring the more arcane aspects of fractions.

Nevertheless, when you got the drawing a couple of days later, it was exactly what you wanted. If you were writing for a two-column paper, the drawing would fit neatly into one column. If it was for a report or a viewgraph, it was bigger, but it was still the right size. When you got back the reviews on your paper, the reviewers might have made helpful comments about the material ("This model is an accurate representation of the system, as it was when I described it 20 years ago in an unpublished report") or the text ("The author would do well to review the rules of grammar, and try to simplify the task of the reader by producing a paragraph break at least once a page"), but they would not criticize the diagrams.

Things have changed. Your graphics department closed their doors years ago, unable to compete with the snappy output of PowerPoint® produced at your very desk, and the desks of their other erstwhile customers. They probably followed your secretary down to the unemployment line. When you started doing all your own typing, her job became simplified out of existence. Bored with nothing to do except spend the day gossiping with her buddies, she walked out one day. It was a week before you noticed.

Something else changed at the same time. Now, about half the papers you write are rejected because the graphics are not considered acceptable by the simple idiots they get to review papers these days. Where do they find these people? They couldn't *tear* a paper, let alone review one! Reviewers?! They must be half blind if they can't understand ... Sorry about that. It won't happen again.

Where were we? Oh yes. It is true that about half the time papers are rejected these days, the reason is poor graphics. The reason for this—even with the help of the latest graphics package, most scientists and engineers still don't know all the tricks of the trade that the graphics department used to know.

That's where we come in. Though we are both now senior staff at the Jet Propulsion Laboratory, our backgrounds are relevant, including time spent in the graphics department, engineering, and IT. It is our intent in this book to show you, at the very least, how to do graphics that will not get your paper rejected, and perhaps will even help you make your point. We intend to address questions of content and style for a number of publication media: papers, books, viewgraphs, and even patents.

These ideas for improvement are aimed primarily at the authors, technically adept, but not trained in document design. We will give guidelines and rules, ideas and examples. We will show both bad technique and good, and explain what is good and what is bad. Throughout, we will give tips on how to get the best result, and how to save time and energy doing it.

This book is aimed at providing guidance to the technical author on aspects of content and style on the one hand, and process on the other. These various kinds of information are not completely segregated in the book, which proceeds from a basic level, assuming the reader has some idea of the purpose of graphics and some idea of what software to use, and advances along the fronts of material and mechanics at the same time. The reader can dip into the book at any point, and pick up this dual narrative.

We have chosen examples from the real world to illustrate the various aspects of graphics, in the hope you find the book entertaining as well as instructive. When we are critical of the work, we usually try to present it in such a way that the authorship cannot easily be determined. This saves us having to seek anonymity ourselves and joining the witness protection program. For the most part, we have redrawn the graphs that we use, to maximize the image quality. We apologize now for any errors of transcription that we may have made.

We have used a variety of software in the production of this book, and we would like to acknowledge the hard work and talents of the people who wrote that software. Without the tremendous capabilities of the word processor, the spreadsheet, and the graphics program, this book would not have been possible. Without the features (and the bugs) in those programs, it might have been less fun to write.

Writing a large and complex piece of software has been likened to building a house. As things approach completion, a "punch list" is drawn up—items that must be fixed before handing over to the customer. The larger and more complex projects naturally have longer punch lists, and the list may not stop growing until after the customer takes over. With the house, it may suddenly be discovered that the air handling system cannot cope with some particular combination of doors and windows being open. With the software, it may be found that some particular combination of operating system and hardware driver does not perform as expected, and a line on the screen becomes invisible.

Here and there in what follows, we may have occasion to note a software issue. There may be a limited capability in some seemingly important area, or there may be a bug. It may also be that by the time you are reading this, the particular issue has gone away, fixed by a new release of the OS or the application. Fear not, you will always be able to think of new features you would like, and there will always be other bugs to keep us on our toes.

Meanwhile, we would like to express our sincere appreciation of the efforts of the folks at Adobe, Corel, and Microsoft, and acknowledge the inestimable contribution of their products. In particular, we have made extensive use of the following applications: Adobe Illustrator®, Adobe Acrobat®, Corel PHOTO-PAINT®, Corel QuattroPro®, Corel WordPerfect®, Corel WordPerfect Presentations™, CorelDRAW®, Microsoft Excel®, Microsoft PowerPoint®, Microsoft Word®, and Microsoft Windows XP®.

Screen shots from Corel products are copyright Corel Corporation and Corel Corporation Limited, reprinted by permission. Microsoft product screen shots are reprinted with permission from Microsoft Corporation.

HAROLD KIRKHAM
ROBIN C. DUMAS

West Hills, California
Monrovia, California
June 2009

Basics

WHAT THE BOOK IS ABOUT

Here is a short list of reasons you might want to use graphics in a technical document:

- Graphics can make it look as if you actually know what you're talking about.
- Graphics can make a plain document look more attractive.
- Graphics can be used to deceive the unwary reader.
- Graphics can make a thin report fatter, or a short paper longer.

However, if they catch you using graphics for just these reasons alone, it won't do your career any good!

As it happens, if they are done right, technical graphics can make things easier to understand. They can clarify relationships. They can allow straightforward extrapolation. They can present a lot of data concisely. For these reasons, technical documents almost always contain graphics. Why so many technical documents contain lousy graphics that make nothing seem simple is another question. Teaching you how to get the point across, and how to avoid the pitfalls, is the purpose of this book.

Along the way, we will look at the questions of style that make some graphics attractive as well as functional (and others, neither one), and we will see examples of graphics so bad they deceived even their famous authors. We will also see what kind of graphic suits what purpose, and what you can do with your software to produce the desired result. We will also note some shortcuts that can save serious amounts of time.

In brief, we assume that you are using a computer (actually, a PC or a Mac®) to produce a technical document of some kind: a report or a paper, viewgraphs for a presentation, or perhaps a drawing for a patent. We aim to help you get the best result possible with your resources, with the minimum effort on your part.

SOME BASICS

Let's get started by looking at some of the terms we will use. First, the word *graphic*. For the purposes of this book, graphic is a general term that includes graphs, bar charts, diagrams, and drawings. It does not include *tables* or *equations*. There is a clue to a generalization here. For our purposes, graphics are produced by software other than your word processing software. The software you use to produce graphics therefore needs to be aware of the word processing software you use—a feature called *integration*. It is a sad fact that graphics software is usually not well integrated with word-processing software, even if they have the same manufacturer's label. We will show you some ways around the problems.

Now, what were those other terms? Graphs, bar charts, diagrams, and drawings.

A *graph* (Fig. 1.1) is a device to show how one parameter varies as a function of another. Often, but not always, one of the parameters is *time.* For example, one might plot the intensity of the sunlight at the surface of the Earth at some particular place. Usually, but not always, the parameter of interest is arranged on the vertical axis, called the *ordinate.* Exceptions occur in a number of fields. In oceanography, for example, the depth is—sensibly—vertical, so that a graph showing temperature as a function of depth would have temperature (the parameter of interest) horizontal. The points in a graph may be connected by a line.

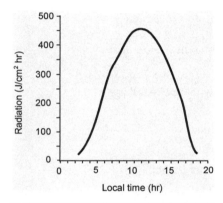

Figure 1.1

A *bar chart* (Fig. 1.2) is a device to show the differences or similarities of a number of separate things. Usually, the things themselves are separated horizontally, and their size is shown by their vertical extent. The items shown in a bar chart are not connected by a line, as there is no meaning to a point part-way between one entry in the bar chart and the next.

Some software uses the term *chart* when it means to say *graph.* Originally, chart meant the same as *map.* Nowadays, the typical chart is the *organization chart,* as in Figure 1.3. Here, a structure is shown where the vertical extent is an indication of seniority (lower levels report to higher levels) and the horizontal separation has no particular significance. In a similar way, a chart can be drawn of the branching of one species into two, and the entire history of a line of evolution can be captured. In this case, the vertical axis can represent time (usually with the more recent at

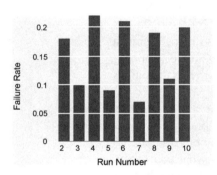

Figure 1.2

And so you see, instead of reporting to *me,* you'll report to *him* and *he'll* report to me. It's what we call a lateral move.

Figure 1.3

the top) and the horizontal spacing represents movement in some undefined gene space. Pie charts and flow charts are other members of the chart species.

A *drawing* is something that, before the availability of low-priced hardware and easy-to-use software, would have been produced by a draughtsman (or a draftsman) working with pen and ink. A drawing can show a machine or a landscape, or a mechanism that can be patented.

A *diagram* is literally anything marked out by lines. Sometimes, we will use the word diagram to mean any of the graphics discussed previously. Just to confuse things, we will also use *diagram* to mean graphics that have not already been described. Electrical schematics are examples of diagrams. Perhaps in this sense, the thing that distinguished a drawing from a diagram is that a diagram uses symbols.

Bearing in mind the many kinds of graphic, our aim with this book is to show you three things:

- Which form of graphic belongs with which type of information. You don't want to be using a graph if a bar chart is appropriate.

- What the particular graphic should look like, for its intended use. A diagram produced for a viewgraph may not be right for a technical paper, and vice versa.

- How to achieve the result you want most easily with the software you have, including converting files from one format to another.

We will show you some shortcuts, to save time and energy. The production of graphics should be approached with a different mind-set from the production of text, and the difference manifests itself largely in the time taken to do the job.

At the end of the book we will examine some good graphics and some bad, some graphics that didn't make it, and some that did but shouldn't have, in a series of case studies.

In the rest of this chapter, we'll continue by looking at a few *definitions*. These won't take long, but it will help if we all know the names of the things we're going to talk about. After that, we'll give some general *guidelines* on what to strive for in terms of appearance.

The next chapter shows you what kind of graph to choose for what purpose, and the one after that deals with the complicated question of joining the dots in a graph.

In a later chapter, we'll have some suggestions to help you get the most out of the software you have. If you are already familiar with all the software you use, you can skip the first part of this. However, it might be worth checking to be sure you didn't miss anything important. Other chapters will deal with using spreadsheets, making patent drawings, perspective (even if you don't consider yourself an artist), presentations, and matters of style.

DEFINITIONS

The ultimate purpose of a graph is to persuade someone of something. The graph accomplishes this by showing the relationship between two (or more) quantities; or sometimes by showing the lack of such relationship, for example, the temperature independence of a measurement device. In either case, a lot of the essentials are the same. You will end up with something like Figure 1.4.

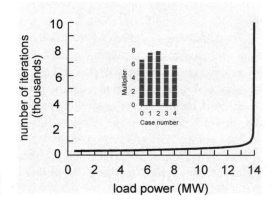

Figure 1.4

Of course, normally there would be a caption under (sometimes over) the figure. We'll discuss captions in Chapter 4. For the time being, accept that this is a curve showing the number of iterations required for a solution of an amazingly inefficient load-flow program for some particular electric power system. The graph shows how the number of iterations varies as a quantity called the load power varies.

First, let's use this example to identify all the parts of a graph; see Figure 1.5.

By the way, the labels added to the graph in Figure 1.5 are called *annotations,* and the lines with arrowheads from the annotations to the thing annotated

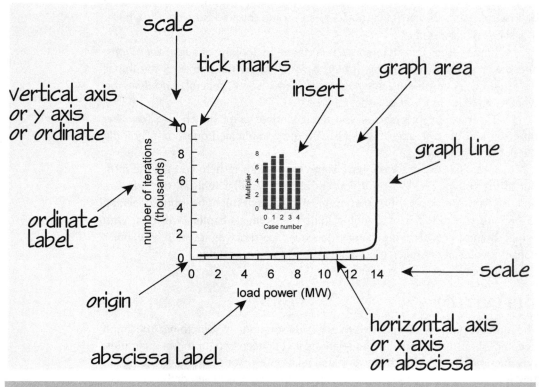

Figure 1.5

are called *leader-lines* or *leaders*. Collectively, annotations and leaders are sometimes referred to as *callouts*.

As well as identifying the major parts of a graph, what we have tried to do here is present an example of what a graph of this kind should look like; in terms of its style rather than its content. More on style in Chapter 13.

Strictly speaking, the *y*-axis need not be vertical. The *x*- and *y*-axes refer to the independent and dependent variables, and as we shall see in the next chapter, there are occasions when it makes sense to plot the independent variable vertically.

GUIDELINES

The ideas on the following pages are offered as guidelines.

Strive for a Clean, Neat Appearance. This means limiting the amount of "things" there are in the graph area. Don't clutter it up with too many labels or callouts to every single item in there. Sometimes it is convenient to put a label next to a curve, but sometimes it gives a neater appearance to put the label outside the graph area and use a leader line.

Think About Copying or Faxing. Your work is going to be reproduced, adding to your fame if not your fortune. If your work is full of fine lines and tiny symbols, it may not copy well or fax well. The symbols seen in Figure 1.6 were used in a graph generated by software that evidently tried to reproduce on a laser printer the effect of a dot-matrix printer.

Figure 1.6

Even this original is hard to make out. Imagine, if you will, what it would look like (Fig. 1.7) after it has been copied a few times! Thus, if you use "dots" for data, make them as different as you can for different data sets, and make them big enough that they will survive even a noisy copying process.

While we are thinking about copying, a word about color is in order. While color printers are becoming fairly commonplace, for some obscure reason color copiers are not. If there is some compelling reason to use color, go ahead and do it. But it might be good to think about how

Figure 1.7

your color graphic will reproduce on a monochrome copier. Convert your graph to monochrome, and make sure it still "works."

Here's an example of the sort of thing that should *never* happen. In 1975, IEEE Press reprinted an article that had appeared in *Scientific American*.[1] Like most articles in *Scientific American*, the article had made use of color in its graphics. IEEE, however, did not. When one of the graphs was reproduced, the reference to color in the caption did not make sense, as it was not possible to say which black line had been which color.

---

[1] Earl Cook, The flow of energy in an industrial society, in *Energy and Man: Technical and Social Aspects of Energy*, IEEE Press, 1975, New York.

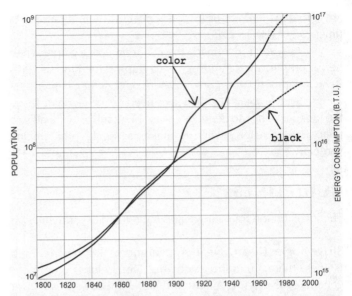

Figure 1.8

U.S. ENERGY-CONSUMPTION GROWTH (*curve in color*) has outpaced the growth in population (*black*) since 1900, except during the energy cutback of the depression years.

Some genius (presumably at IEEE Press) solved the problem by adding labels to the graph. We have reproduced it here (Fig. 1.8) complete with caption so you can see how dumb some solutions can look.

OK, OK, now you can reconstruct what the original must have looked like. But wouldn't it have made more sense to make the changes shown in Figure 1.9?

Here we have reversed the ordinate scales to simplify the placing of the big arrows that identify the curves, and rewritten the caption to remove any reference to color. An alternative approach would have been to use a dotted line for one of the curves, but this might have led to confusion with the (unlabeled) dashed section where the post 1970s estimates are shown.

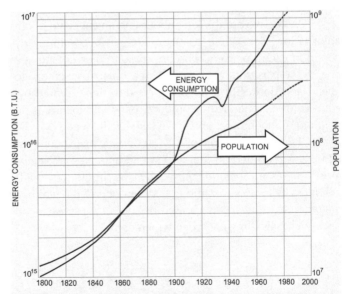

Figure 1.9

U.S. ENERGY-CONSUMPTION GROWTH (*left scale*) has outpaced the growth in population (*right*) since 1900, except during the energy cutback of the depression years.

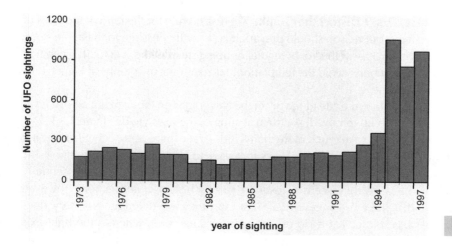

year of sighting

Figure 1.10

Be Careful to Limit the Number of Tick Marks on Each Axis, and Don't Feel Obliged to Put a Label on Every Tick Mark.[2] In most cases, there should be at least 5 ticks on the vertical axis, and at most 11. If there are 9 or more, label only every other one. The horizontal axis is frequently longer than the vertical one, so it can have a few more tick marks.

If you are going to skip some numbers on the axes, label every other one, or one out of five or one out of ten. *Under no circumstances, label just every third tick mark.* For some reason, we do not do well mentally interpolating if the axis is divided this way. Look at the example in Figure 1.10.[3] Where is 1000 on the vertical axis? Or 1990 on the horizontal? Clutter control is a matter of judgment and style, so the details are up to you. But don't do the thing with the threes.

Take the example of the power system, shown with two different axes in Figure 1.11. When all the tick marks are labeled, the legibility is actually lower. But using just the even numbers works.

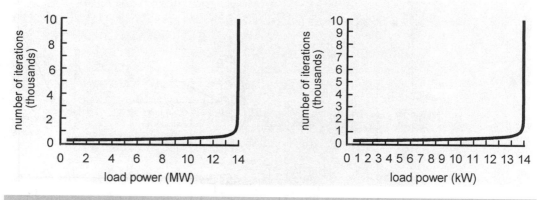

Figure 1.11

[2]By the way, the word is "tick." We have seem some authors wrongly use the word "tic." A tic is a sort of nervous twitch. The spelling comes from the French *tic douloureux*, literally *irritating (or painful) tic.* "Tick" on the other hand seems to have a Teutonic origin and means a small dot or dash made with a pen or pencil, to draw attention to something or to mark a name, figure, and so on, in a list as having been noted or checked.

[3]The graph is not an accurate representation of the numbers: both the largest two numbers were actually slightly larger than shown here, according to the website of the National UFO Reporting Center. We reduced them so that the vertical scale would stop at 1200 rather than 1500.

Don't Lie, Don't Distort the Graph. There is no rule for the ratio of the sizes of the two axes, but reason should prevail. It is perfectly possible to make the same curve give the appearance of being flat or being steep without actually changing any of the numbers: avoid the temptation! For example, the graphs in Figure 1.12 all have a slope of 1.

Of course, the "dead space" in the two graphs on the right is a bit of a giveaway, so it can be removed. A little rescaling of the axes yields Figure 1.13.

It doesn't take much more than a look to see that some deception is going on here, but deception isn't always that easy to spot. How about Figure 1.14? Here, a glance at the graph gives the impression that the two curves are essentially similar, because they seem to overlap so much. Closer inspection, however, reveals that there are two vertical axes, and they are different. Not only does the left axis start at 1 instead of 0, it spans a range of 3, whereas the right axis starts at 0 and spans only a range of 2. The difference is undermined by the horizontal line at the top of the graph that repeats the tick marks of the lower horizontal axis.

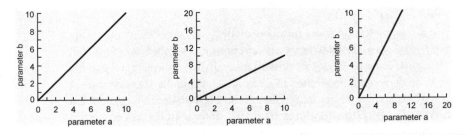

Figure 1.12

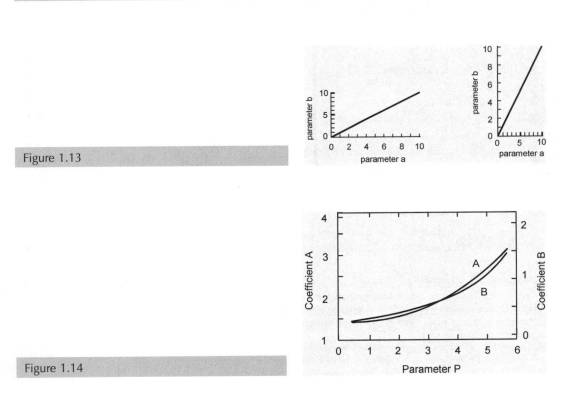

Figure 1.13

Figure 1.14

It might help to remove this line, but still the true dissimilarity of the axes would be buried under false impressions. A fairer presentation would be the version shown in Figure 1.15. Here the two curves are seen to overlap but slightly. They do appear to have generally the same shape, and it still takes a close inspection to see that they span a different amount. At least the presentation itself adds no bias.

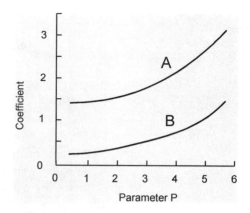

Figure 1.15

If There are Two Axes, Label Them Both! We've all seen the TV advertisements for analgesics, where the brands are compared in a bar chart. The brands associated with each bar are identified, but the performance measure (presumably the length of the bar) is not! The voice-over talks soothingly about the product being recommended by 3 doctors out of 4. In fact, one wonders if they lined up just 4 "tame" doctors and, with a consulting agreement in hand, asked them something like "For $5000, which brand would you recommend?"

But even supposed scientific writers can be guilty of this kind of error. In his 1997 book, *Why People Believe Weird Things*,[4] the professional skeptic Michael Shermer illustrates the fact that various altered states of mind exist with the curves shown in Figure 1.16.

The curves are said to be from EEG measurements, but we are not told either the horizontal scale or the vertical, despite the fact that there is a marker at the end of each of the supposed recordings. Perhaps each vertical marker

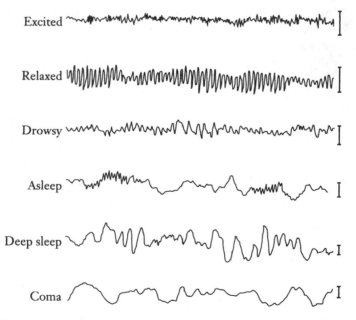

FIGURE 7:
EEG recordings for six different states of consciousness.

Figure 1.16

[4]MJF Press, New York. The figure is on page 76.

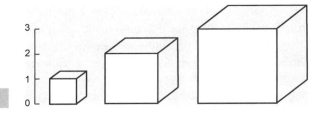

Figure 1.17

indicates the same quantity—in which case the curves would be more easily compared if the scales were such that equal magnitudes were represented by equal lengths. But we are not told. The lack of such information undermines the case being made. A reasonable observer could be forgiven for wondering if, for example, the slow undulations labeled "coma" are simply the fast undulations of "relaxed" at a different time scale.

Don't Use Areas or Volumes in Simple Numerical Comparisons. This advice applies mostly to bar charts. Since some bar charts produced by spreadsheet programs will often gratuitously add a "depth" dimension, you should pay attention to the output of such programs.

Worse, however, is the "bar" chart that, instead of having a *bar* whose height represents the magnitudes in question, has a three-dimensional box, as in Figure 1.17, each of whose sides varies with the magnitude. Suppose we have something—a single parameter—with magnitudes 1, 2, and 3 sampled at some interval.

The addition of a couple of lines, as in Figure 1.18, reveals that the linear dimensions of the cubes are indeed in the ratio 1 : 2 : 3, but there is no doubt that the much larger volume of the right cube, and the fact that it protrudes so far above the labeled axis, helps it create the impression of much larger size.

The proper way to show the relative sizes, of course, is to use the *height* of the bar only, as in Figure 1.19. For some reason, this approach creates the impression that the taller boxes are thinner than the shorter ones. In fact, they are the same. At least, there is no deception here.

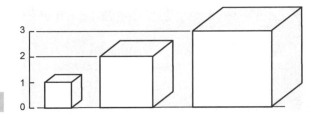

Figure 1.18

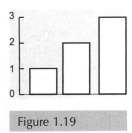

Figure 1.19

As we will see throughout the book, there are several ways graphs can deceive the reader. Deliberate deception is usually detectable—the blank space or the suppressed zero in a graph can be a signal—but some deceptions may not be noticed. These are caused by the almost accidental impressions created by the graph, the optical "illusions."

It turns out that if we have two bars in a bar chart, for example, we may make a reasonable judgment about their relative size. (If the bars are close enough to the axis that we can estimate their values, we can

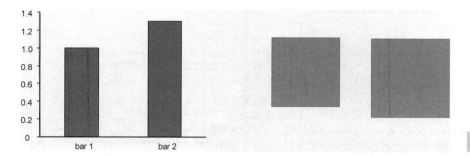

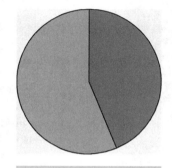

Figure 1.20

confirm our estimate.) However, we do not do so well comparing two *areas*, and it is hard to have an axis for comparison.

In Figure 1.20, the lengths of the bars on the left are in the ratio 1 : 1.3. That is, the large one is 30% larger. The same ratio applies to the *areas* on the right. Most viewers would judge the rectangles reasonably well, because the widths are the same, and only the length has changed between bars. Most of us would imagine the squares to be much closer in relative size.

We do not do well with angles, either. In the pie chart of Figure 1.21, the ratio of the areas of the two pieces of the pie is also 1 : 1.3.

Figure 1.21

Some things we can judge accurately. One of use (HK) is a twin. His parents brought up the twins in a "You divide, and you choose" environment. As a result, the accuracy with which he can divide a pie into two equal parts is truly incredible (typically a few ppm, he says). But even he does not do well with unequal quantities and unusual shapes.

Keep the Graph Line Within the Rectangle Formed by the Axes. Any automatic graphing software will have this feature, and it seems such an obvious thing to say, but once in a while you can see a graph that doesn't obey this rule.

A graph very much like the one in Figure 1.22 appeared on page 993 of the second edition of a well-known electronics book, *The Art of Electronics*, by

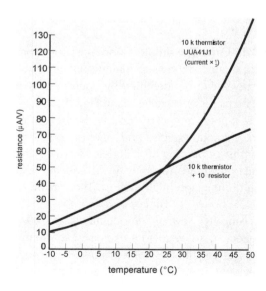

Figure 1.22

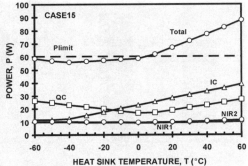

Figure 1.23

Horowitz and Hill (Cambridge University Press, 1989). For some reason the curved line of the uncorrected thermistor characteristic was drawn as if the vertical axis went as far as 140 or 150, instead of the 130 that it actually spans. Perhaps we are used to graphs that look as if they were drawn on graph paper. Whatever the reason, it looks odd to see the graph line "outside the box."

We will return to this graph for further discussion in Chapter 4.

The "opposite" of this problem might be the graph that occupies too little of the space indicated by the scales. Normally, the scale should have the same range as the data, or a little more, in both the vertical and horizontal directions. Given that you should have "nice" numbers on the tick marks (1-2-3-4 or 5-10-15-20 and so on), you sometimes will have a little space left over at the end of the data. This blank space should not be large. The exception to that rule is that if you have a series of graphs (perhaps even on separate pages) that you wish to compare and contrast, they could have the same scale as one another *even if it meant creating blank space.*

Figure 1.23 is an example of this situation. Only the graph of Case 15 uses anything like all the vertical space. Case 10, typical of most of the graphs in this particular series, has a lot of empty space at the top. But since at least part of the idea of presenting the data like this is to show that the other cases remain below the threshold temperature of 60 °C, the fact that the axes are consistent from case to case makes sense.

Avoid Picture-Coding Schemes If You Can. The simplest (and most common) coding scheme is the use of different shaped symbols to identify different parameters on a graph. For example, curves can be identified by open or closed circles or squares or triangles, as in Figure 1.24.

This graph purports to show the benefit of short but vigorous exercise and alcohol consumption in terms of burning off calories. However, the little squares would not copy well, and the real meaning of the graph might be lost. The version in Figure 1.25, however, will copy well.

Even after several reproductions, the beneficial effect of a beer before exercise can clearly be seen. The pointlessness of a prolonged workout is also evident. (OK, this is a book about graphics.

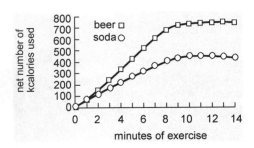

Figure 1.24

Thermodynamic truth can sometimes be sacrificed to make a graphics point!)

While we're on the subject of coding schemes, don't use cute little images, as in Figure 1.26, to emphasize the kind of quantity being represented. You are not writing for your high school magazine, but for an audience of your professional peers.

Come to think of it though, that might not make a bad viewgraph. We can all think of meetings where a round of beers at the beginning would have had a beneficial effect!

Be Concise. We do not want to make the reader spend a lot of time working out what it is we are trying to say. Don't let the design get in the way. Graphs don't necessarily speak for themselves, especially if you clutter them up. So it pays to be simple, and visually obvious. It's OK to hit the reader in the face with your message.

Figure 1.27 is an example of what you might have expected if you were doing graphics in the middle of the 20th century. A graph similar to this

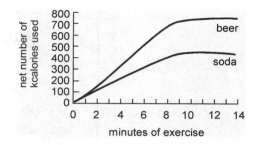

Figure 1.25

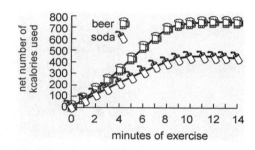

Figure 1.26

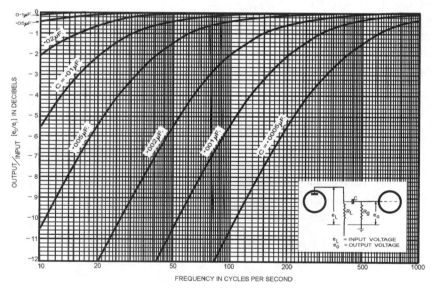

Figure 8. Frequency characteristics due to grid coupling condenser with a grid resistor (Rg) 1.0 megohm. It is assumed that the plate resistance and load resistance in parallel of the preceding valve are negligibly small in comparison with Rg. When this does not hold (as with a pentode valve in the preceding stage), the attenuation is less than is shown by the curves, but may be obtained accurately by making Rg represent the grid resistor in series with r_p and R_L in parallel. These curves may be applied to any value of Rg by multiplying the values of C shown on the curves by a factor equal to that by which Rg is reduced (e.g., for Rg = 0.5 megohm multiply values of C by 2.).

Figure 1.27

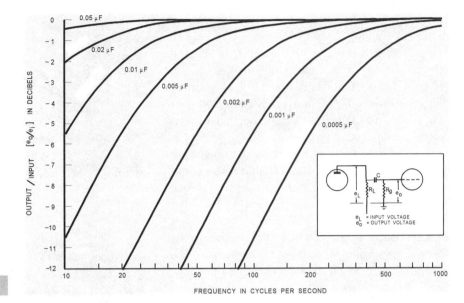

Figure 1.28

appeared in the first British edition, ca 1940, of *Radio Designers Handbook*, originally published in Australia as *Radiotron Designers Handbook*. (Ever wonder what a Radiotron was?) Let's look at how things have changed.

First, the background of this graph is so *busy* it looks a bit like a pattern that might have been rejected by Hart Schaffner Marx, or Moss Bros. That is because it was originally produced on real graph paper. There was a time when graph paper was in common use. The lines of the graph paper were useful when the graph was being drawn by hand, and it was not the style of the times to remove them afterwards. The labels on each curve, and the insert, look as if they were typed on slips of paper that were pasted on top of the graph—a simple but effective approach.

Some graphs—including this one to judge by the caption—were drawn so that numerical values of the quantities being plotted could be extracted. That would not be common these days: the point of a graph is more often to show the variation in the data, or the correlation, but not to indicate the exact numbers.

If we remove the background lines, and relabel the curves, we get a much simpler appearance, as in Figure 1.28.

It may be that the text should all be reproduced larger, and the circuit diagram in the insert on the right removed. If these changes are made, the result is really quite presentable—see Figure 1.29.

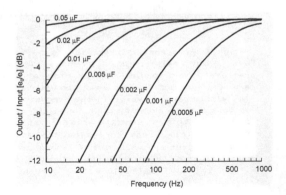

Figure 1.29

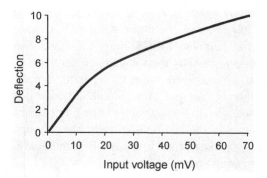

Figure 1.30

Note also that since the text is larger relative to the graph, the whole thing can be reproduced at a smaller scale, so it takes up less space.

Try to avoid the lines across the graph area. Generally, the result will be a neater, cleaner graph, with adequate detail. Figure 1.30 is another example; it represents the deflection obtained on an analog meter as the input drive is changed.

The meter deflection is quite nonlinear. (In fact, it was designed this way. The meter is the indicator for a bridge, and higher sensitivity near zero deflection makes good sense.) Perhaps if you were designing a replacement, and needed to understand the exact levels at which things changed, a few lines across the graph area, as in Figure 1.31, would help. For most purposes, that level of detail should suffice. More would be clutter, as in the "Radiotron" graph.

For some reason, the designers of spreadsheets seem to think clutter is a Good Thing, and it can fairly be said that a more up-to-date version of graphical clutter is readily obtained by using a spreadsheet. For example, the graph of Figure 1.32 was obtained from a spreadsheet without any hard work at all.

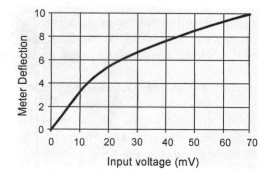

Figure 1.31

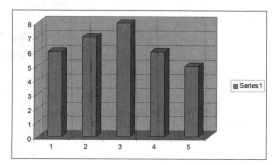

Figure 1.32

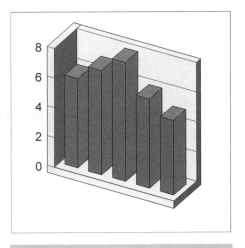

Figure 1.33

The numbers 6, 7, 8, 6, 5 were entered into the spreadsheet, and one of the standard bar charts drawn. The numbers 1 to 5 along the horizontal axis were generated by the software, evidently in an attempt to help us keep one entry separate from the others. The tag "Series 1" was added, although it is hard to see why, in the absence of any other series. The columns of the chart are a lovely blue, though the shading makes it hard to guess where the illumination is coming from.

The columns of the bar chart were then corralled by a wall with lines on it, so we could see the sizes of our data entries. However, since the whole thing has been drawn with perspective, it is not obvious that the data are actually integers.

A different spreadsheet does not come off much differently; see Figure 1.33.

Now the corral has thickness! Here the data are not separately identified, and the unnecessary series number label does not appear. However, the perspective problem is no better, and the whole thing has acquired a distortion (it is narrower here than it is in the spreadsheet) in the process of exporting to a file. (More on this in the chapters on spreadsheets.) The columns are still blue, though it is now clear that they are illuminated from above.

It seems that the software for spreadsheet graphics is being written by the Marketing Departments. Look, it's a job. Just because these people did not graduate top of their class, does not mean they don't need work! There are just not that many jobs for people who can decide that bar charts should be red. (As a winemaker friend once said, "The first duty of any wine is to be red!" He had no opinion on the color of bar charts.)

You *can* get what you want out of these programs, it just won't be the default. Spreadsheets are such an interesting and useful way of graphing data that we've dedicated several chapters to them. Meanwhile, Figure 1.34 is an example of how the same data could be presented in a clean, uncluttered way.

Note that the two axes are labeled, and that there is no perspective. The lines from the ordinate numbers are in white, and on top of the columns of the chart. We have chosen to have the columns presented as gray bars with no line around them. This is purely a matter of style. Almost equally acceptable (in our view) is the version shown in Figure 1.35.

The main "problem" with the graph of Figure 1.35 is the need to decide how long to make the horizontal tick marks. The white lines of the graph of Figure 1.35 do not show against the paper, and so can be any length greater than the width of the space occupied by the columns.

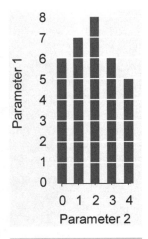

Figure 1.34

Let the Data Make the Case. Focus on values, and variations, but don't feel that you always have to smooth or otherwise reduce the amount of data in order to make a point. Most software will handle large quantities of data, so show the data by all means, if that would help support the case you are trying to make.

For example, back in the late 1970s it was well known that a wet high-voltage power transmission line produced an audible noise. The exact mechanism was not well understood, and while the factors that contributed to changing the noise level were known (the voltage, the conductor surface condition, the rain intensity), the relationship between them was not known. A paper was written that showed measured noise data as a function of rainfall rate. A graph from it is redrawn here as Figure 1.36.

This version differs from the original[5] only in that the dots are bigger. Bearing in mind the relatively primitive state of interactive computer graphics when the original work was done, the result is quite satisfactory. There are nearly 500 data points shown here. (The conditions for selecting them were explained in the paper.) Because each point is no more than a dot, the density of dots can be inferred quite well from the presentation.

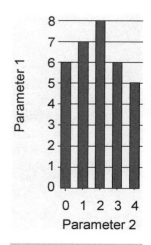

Figure 1.35

One can make some interesting observations, based on this presentation. At rainfall rates above about 1 mm/hr, the noise is relatively high, and somewhat independent of the rainfall rate. At low rain rates, the noise is lower, more variable, and quite dependent on the rain rate. That the noise was so complex a function was a new finding and prompted further analytical work.

Strive for Artistic Balance, and Reasonable Proportions. What are reasonable proportions? You probably never thought of television as having anything to do with balance or reasonableness but, roughly speaking, the aspect ratio of a graph should be the same as the aspect ratio of a TV screen. That may be a bit vague.

Ours is a changing time. For many decades the aspect ratio of TV screens was 4 : 3 (width to height). Then, when it seemed everyone who was going to get a TV had already got one, or two, or more, and solid-state technology and quality control techniques meant that TVs were not breaking down all the time,

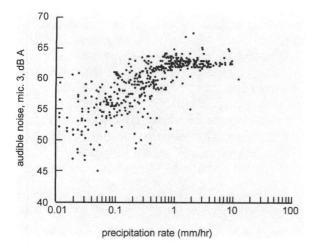

Figure 1.36

[5] H. Kirkham, instantaneous rainfall rate: its measurement and its influence on high voltage transmission lines, *Journal of Applied Meteorology* **19**(1), 35–40 (January 1980).

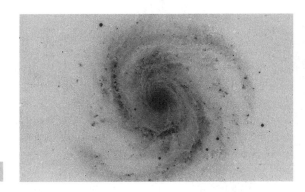

Figure 1.37

high-definition digital TV came along. Well, it had to, or Sony and Panasonic and all the rest of them were going down the tubes. And guess what, you couldn't watch HDTV on your regular screen, because the aspect ratio was different. Now it's 16 : 9. (Somebody was into squares.)

In England, in the first part of this century, you could buy a TV that would let you watch any of the current standards. That's pretty useful. It would also let you (separately) choose the aspect ratio of the image. So if you wanted you could watch a movie at a 16 : 9 ratio that was shot with a 4 : 3 ratio. OK, so the engineers said they could do it, and there was nobody in the marketing department smart enough to see how stupid this would look. Which makes more sense, a letterbox or a stretch?

Don't be doing the same thing to your graphics, just to get them to have some particular aspect ratio! You really don't want to convert the image of galaxy M100 shown in Figure 1.37 into the distorted version of Figure 1.38 even if it *is* technically possible! An error like that would be hard to spot, too, since galaxies are at various angles to the Earth, including edge-on.

Exceptions to this guideline should be made, of course, as necessitated by the page layout (of which there is more in Chapter 5). A two-column format can accept a graph one-column wide with the 5 : 4 aspect ratio, or it can accept a tall, thin graph without looking odd. Tall thin bar charts are often the best way to

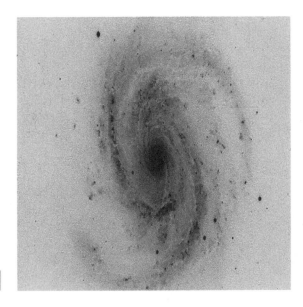

Figure 1.38

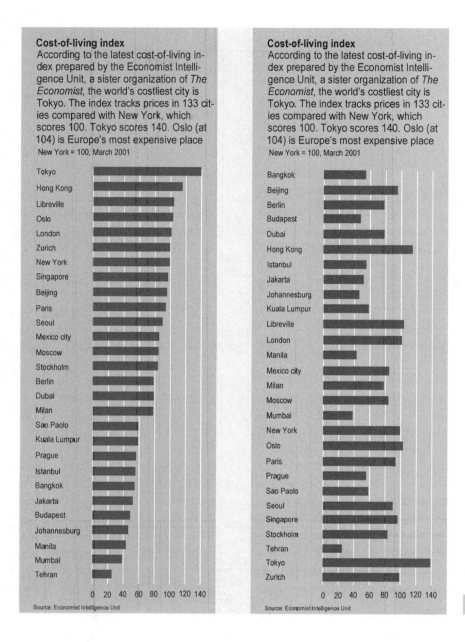

Cost-of-living index
According to the latest cost-of-living index prepared by the Economist Intelligence Unit, a sister organization of *The Economist*, the world's costliest city is Tokyo. The index tracks prices in 133 cities compared with New York, which scores 100. Tokyo scores 140. Oslo (at 104) is Europe's most expensive place
New York = 100, March 2001

Figure 1.39

present a lot of data, as in the example in Figure 1.39 from *The Economist* (July 7 – 13, 2001).

Here the individual entries have reasonable size and proportions: it's just the overall collection that takes up a lot of vertical space.

Note that if you are going to do a really unusual bar chart such as this, it makes sense to spend a moment considering the order in which the entries will appear. *The Economist* always puts such things in rank order, as on the left of Figure 1.39, rather than alphabetical order. See how it looks organized by the coincidence of the English name for the city and you can see why.

Well, the second presentation brings out the absence of an alphabetical bias in the cost of living, but it hides everything else! It is no longer obvious which is the most expensive city, or the least. How does Moscow compare with Stockholm, or Berlin with Milan?

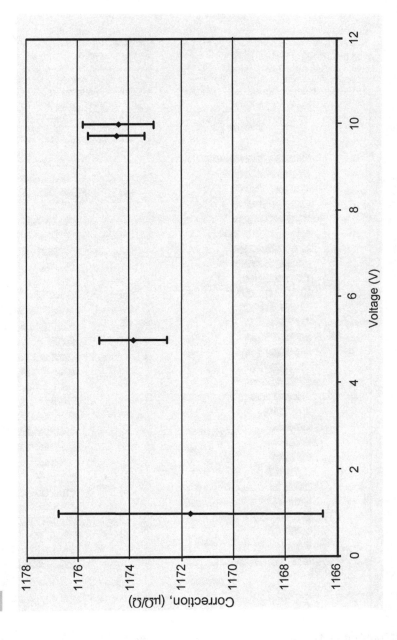

Figure 1.40

With the Aspect Ratio of Your Graph Fixed, *Size* it to Suit the Content. Just for the fun of it, see how small you can make it, before things start to look bad. Seriously!

Here's an example of an actual graph from a journal, reproduced here (Fig. 1.40) actual size. Considering the information content of the graph, the presentation is a waste of space. Figure 1.41 gives the same information, packed a little more densely, and with the horizontal lines removed. Just as useful, and only one fifth the area.

Consider an Alternative to a Graph. Sometimes, we get so accustomed to producing graphs that we don't consider whether a graph is the best way to present the data. In fact, for the graph in Figure 1.41, with only four data points,

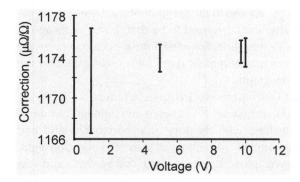

Figure 1.41

showing just one thing as a function of another, it hardly seems worth using a graph. The following is a table that presents all the information in the graph:

Voltage (V)	Correction ($\mu\Omega/\Omega$)	Uncertainty ($\mu\Omega/\Omega$)
1	1171.70	± 5.05
5	1172.85	± 1.25
9.7	1174.50	± 1.55
10	1174.45	± 1.35

Without the addition of the uncertainty numbers, the table would be even smaller, whereas the graph would be about the same size.

For a large number of points, or a more involved set of relationships, a graph is usually a clear choice. For a small number of points, and a simple interrelationship, a table may be a more compact and effective way to present the data. Presumably, somewhere in the middle—a modestly large data set and a fairly complex interrelation—there is a number of points and a degree of complexity that would be equally well served either way.

You will have to decide for yourself where that dividing line is. Don't let the habit of making graphs win every time!

Adjust the Line Weights in Graphs. As with font sizes, with most graphics packages line weights (the "official" term for line widths) change when you scale an entire image. With some, you have the option to fix the width or let it scale. Either way, the line width should be adjusted carefully, so as to preserve a reasonable balance to the graph.

The table below summarizes our suggestions.

	Two Column		Single Column		Viewgraph	
	thousandths	points	thousandths	points	thousandths	points
Axis lines	8–10	0.6–0.7	12–15	0.8–1.1	15–25	1–2
Tick marks	6–8	0.4–0.6	8–10	0.6–0.7	12–20	0.8–1.5
Graph lines	10–20	0.7–1.4	15–30	1.1–2.2	20–40	1.5–3
Leader-lines	7–10	0.5–0.7	10–12	0.7–0.9	15–20	0.8–1.4

Viewgraph, in this context, applies to the computer-based presentation as well as the one based on a transparency. It used to be that, a few years ago, a formal presentation required a 35-mm slide. For a less formal occasion everyone used viewgraphs. (There was once a retirement party for a colleague, and the speakers honoring him used viewgraphs.)

The original viewgraph was based on a transparent film about $8\frac{1}{2}$ by 11 inches, sometimes with a cardboard frame. If you were an engineer, the viewgraph was arranged landscape (i.e., with the long side horizontal) and had been done by the graphics department, sometimes at great expense overnight. If you were a scientist, the viewgraph was portrait (long side vertical) and was hand drawn, often during the presentation of the previous speaker, and sometime in real time during the presentation.

These days, things are much better. Almost everyone is using the computer for presentations, formal or informal, and almost everyone prepares the "slides" ahead of time. Nobody has figured out how to turn the screen sideways, so even scientists now make landscape presentations.

Chapter 6 discusses presentations in more detail.

Think About Not Connecting the Axes. Because graphs were historically drawn on graph paper, the axes were always joined. Now that graphs are drawn on a computer, that little implementation detail is no longer a requirement. The two graphs in Figure 1.42 convey the same information, after all.

If one or both of the axes do not go to zero, it might make good sense to ensure the lines were not connected, or even aligned. In essence, the separation of the axes could be thought of as a special case of a scale break. Which reminds us...

Avoid Scale Breaks If You Can. Scale breaks, which can involve changes in scale factor as well as simple offsets, tend to give the reader the wrong impression about the data. Fortunately, most software does not readily produce scales with breaks, so you would have to make some effort in order to create a graph that

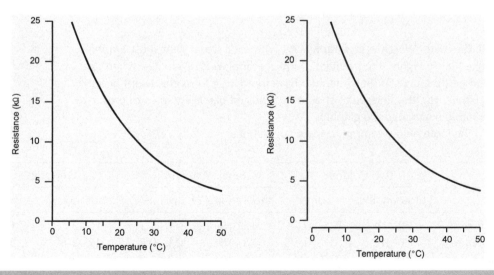

Figure 1.42

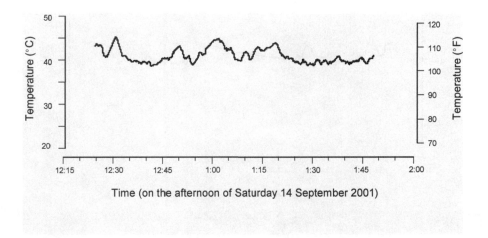

Figure 1.43

uses one. If you decide you absolutely cannot avoid using one, be sure the break is obvious, and under no circumstances "join the dots" across either side of the break. Consider the graph shown in Figure 1.43.

These data were collected in the Mojave Desert one hot and dusty afternoon. There is a continuous record lasting about 90 minutes. In fact, it is most unlikely that the temperature fluctuated this rapidly—the reading seems highly correlated with the reading from the anemometer, as shown in Figure 1.44, implying that the sun was heating the thermometer and the wind was cooling it.

Be that as it may, suppose there had been a break in the record, with no data collected for some of the period. It would certainly be incorrect to show the data as in Figure 1.45 to correct the omission. Far better to leave out the line for the period where data are missing, as in Figure 1.46. (The question of whether or not to "join the dots" is surprisingly difficult to answer. It is addressed in Chapter 3, where some generally applicable rules are given.)

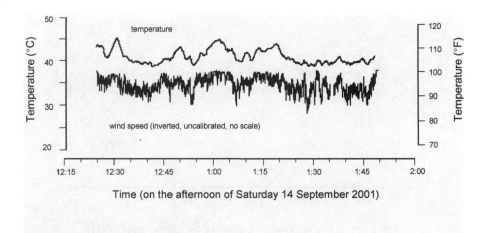

Figure 1.44

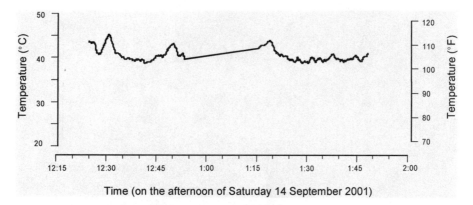

Figure 1.45

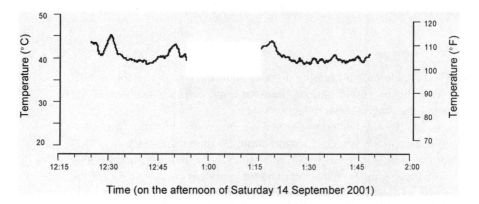

Figure 1.46

If the data gap were longer, you might be tempted to show the scale break as in Figure 1.47. Resist the temptation! It is too easy with this presentation to imagine that there is only a *small* period of missing data. If there must be a break, but you must save space, be sure that the reader has no doubt that data are missing. You might even consider the method of Figure 1.48.

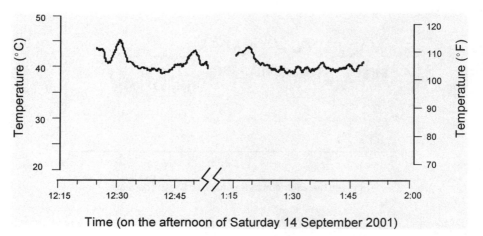

Figure 1.47

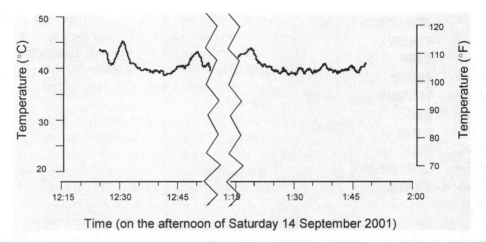

Time (on the afternoon of Saturday 14 September 2001)

Figure 1.48

A scale break is sometimes a possibility in the vertical axis, too, again because of missing data, although the data may be missing in a sense other than that you stopped collecting data for a while. The only excuse for using a scale break is that it is essential to show two distinctly separate parts of a graph in similar detail. Instead, consider using a logarithmic axis, and consider using two separate graphs.

The same argument applies to bar graphs, too. A bar graph was used by *The Economist* to show the effect of the Chinese astrological calendar on birth rates, in particular, that in certain years, viewed as auspicious, there were more births.[6] The actual number of births in Huludao, a port city in northeastern China, were plotted. The obvious way to plot the data would have been as in Figure 1.49.

However, the idea was to show the dramatic increase in the Year of the Dragon, and so the axis was changed as in Figure 1.50. Note that the left end of the bars has a very pronounced raggedness, so that even the casual reader must see that the bars were "torn" at this end.

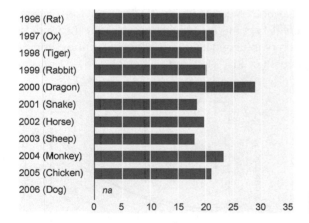

Figure 1.49

[6] The Golden Pig cohort, *The Economist*, page 44 (February 10, 2007).

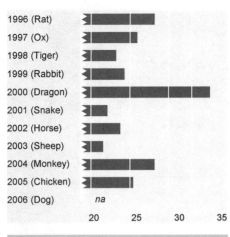

Figure 1.50

You Don't Have to Put Lines at the Top or Right. Since you are not cutting the graph out of a printed page, and you are not drawing it on actual graph paper, there is no need to put the graph area inside a four-sided box. A line at the bottom and one on the left will do.

This is just a matter of style, and that topic is addressed in Chapter 13. The modern style of things is for a clean, uncluttered appearance, and an approach that says, in essence, "justify every bit of ink on the graph" is in keeping with that.

If you have multiple parameters that require multiple ordinates, you can put them all on the left, or (for the special case of two axes) put one each side (like we did above with the Fahrenheit and Celsius scales). If you follow this rule, the scale on the right will be different from the one on the left, and so a line joining them at the top will not intersect at a tick mark. Ugly, very ugly! Remember Figure 1.14? It is reproduced here as Figure 1.51 for convenience.

Close examination reveals that the right scale was adjusted so that the two curves lay on top of one another. Neither horizontal line intersects the right axis at a tick mark.

You Don't Have to Put Lines Across the Graph Area. Except for lines of longitude and latitude, it is no longer considered good practice to put lines across the graph area. (Longitude and latitude are different, because it may be that on flat paper they are not straight lines, depending on the projection being used.) Several examples of graphs with lines on them have already been presented in this chapter, so no more will be added here.

It seems that the default appearance of graphs produced by spreadsheets includes lines. However, these can usually be removed, either by the spreadsheet software or in your graphics package. More on this in the chapters on spreadsheets.

Look Out for Step Functions. Another almost accidental feature of spreadsheet graphs is the joining of adjacent data points by straight lines. Usually the result is acceptable, though the question of joining data dots is actually far from simple, and we have devoted a fair amount of space to it in Chapter 3.

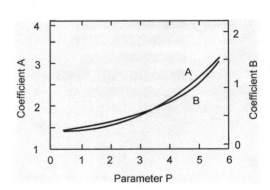

Figure 1.51

One particular aspect of this feature of spreadsheet graphics is apparent in graphs of funding and expenditures, exactly the kind of thing spreadsheets were originally designed for. For example, suppose we have a project that is spending money at a reasonably constant rate, and receiving funding on an annual basis. One way to present to the manager her financial information is to plot the accumulated expenditures and the funding received as a function of time. The result might look like the graph in Figure 1.52. The light gray line is the available funds.

This is the sort of thing the default graph from a spreadsheet would give you. New funding is expected in January, so the manager may glance at this graph and conclude all is well. It isn't. By joining the adjacent data points by a straight line, the spreadsheet obscures the fact that the project will run out of money late in December. Happy Christmas, indeed!

The problem lies in the way the available funding line appears to ramp up from one value to another. It would be a more accurate representation if it looked like a step function.

One way to achieve the right effect is to add another data point, just before the value changes. It may be possible to do this in the spreadsheet, or it may be necessary to import the graph into your graphics package. The desired result is shown in Figure 1.53.

This is closer to what you might want.

Actually, some spreadsheets offer a step-function graph, but only if you make it three-dimensional, at which point it looks like the diagram for the air-conditioner ducts above the office. See Figure 1.54.

More on peculiar defaults in Chapters 8 and 9 on Spreadsheets.

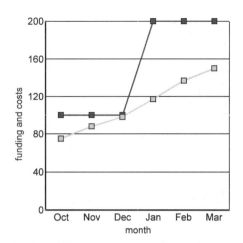

Figure 1.52

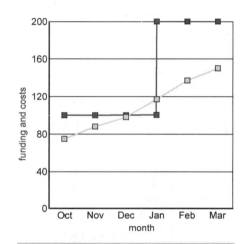

Figure 1.53

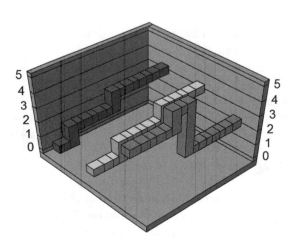

Figure 1.54

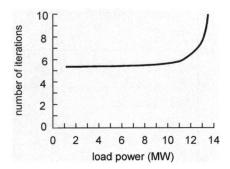

Figure 1.55

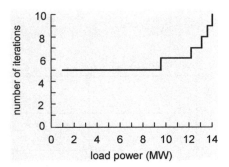

Figure 1.56

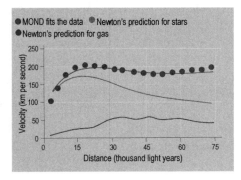

Figure 1-57   See insert for color representation of this figure.

Step-function graphs like the one in Figure 1.53 are reasonably common. For example, when interest rates and exchange rates are changed, they change abruptly from one value to another. If your data produce something that looks almost like a step on your graph, it might be wise to think about whether it actually *should be* a step.

For example, suppose the power system load flow that we showed at the beginning of the chapter was made much more efficient, so that only a few iterations were needed instead of a few thousand. The graph of Figure 1.55 would certainly not be correct. The number of iterations must be an integer, so Figure 1.56 shows what you might expect.

The Key. The *key* used to be a common part of a graphic. It was the little box in the corner that explained what the various colors signified on a map, for example, or explained the meaning of the solid and dotted lines on a graph. This way you could tell the difference between the route of the treacherous Pizarro and the way to Piglet's house.

In the latter part of the 20th century, the key fell out of favor, and other ways were found to get the information across. Now the key seems to be making a comeback, largely because of the increasingly gratuitous use of color in graphics. Resist the trend. Avoid using a key, if possible. *The key is an obstacle the reader has to get past before he can understand the graph.*

In the graph in Figure 1.57, which is similar to one that appeared in *New Scientist* (January 25, 2003), conventional Newtonian gravity is compared with an altered version (MOND—modified Newtonian dynamics). MOND is a possible alternative to postulating large amounts of cold dark matter in the universe as a way to explain the motion of stars in the outer reaches of galaxies. The text makes it clear that the distance referred to in the graph is the distance of a star from the galactic core, though the galaxy whose data are being presented is not identified.

The *key* in the graph of Figure 1.57 consists of the top two lines.

Well, actually, it's more of an intelligence test than a key. Remember intelligence tests? They were used to find out whether twins or tall people were smarter. "Find two symbols that have something in common in the following set":

See how the MOND graph fits this pattern? There are four colors in the graph (not counting background colors), and only three in the key. There is one set of round blobs in the graph, but there are three in the key. After a moment of study, you see that the round blobness of the items in the key is irrelevant. The key applies to the color of the *lines*. The round blobs on the graph are not described anywhere, but it seems likely they represent data.

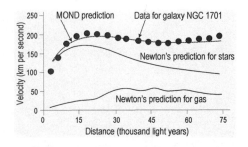

Figure 1.58

The graph highlights some of the problems with producing a key. If the purpose of the key is to explain color, then the colored part of the key must have a shape. If the shape of the object in the key is different from the object in the graph, the reader is faced with an intelligence test: Is the important part of the key the shape of the objects or the color?

This conundrum is exacerbated by the fact that lines act somehow to make color difficult to see, and colors appear to change when surrounded by other colors. Since the key and the graph have different background colors, there is a potential problem.

The MOND graph could just as well have been presented in black and white, as in Figure 1.58, with the dots and curves identified conventionally.

Note that there are now four things identified in the graph, instead of the three of the original. Perhaps the presentation is not quite as spiffy as the colored version, but the point is made much more clearly. The MOND prediction is a reasonable fit to observation, Newtonian dynamics for stars is not. (It is to be noted, however, that the "observations" shown in this graph are just a shade suspicious. The spacing between dots is remarkably regular where the curve is flat—above a distance of 5 kparsec, say—and this spacing decreases where the curve is steeper. Although we have no basis in evidence, we suspect that the dots are merely an artist's impression of how data might look superimposed on a curve.)

Here's a similar example from *Science News.* Under the heading "Our big fat cancer statistics" (November 19, 2005, Vol. 168, page 334), the argument was made that obesity was the second largest cause of cancer in the United States (responsible for 10% of all cancers, compared to smoking's 30%). The data were presented in a graph like the one reproduced in Figure 1.59 (in which the background color of the original has been removed).

In this example, the key seems to be entirely unnecessary, and is again more of an annoyance than an added value. Suppose you are curious which is the most common type of cancer attributed to obesity. You look at the chart, and decide it is the blue one (although you are not sure what the percentage is, since the bar is taller than the axis). You search through the colors in the key area below the chart, and you find "endometrial."

Why should "endometrial" be blue and "colon" beige? No reason. Why should the types be written in the order Colon, Breast, Endometrial, Kidney, and so forth? No reason. The order is not the order of occurrence or the order of the alphabet. Nor do the initial letters spell anything pronounceable: C-B-E-K-E-P-L-M.

Compare the graph of Figure 1.59 with the one in Figure 1.60, which contains all the same information.

In directly labeling the bars of the bar chart, we have avoided the need to use a key. This means that one step has been removed from the process of reading the graph. As a result, the presentation is simpler to understand. As a side benefit,

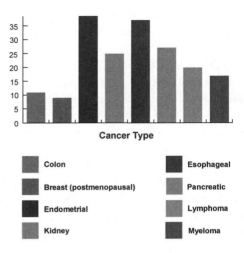

ADDED CANCER RISK FROM OBESITY*

Cancer Type

Colon Esophageal

Breast (postmenopausal) Pancreatic

Endometrial Lymphoma

Kidney Myeloma

* Extra percentage of cancer risk, by type, attributable to obesity in the U.S. population

Figure 1-59 See insert for color representation of this figure.

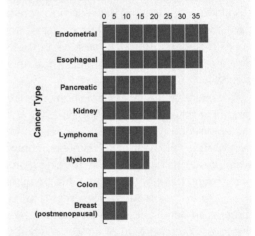

ADDED CANCER RISK FROM OBESITY*

* Extra percentage of cancer risk, by type, attributable to obesity in the U.S. population

Figure 1.60

it takes up slightly less space. If *Science News* had done this, they could have retained the use of color simply by coloring the bars—but they could all have been the same color.

The presentation is also improved by ordering the types by their relative contribution, in the manner of the graph from *The Economist* presented earlier in the chapter.

Finally, comparison of the types is aided by running the lines for the percentages over the top of the bars.

Sometimes a key makes sense, however. In the graph shown in Figure 1.61, based on one in *The Economist* of February 1, 1986, the key is a good way to explain the difference between the light and dark shaded parts of the bar chart. Really, this is two charts combined, one showing the performance of Singapore, the other that of South Korea, between 1979 and 1984. There is a lot of information in this graph!

Note that the key contains objects that are the same shape as the objects they explain in the graph: rectangles. Note also that the graph would copy well. It may not be as visually appealing as a graph done in color, but from the point of view of usefulness, it is hard to beat. Usefulness is surely a primary goal in works of science and engineering, with visual appeal coming second.

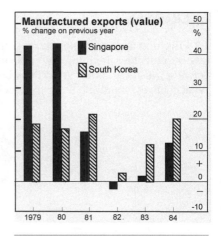

Figure 1.61

SUMMARY

We have looked at several kinds of graph and examined some of their peculiarities. The next chapter will look at the applications of various graphs in more detail. For all graphs, the basic rule for the author is to strive for clarity and avoid ambiguity.

- Think about what it is you want your graphic to say (and even whether a graph is the best way to go), and how best to arrange things to that end. (This question is addressed further in the next chapter.)

- Let the data make the case. Don't distort the graph to make your point. Don't use areas or volumes in simple numerical comparisons.

- Aim for a neat, clean appearance, with balance and good proportions. Varying line widths, making the graph line heavier than the axes, is a good way to improve appearance.

- Minimize the size of the graph, resizing the words if necessary.

- Control the amount of clutter in the graph. For example, use tick marks sparingly, and don't feel obliged to label all of them.

- Don't feel obliged to have lines around the graph on all sides, or all the way across the area.

- Clutter control is a matter of judgment and style, so the details are up to you.

- Think about what happens when the graph is copied or converted to microfiche.

- If there are two axes, label them both.

- Keep the graph within an imaginary rectangle, but don't feel you have to draw the rectangle.

- More or less fill the rectangle range with data, unless you have several similar data sets to compare.

- It is not essential to join axes—but avoid midscale breaks.

- Look out for step functions.

- If you can avoid using a key, do so. If you must use a key, make sure it is clear.

EXERCISES

1.1. The most important thing before you start making a graph is to figure out your objective, and choose a way of presenting the information. Something like the graph shown in Figure 1.62 appeared in *New Scientist* (May 31, 2008). Would this be appropriate, assuming you could guess the writer's purpose? What changes would you make for a technical paper?

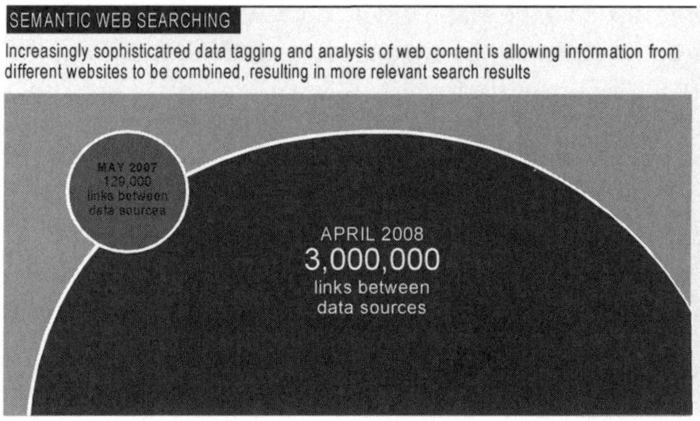

SEMANTIC WEB SEARCHING

Increasingly sophisticatred data tagging and analysis of web content is allowing information from different websites to be combined, resulting in more relevant search results

MAY 2007
120,000
links between
data sources

APRIL 2008
3,000,000
links between
data sources

Figure 1.62

1.2. The following are some ranges of data. For each range, generate a sample scale for a vertical axis. Consider carefully how many tick marks, and how many labels there should be.

	Range 1	Range 2	Range 3	Range 4
Start	5:48 am	−3.1	0.1	1979
End	6:17 pm	17.6	100	2008
Comment			Logarithmic	Years

Would the scale be different if the data were used for the horizontal axis?

1.3. What is wrong with the graph of Figure 1.63, which is similar to a graph in *Science News* (May 24, 2008), that was aimed at showing that buying local produce would not have much impact on one's carbon footprint?

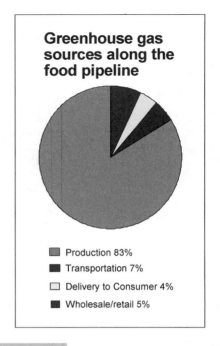

Figure 1-63 See insert for color representation of this figure.

1.4. What changes would you make to the graph seen in Figure 1.64, which is similar to a graph that was in an *IEEE Transactions* paper?

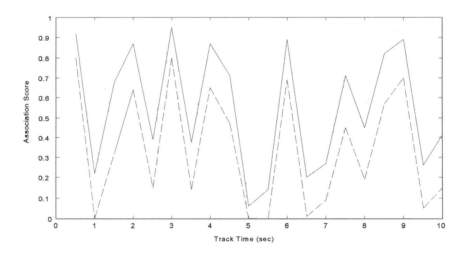

Figure 1.64

Which kind of graph? 2

We mentioned in the previous chapter that the ultimate purpose of a graph was to persuade someone of something. The graph accomplishes this aim by showing the relationship (or lack thereof) between variables. Honesty and integrity dictate that it is not fair to modify the data to make your point—unless you particularly *want* to appear on *60 minutes*, denying everything. However, there is nothing wrong with presenting unadulterated data in a way that supports your argument rather than a way that does not. Given that you won't actually *cheat*, there are still some surprising choices to be made.

Each kind of graph is best suited to one use. Really! The wrong graph in a paper may not only fail to make the point, it could undermine the work. You can actually get a paper rejected because of the wrong choice of graph type, or the use of the wrong data.

This chapter will examine your choice of graph type. Here and there, we will make use of data from actual papers, and here and there we will use artificially created data.[1] This way, if you happen to see a graph that looks familiar—fear not! You can always pretend it's artificial data. If you *did* happen to produce a graph like one of our bad examples, or you did make a wrong choice, be assured that you remain anonymous. Honest!

As we began writing this chapter, we did what any self-respecting researcher would do—we looked at what other writers had to say on the topic. We found one book that used the same data set to demonstrate several different kinds of graph. The reader was therefore given the impression that, faced with plotting one parameter against another, he could choose freely among any of the presentations in Figure 2.1.

Well, you really can't just pick and choose. Even though these graphs all look similar, they are not interchangeable. What's more, it's not even as if this is the complete selection of ways the same data can be presented. Consider the options shown in Figure 2.2. You can't just pick and choose from these either!

---

[1] The artificial data sets were created using a spreadsheet to generate "perfect" data, and to add to it biases and random errors as appropriate.

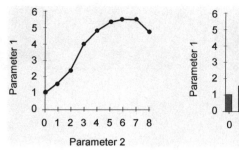 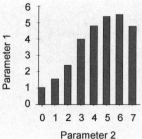 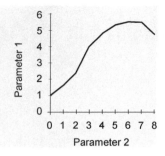

Figure 2.1

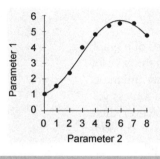 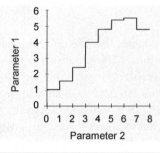 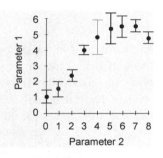

Figure 2.2

Another book we found said that the "histogram (or bar chart) is used in presenting material to the general public." That may or may not be true, but it doesn't help to know it. Consider Figure 2.3, adapted from the *Wall Street Journal*, July 31, 2001, page B1.

Without changing the numbers, the information could have been presented as shown in Figure 2.4.

Which is better? Is one way right and one wrong? Is there a third (and better) way to present the data? All sorts of stuff is used in presenting material to a general audience. Would John Q. Public understand the second graph?

In fact, this business of choosing the appropriate graph is a bit of a minefield. There are many ways that data *can* be graphed, as we've shown, but there are some restrictions. Sometimes, the choice of graph type *can* be a matter of style. Sometimes, however, it really is a matter of right and wrong. Our hope in this chapter is to find a way through the minefield.

Changing Channels

Growth and decline in household audiences, in thousands of households

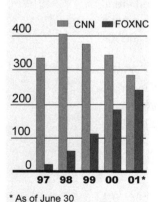

* As of June 30
Source: Nielsen Media Research

Figure 2.3

Changing Channels

Growth and decline in household audiences, in thousands of households

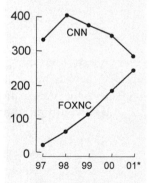

* As of June 30
Source: Nielsen Media Research

Figure 2.4

CHOOSING A GRAPH

We summarize here the appropriate applications for some of the more important types of graph. We will begin by looking briefly at some of the different kinds of graph you can choose from. It may be that there is no limit to the number of choices that you have: people are capable of inventing new graphs, and why not? However, some graphs are relatively common, and some rarely seen.[2]

In what follows, the graphs are ordered approximately according to their frequency of use. We will look at the following types:

- Trend plot
- Stacked trend
- Pie chart
- Bar chart and histogram
- Scatter graph

Later in the chapter, we will look at some unusual graphs developed for specific applications:

- The coxcomb, invented by Florence Nightingale
- The petal plot, used in space navigation
- The subway map, applied to cancer

TREND PLOT

If you are recording data from an experiment, it is almost certain that you will have the data stored on your data acquisition system, or in your notebook, in the order in which they occurred. That is, the data are all time-tagged, and stored in time sequence. When you want to have a quick look at the data, the chances are good that you will begin with a look at the data as a function of time.

The trend plot is a useful way for you to scan a lot of data. Any tendency for the data to show correlation may be evident, and any drift problems or noise can often be spotted. If you have real data, and the time[3] is available, there is never anything *wrong* with plotting data against it, though the result may not always be as revealing as you hope.

Really, the trend plot is a special case of the more general *scatter graph* or *scatter plot*, in which one parameter is plotted against another. In the case of the trend plot, the time is the second parameter, and it is almost always plotted on the horizontal axis.

You can draw a trend plot for a single variable as a function of time, as in Figure 2.5. Here the graph is the output of a *sampled* pulse system. The plotting scheme makes a zero-order assumption, that between the samples, the magnitude

---

[2] There is a summary of graph types on the website of the National Institute of Standards and Technology (NIST) at `http://www.itl.nist.gov/div898/handbook/eda/section3/eda34.htm`. These are largely statistical in nature, to do with exploratory data analysis.

[3] One of the most difficult aspects of the trend plot is the great variability that the time axis can have: it can deal with nanoseconds or years. To add to the complexity, the way the axis is divided changes from decimal to odd-ball once the time gets into the range that humans are familiar with: from seconds on up. Sadly, spreadsheets are generally quite bad at handling this large range of time. Some aspects of a solution are presented in Chapter 7, which introduces the use of spreadsheets.

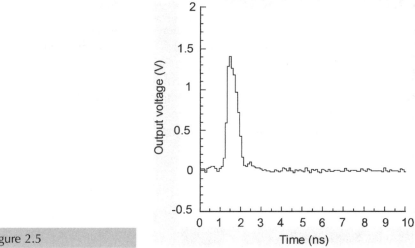

Figure 2.5

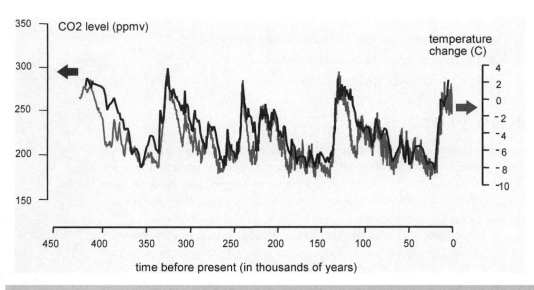

Figure 2.6

of the output does not change. The presentation came from some simple software that allows the approximate width and height of the pulse to be seen at a glance.

You can also draw a trend plot for multiple parameters varying as a function of the time, and use more than one ordinate scale if necessary. The graph in Figure 2.6 is sure to stir someone up. This graph was constructed[4] using data from the Web, surely the source of all wisdom and knowledge. This kind of presentation is usually intended to support the hypothesis that atmospheric carbon dioxide is correlated with (and therefore causes?) planetary warming.

From the graphics point of view, there are a couple of things to note. First, the horizontal axis has numbers on it that decrease from left to right. This is because we are so used to seeing graphs with time going from left to right that it would be misleading to present the data the other way round (as a function of

[4]We will look later at how it was done, in Chapter 10 (Fixes Using Graphics Programs).

increasing ice age) without a good explanation in the text. (As a matter of fact, the website from which the data were obtained does contain a "backward" graph, with no warning comment in the text. See Chapter 10 for a link.)

Second, note the way the appearance of correlation is strengthened by the choice of scales. The CO_2 scale has a suppressed zero (because the numbers are all between about 150 and 300) and the temperature scale is adjusted so that the data more or less overlap. There is nothing dishonest in this, nor in the fact that the vertical axes are separated from the time axis. (We shall see in the last chapter, however, that honesty limits this kind of thing, and some distortions are the equivalent of cheating. This presentation, however, is honest.)

It is an oddity of the process by which our brains interpret such graphs that when we see two lines climbing or falling side by side, such as we see here at the left end of the graph, we tend to estimate their separation by mentally estimating their separation along imaginary lines at right angles to the original ones. The process greatly underestimates the separation. While the sharp peaks are aligned, we see, here and there, that the temperature (the generally lower of the two curves) often appears to be thousands of years ahead of the carbon dioxide level in falling. Curious, and perhaps an artifact of the presentation.

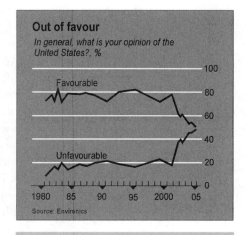

Figure 2.7

You should be aware of this human tendency to underestimate the separation of sloping lines. Be cautious yourself when drawing inferences from graphs that have this kind of feature. Watch out when others have used graphs that include this kind of data.

The trend plot can also be used to suggest cause and effect when the cause is an event that occurs at some particular time rather than a continuing function of time. For example, according to *The Economist* (December 3–9, 2005), various of the Bush administration's trade actions and inactions led to the United States falling out of favor with the Canadian public. A graph much like the one in Figure 2.7 is was offered as evidence (*Survey*, page 12).

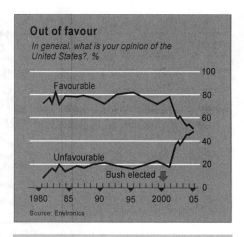

Figure 2.8

The point might have been strengthened by the addition of a symbol indicating the timing. For example, an arrow indicating the Bush administration's coming to office could have been added, as in Figure 2.8.

But a word of warning is in order. While a trend plot such as shown here can *suggest* a correlation, it does not *prove* one. Whole chapters of books have been written on this topic, for example, in Diggle's book (*Time Series: A Biostatistical Introduction*, Oxford University Press, 1990), which contains much guidance on the mathematics of time series. However, this is not a book with any interest or advice when it comes to graphs: the words "graph," "chart," "plot," "histogram," "bar chart," and "scatter plot" do not appear in the index.

STACKED TREND

You can also *stack* the information in a trend plot. This approach was taken in an article in an IEEE book[5] with a graph like that shown in Figure 2.9. Either this graph was originally made for presentation on the big screen, or here is definite proof that the Neanderthals did not die out 30,000 years ago, as is commonly thought. In general, we have reserved discussion of style for Chapter 13, but we cannot let this graph go by without at least commenting on the ugliness achieved by the choice of line weights and the use of only uppercase lettering.

This kind of curve has become particularly associated with presentations about energy (Figures 2.9, 2.10, and 2.11), largely because so much information can be contained in one graph. With this sort of graph, it is important that the reader know whether the "height" of the various lines should be viewed as height from the baseline or as the height on top of the line below. In this example, the clue is given by the fact that the areas are shaded, and each area is shaded differently. Evidently, these shaded areas represent the energy use *on top of* the areas beneath. This conclusion is borne out by the magnitude of the vertical scale: U.S. annual energy consumption at the turn of the century was around 100×10^{15} BTU, never having quite reached the level predicted here in 1975.[6]

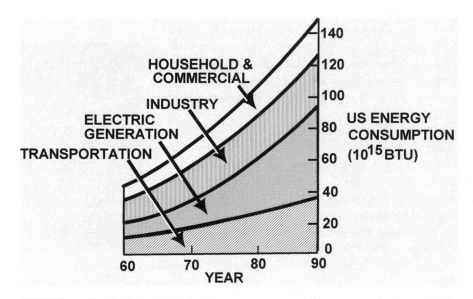

Figure 2.9

[5]Charles A. Zraket, Energy, resources and the environment reprinted in *Energy and Man: Technical and Social Aspects of Energy,* IEEE Press, New York, 1975.
[6]Which shows that the predictive ability of a trend graph cannot be counted on. In the case of this graph, which was generated around 1973, the data were a good fit to a model that assumed exponential growth. But the fact that a curve fit could be done does not mean that there is any underlying causal mechanism. According to data from the Energy Information Administration (http://www.eia.doe. gov/), the growth slowed down soon after this curve was drawn, likely in response to the oil price "shocks" of 1973 and 1979.

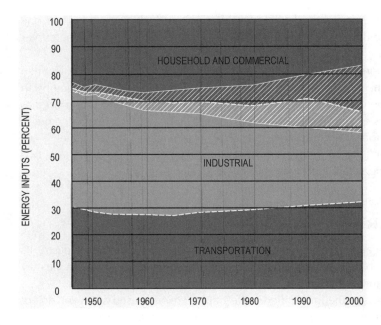

Figure 2.10

This type of presentation is a sort of multiyear pie chart, except that the pie chart always adds up to 360 degrees. A vertical slice through this diagram would add up to 100%, but the value at 100% would be expressed directly. If you "normalize" the data, you get the sort of graph shown in Figure 2.10 (this one is also from the same IEEE book). (Here the cross-hatched area is the part supplied by electrical energy.)

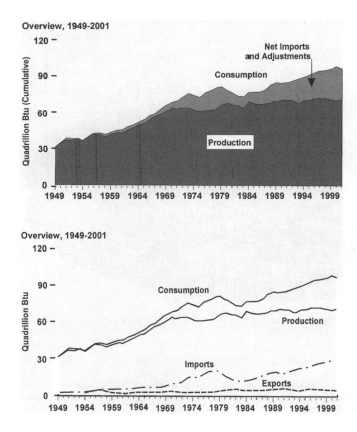

Figure 2.11

Do not treat lightly the question of shading. If the curves add to one another so that the top curve is the total, shade the different areas. If the curves are each separately referenced to the abscissa, do not use shading.

This approach was *not* followed by the Department of Energy (DoE) when it produced the graphs shown in Figure 2.11. Originally, the two graphs in Figure 2.11 were side-by-side. Because of the shading, the top graph *implies* that the curves are cumulative, and the ordinate label says so explicitly, but—since production and consumption must at least balance approximately—we know that can't be the case. Why the DoE failed to notice that these two graphs were identical, apart from the shading, is anybody's guess.[7]

PIE CHART

The pie chart seems to be a particularly useful thing for the persuasion industry. Its application is obvious: it shows how things are *divided*. The pieces of the pie may be identified numerically or as a percentage of the total.

Pie charts, sometimes called circle graphs, are often used when you do not wish to give information about the total quantity of something, but do wish to give information about how the total is broken down into its constituent parts. For some reason, pie charts are often associated with *money*. "Here's how your tax dollar is spent," and so on.

The pie chart is properly viewed from above (plan view) and appears to be circular. However, for maximum impact, some pie charts are given perspective. For example, the chart in Figure 2.12, by having a thickness and by putting it in the foreground, gives added weight to the fraction of the whole that is spent on the military. We found this one on the Web, by searching for "tax dollar."

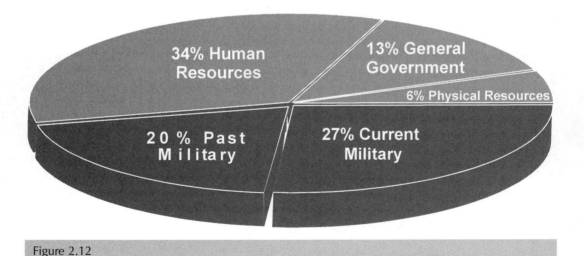

Figure 2.12

[7] Note also the unusual choice of years for the labels on the abscissa. Oh, well, nobody's perfect. Most of the graphs on the website (start at www.eia.doe.gov/emeu/iea/contents.html) are excellent examples of clear presentations and consistent style.

In general, it is not good practice to add a third dimension to graphs unless there is a third parameter involved. The gratuitous use of "3-D" to create depth on a graph creates the impression of a less-than-serious piece of work. This comment applies to any kind of graph, not just pie charts.

For emphasis, you may want to slide one piece of the pie out of the circle. Figure 2.13 is an example. This works, sort of, if the segment being removed is small, and if it isn't removed very far. You may not think it looks good when the segment is large, however, because the outer edge appears to be part of a circle that has a different radius.

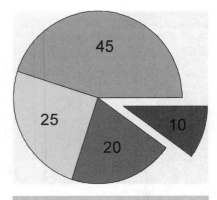

Figure 2.13

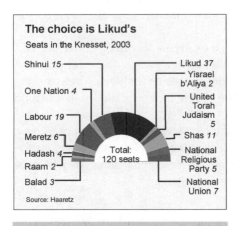

Figure 2.14

The pie chart is the most used of what might be called an *area chart*. A variation of the pie chart is what might be called the *donut chart*, which is really no more than a pie chart with a hole in the middle. This device seems to be a relative newcomer on the scene. As far as we can tell, it usually has little to commend it.

One application where the donut chart makes at least a little sense is the presentation of the composition of governing bodies, such as the U.S. Congress or the Israeli Knesset. Look at the example of Figure 2.14, adapted from *The Economist* of January 30, 2003, which shows the party composition of the Knesset after an election in early 2003. This presentation makes a sort of sense because the Knesset, like many legislative bodies, is arranged in approximately the shape shown, with the seats of the legislators in a semicircle centered on the podium. It may well be that the parties are ordered in the graph according to their politics, with the left wing parties on the left and the right wing ones on the right of the diagram. *The Economist* has made use of the space in the middle to insert the total number of seats.

Altogether, this is a very effective presentation. Figure 2.15 is an example of the donut chart that is much less effective. This donut chart was used in the *New Scientist* (March 15, 2003) to show the contribution of various sources to the planetary level of the greenhouse gas methane. Perhaps the delicate souls at *New Scientist* were reluctant to make it clear that the biggest methane contributor is cow farts. In any event, the fact that dark blue is 22% is the thing that jumps off the page when you look at the chart, and you have to work on the problem to find out that dark blue is "domestic livestock."

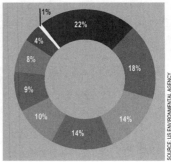

Figure 2-15   See insert for color representation of this figure.

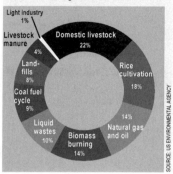

Figure 2-16   See insert for color representation of this figure.

Making the reader work like this is not good practice. Each section of the pie should be labeled, either inside the pie or just outside. Figure 2.16 shows how the previous example might be improved. This is better, but just what the purpose of the hole in the middle could be remains a mystery. Somebody, somewhere, had to make a decision about how big to make the hole, and some editor at *New Scientist* had to approve it. All this work, and it serves no function!

HISTOGRAM AND BAR CHART

The barchart is perhaps the most common of all the statistical graphs. The length of a bar is proportional to the magnitude of the parameter being shown. Several bars are used to represent several parameters. The bars may be vertical or horizontal, and are parallel.

Sometimes two bar charts are shown on either side of a dividing line, so that they can be compared as in a mirror. The most common application of this technique is to show differences or similarities between the aging male and female population. The graph is then called a population pyramid, as in Figure 2.17.

Figure 2.17 is from a website of the Japanese government (http://www.stat.go.jp). Note that, in this graph, a line showing the ends of the bars for the 1935 population distribution has been added for comparison. We will have more on the appropriateness of that in the next chapter.

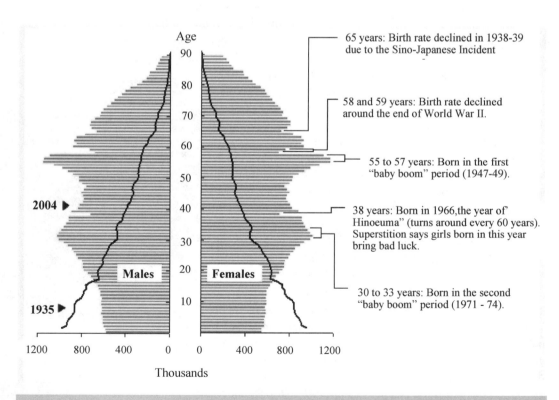

Figure 2.17

Strictly speaking (and we intend to speak strictly), there is a difference between a histogram and a bar chart. A histogram shows the distribution of a single variable. A bar chart may have only one variable, or it may combine several sets of data. This difference means that the two kinds of graph are used for different purposes.

A histogram summarizes the statistics of a single variable: its location (think of the average value), and its spread (as in deviation). It shows whether the data are symmetrical about the center or skewed, whether there is a single mode or multiple modes are present, and perhaps whether there are any outliers.

A bar chart is a much more general kind of graph and can be used to show all sorts of things: how one (or more) variables have changed over time, for example.

This is a *real* difference. The plotted parameter of a histogram should only be the count, the percentage of total population, or some similar parameter, and the "as a function of" parameter should be the "bin," the range of values to which the count applies. The plotted parameter of a bar chart can be anything at all and, come to think of it, so can the other parameter.

The population pyramid is a pair of histograms turned sideways. The plotted parameter is the number of people of a certain age (i.e., born in a given year), and the "bin" is the calendar year. (In Example 6 in the next chapter, we will argue that there is nowadays a new kind of histogram whose appearance resembles a regular bar chart. For now, however, let us stick to the conventional meanings of the terms.)

Figure 2.18 is an example of the "ordinary" kind of histogram. The graph is based on a survey of the years of experience of the members of a professional institute.

The histogram graphically shows the following:

The *location* of the data

The spread (or deviation) of the data

The skewed (nonnormal) nature of the distribution

Had there been any outliers, these might have been evident. Had there been multiple modes, this would have shown up as multiple peaks in the graph. Actually, the ability to show multimodal data depends on the choice of the width of the *bin* chosen. The bin width here is 5 years, chosen for convenience. We can take the

Figure 2.18

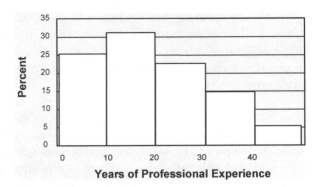

Figure 2.19   **Years of Professional Experience**

same data and choose a bin width of 10 years, with the result shown in Figure 2.19.

Ordinarily, you will have some flexibility about choosing bin width. It could be argued that the width of the bin should be such as to produce "nice" numbers at the edges, as in the two examples just given. However, for a histogram, it is perfectly possible to label the axis with "nice" numbers and let the bin width be more or less separately decided. (This is no different from acquiring data at intervals that are not "nice" or even regular, and plotting them in a trend graph with a convenient and understandable axis.) Figure 2.20 illustrates this with bin widths that are not an integral number of years.

Here we have chosen a bin width of 20 months. Because of that, every 5 years, the edge of the bin coincides with a label on the axis. We could have chosen some other number, even an irrational number, and still had an acceptable presentation.

There are papers on the subject of selecting bin width—if you are interested, we suggest you do a Web search.

The histogram differs from the bar chart in another graphical way. There is ordinarily a space between the bars of a bar chart: no gaps exist between the bins of a histogram. This difference may require you to edit the default product of a spreadsheet.

The typical spreadsheet output has the general appearance of the graph in Figure 2.21, which is an example of a bar chart that could be used to make the case that the United States was lagging the world in an important area of education.

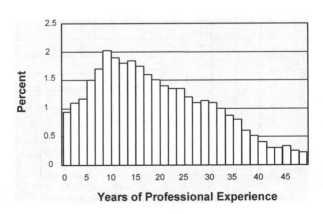

Figure 2.20   **Years of Professional Experience**

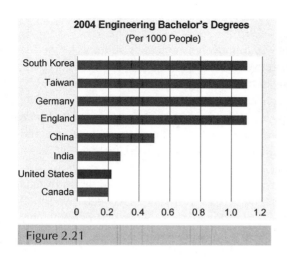

2004 Engineering Bachelor's Degrees
(Per 1000 People)

Figure 2.21

2004 Engineering Bachelor's Degrees (per 1000 people)	
Canada	0.198
China	0.498
England	1.098
Germany	1.100
India	0.278
South Korea	1.101
Taiwan	1.100
United States	0.220

Isn't it easier to see the situation in this graph than in the data[8] from which it was drawn?

In the table to the right the data are ordered according to the alphabetical order of their names in English. In the graph, they were reordered according to the numbers.

It is common to combine data for several bar charts into one when they have a common basis. For example, a certain school has used the presentation shown in Figure 2.22 to illustrate continued steady improvement in language and math abilities, with an increase over time in the fraction of students classified as Proficient and Advanced, and a decrease in the fraction classed as Below Basic or Far Below Basic. Note that, in this example, the two horizontal scales are the same, 0–100%, even though there are no data above 80% in the Language Arts chart (and therefore this section of the graph could have been omitted). This similarity of scales makes for an unbiased presentation.

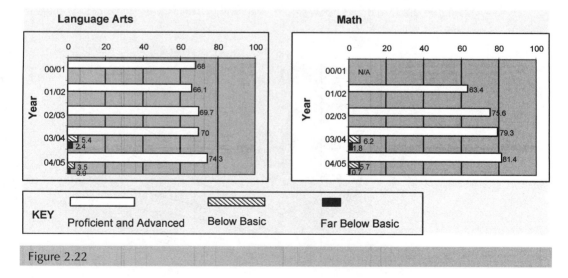

Figure 2.22

[8]Wanda K. Reder, The technical talent challenge, *IEEE Power and Energy*, pp. 32–39 (January/February 2006).

SCATTER GRAPHS

The scatter graph is a convenient way to show how one thing varies as a function of another. It can be used very effectively to illustrate correlation, or lack of correlation. It can show whether some of the results should be considered outliers.

In 1889, Friedrich Paschen presented a paper: *Ueber die zum Funkenübergang in Luft, Wasserstoff und Kohlensaüre bei verscheidenen Drucken erforderliche Potentialdifferenz* (On the flashover in air, hydrogen and carbon dioxide at various required applied potential difference). It is in Wiedemann *Annalen der Physik und Chemie*, 1889, Vol. 37, pages 69–96. This paper has had far-reaching effects. In it, Paschen presented tables of data showing how, for various gases, the flashover distance between two small spheres was a function of potential and distance. The work was painstakingly done, and took long enough that seasonal variations were noted.

Paschen presented only two graphs, though he had information enough for more. He might have started with a graph showing that, at various pressures, the flashover voltage was a fairly simple function of the spacing of the spheres. (As it happens, Paschen did not measure the voltage directly, he had some kind of uncalibrated electrometer, so he recorded its deflection.) Figure 2.23 is a graph showing the results for carbon dioxide.

The graph shows that, for a given pressure, the meter deflection (representing flashover voltage) is without doubt a function of the spacing. (We just joined the dots with straight lines, though we will argue in the next chapter that there may be a better way to draw the graph.) Paschen might then have continued with another graph showing that, at various spacings, the flashover voltage was also a function of the gas pressure, as in Figure 2.24.

But the thing that Paschen is most known for is the realization that the data from both of these graphs could be combined. This result is shown in Figure 2.25. In this graph, the "delta" is the spacing of the spheres (in cm).

Whether or not Paschen presented graphs in his paper, it is clear that he had in mind *exactly this result*, and he must have produced the graphs *somewhere*. And you cannot be surprised that Paschen, seeing the graph shown in Figure 2.25, realized he was on to something! This is true elegance! Workers in

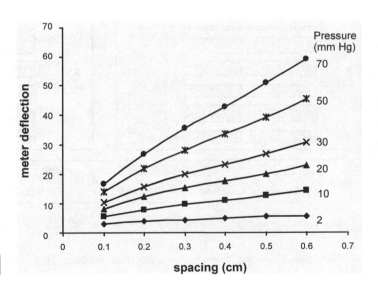

Figure 2.23

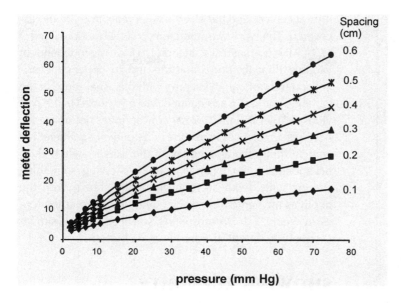

the field of gas breakdown have followed this approach ever since, plotting break-down voltage against the product of pressure and distance.

We can understand much of this behavior today in terms of electric field strength, of ionization by electrons, and of the mean free path.[9] However, in the Germany in which Paschen did his work, the atomic theory of Dalton was not yet

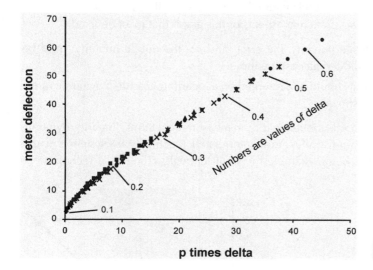

[9]In the region considered by Paschen, the mean free path is smaller than the electrode spacing. The mean free path (mfp) varies as the pressure, the temperature, and as the inverse of the square of the diameter of the molecule (here perhaps 0.2 nm to 2 Å in diameter). For carbon dioxide at 200 K and 1 mm Hg, the mfp is 0.12 mm. The energy gained between collisions can be calculated by dividing the electrode separation by the mean free path and multiplying by the total voltage. If the electrode separation δ is doubled, the energy per collision is halved. If the pressure P is halved, the mean free path is doubled. Thus, provided the product $P\delta$ is constant, the energy per collision is constant. The likelihood of ionization at the collision is therefore unaffected.

It was a piece of good fortune that while the field between two spheres is nonuniform, for the sizes and spacings considered by Paschen, the field was only weakly nonuniform. The average field is indeed simply given by the voltage divided by the spacing.

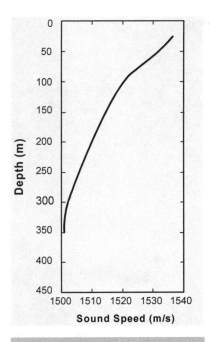

Figure 2.26

fully accepted, and the electron was years from being discovered. The work was, therefore, truly groundbreaking.

It is normal in a scatter plot to have the independent variable along the horizontal axis and the dependent variable on the vertical, in keeping with the usual sense of the variables x and y in an equation. We mentioned in the previous chapter that this normal convention is not universal, however. In sciences where the independent parameter is a vertical thing, the assignment of the axes is switched. In oceanography, for example, curves showing something varying with depth are conventionally shown with the depth as the vertical parameter, and with zero at the top, as in Figure 2.26. Atmospheric scientists have a similar convention, except that zero is at the bottom.

SHOWING LINEARITY

Quite often a graph is used to explore the performance of a system. The graph in Figure 2.27 is based on a graph that was drawn to show the performance of a transducer and was used in a technical paper to justify a claim of good linearity. The transducer being tested may have been linear, but although the data "look" about like a straight line, in fact this figure does not really tell us much at all.

There are two particular aspects of this graph that must be fixed.

■ To indicate linearity, the *error* (and *not* the output quantity) must be shown plotted against the input.

■ The input should be presented on a logarithmic scale if dynamic range is to be shown.

First, let us think about linearity. In measurement terms, linearity, dynamic range, and signal-to-noise ratio are all interrelated. Practically any system is nearly linear if the range being examined is small enough. (There is a technique in

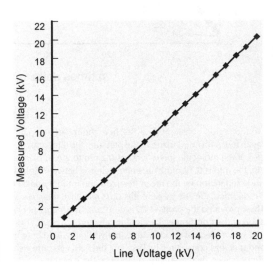

Figure 2.27

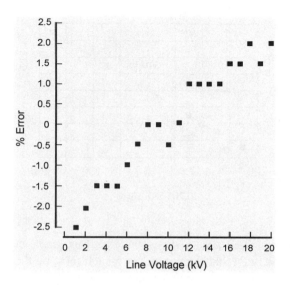

Figure 2.28

control engineering that allows analysis of nonlinear systems by approximating them as piecewise linear.) However, for a small dynamic range, a claim of linearity will not impress.

The graph of Figure 2.28 shows the error in the transducer. This shows *why* it is such a good idea to plot the error, and not the input and output of a transducer. It is pretty clear that this transducer has not been calibrated properly! The error is negative for small signals, becomes small in the midrange, and goes positive at the upper end of the operating range. In fact, because of the bad calibration, the error is as high as 2% or 2.5%. By reducing the gain, and adjusting an offset, it would be possible to keep the error below 0.5% over a good deal of the range—certainly the range of interest (the high end, as this device was a voltage transducer).

But even this improved presentation does not show the limited dynamic range of the device being described. The graph in Figure 2.29 shows the same data but has been plotted with a logarithmic axis. Now we can see that the transducer has not been demonstrated over a particularly wide range.

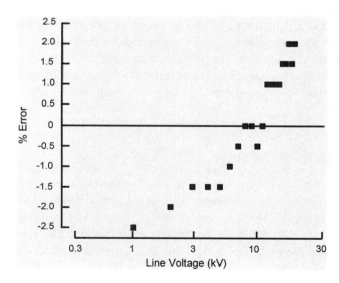

Figure 2.29

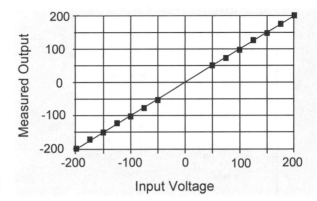

Figure 2.30

Here's another example of how *not* to show linearity. This graph of Figure 2.30 attempts to make the point that the transducer (which must be the brother of the one in the previous example) has a linear transfer curve. The difference is that this one is for direct voltage, and it has to operate with both positive and negative inputs (and outputs).

The graph has the same problems as its predecessor:

- To indicate linearity, the *error* (and *not* the output quantity) must be shown plotted against the input.
- The input should be presented on a logarithmic scale if dynamic range is to be shown.

What is the dynamic range of the device shown (in simulation) in the figure? If we acknowledge the fact there are no data close to zero, we could be forgiven for thinking that the dynamic range was rather poor. Usually defined as something akin to the ratio of the largest signal to the smallest, the dynamic range of this transducer might be 200/50 or 4 : 1.

Thus, a major fault with the graph in Figure 2.30 is that it does not show voltages close to zero. The input was varied between 50 and about 200, both positive and negative. The dynamic range, 4 : 1, is scarcely any range at all!

If the input axes had been logarithmic, this lack of data would have been obvious. Suppose an axis had been labeled as follows: 10^{-1} 10^{0} 10^{1} 10^{2} 10^{3}. Such an axis would allow a dynamic range of four orders of magnitude to be presented. However, all the data presented in the graph above would appear in the last two intervals of such a graph. The smallest input (50) is close (on a log scale) to 10^{2}, and the dynamic range tested was less than one order of magnitude above this. If the dynamic range is this small, claims of good linearity cannot be made.

REGRESSIONS

Figure 2.31 is based on the same data. It represents only the positive side of the tests. The error was calculated by subtracting the output value from the number calculated by a simple linear regression. It can be seen right away that the error varies apparently randomly with the input. The biggest error is about 2%. This presentation shows at a glance that the data in the first figure do not show a very linear system.

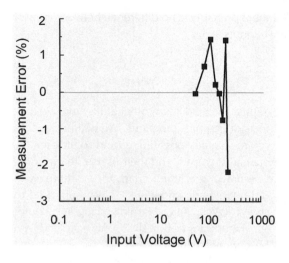

Figure 2.31

Modern spreadsheets can do a linear regression in almost no time at all. Not long ago, it was the custom to enter manually taken data into a spreadsheet to do the graphing. While these days more data-taking is automated, we should not forget the other capabilities of the spreadsheet. Thus, if your data acquisition system does not come with a built-in capability to do a regression, you might spend the time to figure out how to get the numbers into a spreadsheet.

It is fair to acknowledge that some transducers do not *need* a large dynamic range. In electric power systems, for example, adequate voltage measurements can be made with a device with a minimal dynamic range, because power system voltage rarely varies more than a few percent. To present such results, a horizontal axis that is linear (rather than logarithmic) might be acceptable if the region near zero volts is ignored. (An extrapolation through zero is meaningless, especially as we have argued that readings at very low voltage are not needed. Hey, if a power system is operating at more than 10% undervoltage, there is trouble somewhere that a highly accurate voltage reading will *not* be helpful in fixing!) Since in the dc voltmeter graph in Figure 2.31 there are data only between 50 V and 250 V, this range alone can be presented, as in the graph in Figure 2.32. The presentation makes clear that the voltmeter

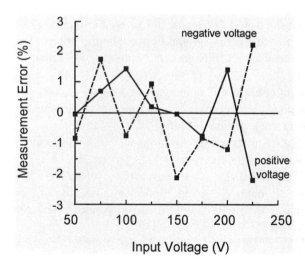

Figure 2.32

is not particularly good, and that there probably is no difference between its performance on positive and negative inputs.

OTHER GRAPHS

We have looked at the most used graphs. There are many other kinds, and you may be producing them with spreadsheets or graphics programs. We will not go into them all here, partly because there are so many possibilities and so little space, but also because they tend to be specialist graphs, and their users already know what they are about. Who could be drawing a Smith chart, for example, who did not already know what it was supposed to show? (Well, OK, Harold's daughter could: see Figure 2.33. But she's only six, and you have to make allowances.)

Figure 2.33

Rather than examine the narrow-focus graphs that will interest only a few readers, we turn instead to the idea that *novel graphs sometimes make good presentations.*

The graphs used in one part of science or technology sometimes make attention-grabbing presentations when adapted to another. There are relatively few graphs that engineers and scientists haven't seen at least a few times, however, so we will not have many examples here. (If the reader has an interesting example to add to our collection, please contact the authors at harold.kirkham@ieee.org.) We present these examples in the hope of stimulating the reader into thinking about something similarly novel.

One of the earliest graphs invented specifically for a single purpose is an example of the *area graph*. The presentation in Figure 2.34 was drawn by Florence Nightingale, as part of her campaign to improve the survival chances of the British soldier, then fighting in the Crimea.

The first impression is that this is a variety of the pie chart, but really it isn't, as here the *diameter* of the pie is the parameter that changes, and the *angle* of the slice is fixed at 30 degrees (representing months). Nightingale included a predecessor of this in a report[10] that she called a *coxcomb*, the ostentatious part of her bigger Royal Commission Report that people would notice. She did not use the word coxcomb to describe the diagram, though it has subsequently come to have that name.

You might have thought of Florence Nightingale as a nurse—in fact she was a remarkably capable woman in many areas. She was enough of a mathematician to be elected a Fellow of the Statistical Society.

Unfortunately, the chart requires a large explanation, but it was (and is) very effective. By far the majority of soldiers were dying of preventable or mitigable disease, and not of battle wounds. What a case she had to make! But the chart doesn't spoonfeed the reader—it makes the reader think! The other side of that coin is, of course, that it makes the author think, too. You cannot easily get

[10] *Notes on Matters Affecting the Health, Efficiency and Hospital Administration of the British Army,* 1858.

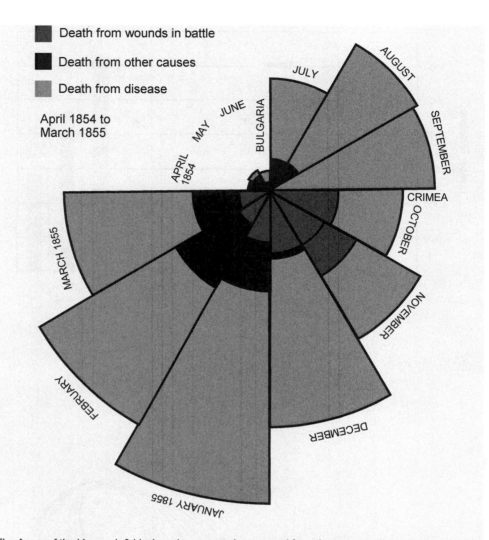

Death from wounds in battle

Death from other causes

Death from disease

April 1854 to
March 1855

The Areas of the blue, red, & black wedges are each measured from the centre as the common vertex.

The blue wedges measured from the centre of the circle represent area for area the deaths from Preventable or Mitigable Zymotic diseases, the red wedges measured from the centre the deaths from wounds, & the black wedges measured from the centre the deaths from all other causes.

The black line across the red triangle in Nov. 1854 marks the boundary of the deaths from all other causes during the month.

In October 1854, & April 1855, the black area coincides with the red, in January & February 1855, the blue coincides with the black.

The entire areas may be compared by following the blue, the red, & the black lines enclosing them.

Figure 2-34   See insert for color representation of this figure.

your spreadsheet program to draw you a graph like this. In fact, given the raw data, you are going to have to work out the radius of the various slices. But maybe the effort is worthwhile.

The same data could be presented as a bar chart, of course, obtained from your spreadsheet software if you like, as in Figure 2.35.

These days, this kind of presentation would probably get nobody's attention the way the coxcomb chart did. Although the worst month (January 1855)

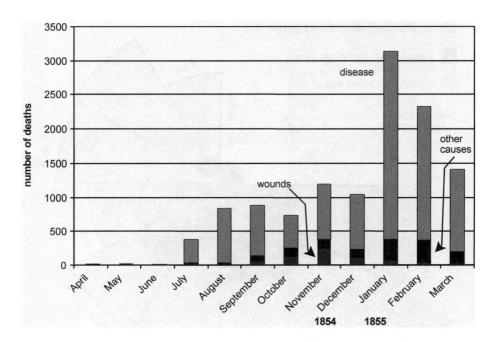

Figure 2-35 See insert for color representation of this figure.

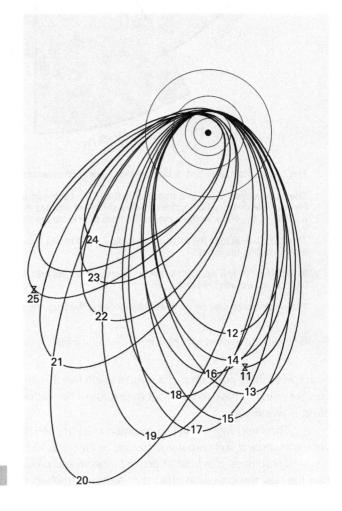

Figure 2.36

looks worse in this graph because the coxcomb diagram compresses the length of the slice according to the square root of the area, the overall impression here is that the problem may already be going away. When more than 85% of the deaths are caused by preventable disease, that is not an impression to foster!

An example of a special-purpose graph from planetary exploration is the *petal plot*. This is a way of viewing the trajectory of a spacecraft in orbit around a body (or around a libration point). The origin of the name is obvious, as seen in Figure 2.36. This chart shows the trajectory of the *Galileo* spacecraft orbiting the Jovian system in the late 1990s and gathering data (and images) on Jupiter and its various moons. (The numbers at apoapsis are the orbit numbers. The symbols on orbits 11 and 25 show where the rocket motor will be used to make adjustments to the orbit.)

The presentation can be adapted as in Figure 2.37 to show the progression of, for example, proposal paperwork through the bureaucracy of a company. It is clear that the planetary system here revolves around the contract department! This graph requires nowhere near the amount of effort to produce as the coxcomb chart, because it is not *quantitative*. On the other hand, it might be just the sort of thing to offer to your management. (You know, not too technical.) You could even add a few dates, to show how slow some of these orbits are.

Figure 2.38 is an example of another unusual graph: it looks at first glance like a subway map, but in fact it shows molecular pathways involved in cancer. The original of the graph reproduced here was in "A Subway Map of Cancer Pathways," by William C. Hahn and Robert A. Weinberg, *Nature Reviews Cancer* **2**(5), May 2002. An online version can be found at http://www.nature.com/nrc/poster/ subpathways/index.html. The instructions with the graph

> Click on any of the stations on the map to go to the LocusLink entry for that gene or group of genes. Click on the main stations (Boxed text) to go to a description of the pathway.

make clear that this is an interactive presentation. This is an excellent example of an unusual application of the commonplace. It is attractive and functional. Would that all graphs were!

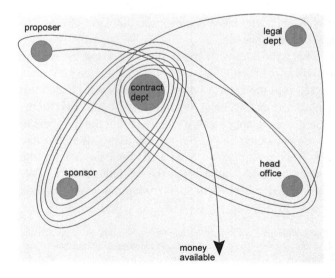

Figure 2.37

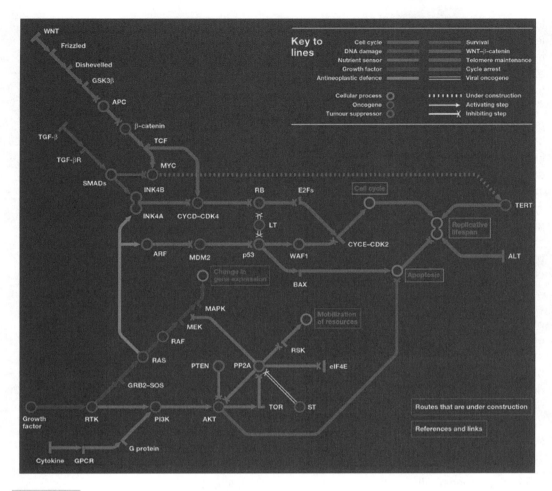

Figure 2-38 See insert for color representation of this figure.

We opened the chapter with an example drawn from the *Wall Street Journal*. We asked whether the original graph, a bar graph, was the appropriate choice. We invited the reader to consider an alternative (shown in Figure 2.39, on the right). We also asked whether a third option might exist that was better. Of course there is! Books, like TV trial lawyers, never ask that kind of question unless the answer is known in advance.

First, ask yourself what is the purpose of the graph. It is to show the decline in the growth rate of the number of households with CNN, a cable news channel, and the simultaneous increase in the growth rate of the number of households with FOX News Channel, a competing news organization.[11] (We say "with" because neither the article nor the graph makes it clear whether the numbers in the graph are the numbers *connected* or the numbers actually *watching*. For our purposes, of course, it doesn't matter.) Note that the data are for the *change* in the number of households, rather than the absolute number. We have no way of saying, from the numbers in the graph, what the relative market

---

[11]In fact, we were a little confused by the title of the graph. While the caption mentions "growth and decline," the data seem to show only growth. None of the numbers are negative. The CNN growth does seem to be declining, however.

Changing Channels

Growth and decline in
household audiences, in
thousands of households

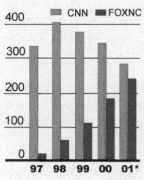

* As of June 30
Source: Nielsen Media Research

Changing Channels

Growth and decline in
household audiences, in
thousands of households

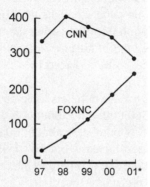

* As of June 30
Source: Nielsen Media Research

Figure 2.39

share of the two channels might be.[12] For our graphical purposes, this, too, is unimportant.

Now let's look at each of the two presentations, starting with the bar chart. Does the bar chart accomplish the presumed goal? It does, fairly well. Mainly, there is the problem that the two channels are obliged to sit side by side, implying (slightly) that there is a time element relating each pair of numbers. On the plus side, the bar chart does carry with it the implicit idea of something accumulating throughout the period of interest. This is fair: the numbers do seem to be the numbers that applied at the *end* of the period. This inference is supported by the statement that the last column presents numbers valid "as of June 30," implying that, had we waited, other numbers would apply.

Also in support of the use of the bar chart is the notion that these data represent new viewers. The *new* viewers in one year are different people from the new viewers of a previous year. Compare with the "Nobel prize problem" of Chapter 3. While a trend may be apparent, it is due to some underlying cause, nothing more.

The bar graph is therefore a fair choice. How does the dot graph do?

First, it lets us present simultaneous data with vertical alignment. This overcomes the side-by-side objection to the bar chart. Second, it lets us put dots to represent the actual numbers. This reinforces the idea that the graph is based on measured data. Though the lines joining the dots don't actually mean much, they do emphasize the rate of change, and can be allowed to stand.

However, somehow the idea of accumulating new customers throughout the year is lost. In particular, consider how the last set of numbers could be

[12] The numbers show that the data are growth numbers and not actual market size. The United States has many tens of millions of households altogether. The biggest numbers here, around 400,000, would represent less than 5% of ten million. One can be sure that considerably more than 5% of households have a cable news service.

Changing Channels

Growth in audiences, in
thousands of households

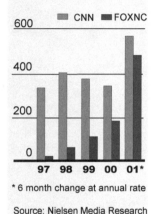

* 6 month change at annual rate

Source: Nielsen Media Research

Figure 2.40

improved. Since each column represents the new viewers for the year, we can readily estimate the end-of-year numbers for the last column from the midyear values—we just double the numbers! (This is the simplest estimate possible and assumes there are no seasonal factors at work.) Figure 2.40 shows what we get.

This rather changes the conclusion, doesn't it? The mental extrapolation from the earlier graphs was one of declining fortunes for CNN in contrast to the steadily rising star of FOXNC. This has now gone, to be replaced by a rebound for CNN, and a jump for FOXNC.

Do these numbers look reasonable? Yes: the CNN number is around 560,000, only about 40% higher than the increase in 1998. The FOXCN number is about 500,000, more than double their previous largest increase, but still smaller than CNN's figure.

Note that we changed the captions. The main title is the same, as an eye-catching headline. The subtitle now does not mention *decline* in audience, as this seemed to be a mistake anyway. The 2001 data are now described as "6-month change at annual rate."

As far as this graph is concerned, the situation with the CNN–FOXNC horse race could be described thus: CNN is out in front. CNN is running faster. But FOXNC is accelerating harder. This hardly justifies the *Wall Street Journal* making a story of it!

SUMMARY

There are three common chart families: histograms and bar charts, *x-y* plots (also called scatter graphs), and area charts (pie charts or donut diagrams). While each of these has several species within it, some generalization is possible.

- A *histogram* is used to present statistical data for a single variable.
- A *bar chart* can be used for numbers that change with time, or to compare two (or more) sets of data. Bar charts are the one kind of graph that can be arbitrarily rotated to be vertical or horizontal
- An *x-y plot*, or *scatter plot*, is used to show how one parameter varies as a function of another. It is usual to have the independent variable horizontal, though it is conventional in some fields (e.g., oceanography) to have an obviously *vertical* parameter (such as depth) shown vertically.
- An *area chart* can be used to show how some entity is divided among its constituent parts.
- Consider using a graph any time you are considering a table of data.
- In general, it is not good practice to add a third dimension to graphs unless there is a third parameter involved. The gratuitous use of "3-D" creates the impression of a less-than-serious piece of work.

After these general guidelines, there are some specifics.

- To show linearity, do not use an *x-y* plot with the output as a function of the input. Show instead the *error* as a function of the input.
- To show a parameter with a large dynamic range, use a logarithmic scale.

EXERCISES

2.1. The following data show the increasing complexity of integrated circuits as time goes by. The data are estimates representing both memory (the first few data) and processors (the rest) from Intel, a chip manufacturer. Show the data as a graph.

Year	Transistors/Die	Comment
1959	1	Transistor
1962	8	
1963	20	
1964	35	
1965	80	
1971	2000	4004
1974	8000	8080
1978	30000	8086
1982	120000	80286
1985	250000	i386™
1989	1000000	i486™
1993	3000000	Pentium®
1997	8500000	Pentium® II
1999	30000000	Pentium® III
2001	70000000	Pentium® 4
2002	200000000	Itanium™

2.2. The following numbers are estimates of the amount of deforestation (in thousands of square kilometers) taking place in Brazil, over the years 1990–2007. Present the data graphically.

Year	Area	Year	Area	Year	Area
1990	13	1996	18	2002	22
1991	11	1997	13	2003	25
1992	13	1998	17	2004	27
1993	15	1999	17	2005	18
1994	15	2000	18	2006	14
1995	28	2001	18	2007	11

2.3. You wish to present data graphically. What kind of graph should you use for the following:

(a) U.S. annual energy consumption from 1949 to 2007?

(b) Height of students in a class?

(c) Sources of information on technology matters used by adults (TV, Internet, newspaper, magazine, books, radio, and so on)?

(d) Number of attendees at an annual conference, 1968–2006?

(e) Percentage overshoot or undershoot in a series of six measurements?

(f) Percentage error in an instrument as input parameter is varied?

(g) Excess number of "heads" in a series of coin flips?

Connecting the dots 3

O n the side of my children's cereal boxes are the dotted outlines of giraffes and elephants, and various other denizens of the veldt whose outlines are, for some mysterious reason, as familiar as that of the cat next door. These outlines come with instructions to "connect the dots" and reveal the hidden creature. So common is this that "connecting the dots" is a widely used metaphor for making an inference.

In the world of graphs, things are not so straightforward. With the help of the computer, the practice of graphical dot-joining has become very common, in spite of the fact that it is often not obvious whether it makes any sense. That is what we will explore and discuss for the remainder of the chapter. And this question of dot-joining can be an important issue. We shall see in Chapter 14 a case study where the wrong choice allowed unfortunate human consequences.

It sometimes takes a bit of digging to see whether or not connecting the dots is appropriate. So let us dig. We begin by going back about a hundred years.

Albert Einstein held that theories fall into two categories. *Constructive* theories are based on *known facts*, but there are no theoretical principles to explain the observations. *Principled* theories are deduced from some basic principles, and may then be used to explain certain facts. Whether or not we consciously knew this, the chances are good that this kind of thinking has been in the back of our minds. Your average physicist would much prefer a principled theory, on the grounds that it leads more directly to *understanding*.

Einstein's own theories of relativity are *principled* theories. Based on the principles (almost assumptions, in fact) that the speed of light is constant, and the laws of physics are the same for everyone, Einstein derived special and general relativity. Quantum theory, which caused Einstein some trouble, is a *constructive* theory based on some known facts. Even today, an underlying principle to explain the weirdness of the quantum world has not been found.

Now, if you are looking to develop a theory, curve fitting to the observations might lead to seeing the principle, so you could produce a principled theory rather than merely a constructive one. Curve fitting is a very useful process.

Figure 3.1 is an example of a curve fit.[1]

---

[1] Shakil A. Awan, Bryan P. Kibble, Ian A. Robinson, and Stephan P. Giblin, A new four terminal-pair bridge for traceable impedance measurements at frequencies up to 1 MHz, *IEEE Transactions on Instrumentation Measurement* **50**(2), 282–285 (April 2, 2001).

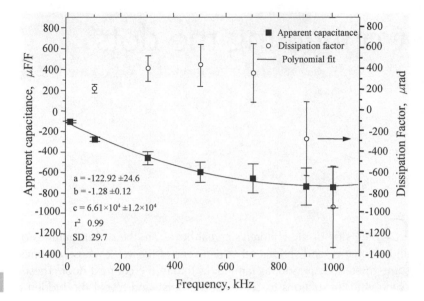

Figure 3.1

The capacitance and dissipation of a 10-nF capacitor are measured. The capacitance is assumed to have a frequency dependence given by $\Delta C(f) = a + bf + cf^2$, where $\Delta C(f) = [1 - C(f)/C_0]$. $C(f)$ is the measured apparent capacitance and C_0 is the 1-kHz value. The values for the coefficients found by the curve fit are given in the graph: they show that the curve is almost a perfect parabola. This observation might support some model of the variation with frequency.

If we remove the curve fit in the graph in Figure 3.1, and substitute a version of the graph with the appropriate dots joined by straight lines, we obtain the graph shown in Figure 3.2. The general appearance is similar to the curve fit of Figure 3.1. And the method is very common. But even so, is dot-joining a useful thing to do? Or fair?

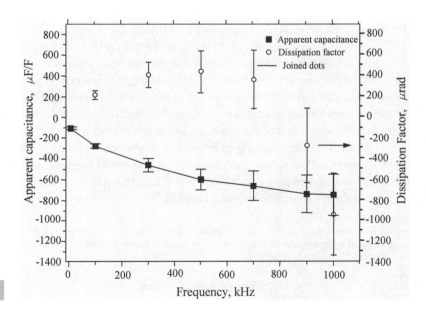

Figure 3.2

We mentioned all the Einstein stuff because principled theories might be advanced from a curve fit to experimental data. Since principled theories are a Good Thing, curve fitting must be a Good Thing too. But similarity of appearance notwithstanding, connecting the dots is not really a Good Thing. It just does not have the same benefit.

A straight line would *not* have been used to connect adjacent data points in the remote past, say, thirty or forty years ago. A smooth line might have been obtained by the artistic use of a French curve, perhaps, if there was some reason to suspect that a fairly simple nonlinear relationship was reasonable, but the only use of a straight line would have been for a linear regression, basically a straight-line fit to the entire data set.

This viewpoint is supported by a few books from the period. A textbook in HK's collection (Forrest I. Barker and Gershon J. Wheeler, *Mathematics for Electronics*, Addison-Wesley, Reading MA, 1968, page 167) instructs us, on the topic of plotting graphs from measured values, to

> Draw the graph of an experimentally determined (measured) function as a *smooth* curve or a straight line. It is almost *never* correct and certainly not in keeping with accepted practice to connect successive points of plotted data with short line segments producing a jagged-line graph .... Using a drafting device such as an *irregular* or *French curve* will help.

(The italics are in the original.)

A little later, Cuthbert Daniel and Fred Wood published *Fitting Equations to Data* (Wiley, New York, 1971). This book contains a fair number of graphs—not a single one shows dots joined by lines. Not one!

Later still, John Chambers, William Cleveland, Beat Kleiner, and Paul Tukey published *Graphical Methods for Data Analysis* (Wadsworth, Belmont, CA, 1983). In this book are several hundred example graphs. Only one of them has adjacent dots joined by straight lines.

So how does it come about that we see dots joined by straight lines so frequently?

Well, it happened because of computers. As the small computer became more used to examine measured data, through the 1970s and into the 1980s, the reasoning behind generating a line changed. Instead of a scientist or engineer deciding that a particular curve made sense, or fitting an equation to verify a theoretical prediction, a computer programmer decided to use a straight line between adjacent points.

We know this to be true. One of us (HK) was one of those programmers.

There were lots of us back then, scientists and engineers writing code to solve particular problems, and we would not claim that our own contribution was in any way particularly important. Typical of what was being done in the 1970s, HK and his colleagues at the American Electric Power Service Corporation wrote data acquisition software with a limited ability to display results. They produced trend plots. (The data were transferred over to another computer for more thorough statistical study.) They had the ability to place a character (such as an X) at a point on the screen where a data point should be. Or they could use a dot.

The monitors they had in those days were not particularly good in terms of contrast, and they decided that the dots alone were not very visible. However, when there were lots of data points, Xs looked klugy. So they put lines joining the data points instead.

Were they breaking any rules? The question did not arise. They assumed that the straight line was just a simple interpolation. Must be OK, right?

The evidence is that until the mid-1980s there was no need to instruct graphics artists on the proper use of the various kinds of graph. However, since the advent of the small computer, we no longer send our graphics to a professional artist before going into print, and the software that we use to replace their services is relatively ignorant of the rules, written or unwritten. That software typically allows us to join the dots—or not—but it does not advise us whether we *should*. We, the scientists and engineers using the software, are ultimately responsible, but if we select the default option, we are delegating to the software writer.

It is time to take back responsibility. To do that, we will need to create some general rules that can cover all situations. No mean feat! Along the way, we will examine graphs that have joined dots, and those that do not. We will make up examples that are clear and examples that are ambiguous. Our purpose is to end up with some *understandable* guidelines so that you, gentle reader, can always decide for yourself.

EXAMPLE 1: GROUPED RESULTS

Consider the following situation. An experiment is performed in which several hundred operations of some system are routinely combined in a way that allows you to record the results on a spectrum of success and failure. (The idea comes from tests on an experimental magnetic recording process, but is surely more widely applicable.) Figure 3.3 is a bar chart of the results. This presentation makes the results look fairly good because the bars in the graph are tall.

Let us spend a few minutes thinking about the use of a bar graph in this case. We could have presented the same data as shown in Figure 3.4. Is this permissible?

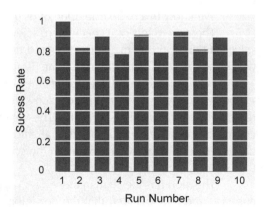

Figure 3.3

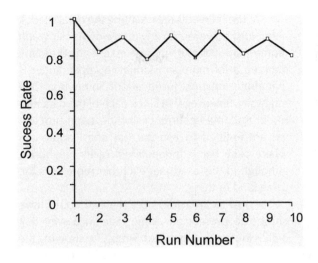

Figure 3.4

Case Against Joining Dots	Case for Joining Dots
The various trials are discrete events and independent. The line joining the dots at least *implies* that one could interpolate the data. But in the present situation, there is no significance to a "Run 1.3."	Time, at least, is increasing between the various runs. That fact might be used to argue that, by and large, things were getting worse at the end! This trend is emphasized by the line.

In the example of the magnetic recording system, it should be clear that there cannot be any data points between the ones in the bar chart. It is therefore not permissible to interpolate, as would be implied by connecting the dots. In this example, we can choose only between a bar chart and a scatter graph with no line.

Be aware that most spreadsheets will automatically add the line in what they sometimes refer to as X-Y plots, but it can be removed, too, either by the spreadsheet or by a graphics program. In the example here, even if a spreadsheet inserts the line, we should omit it. The scatter plot of Figure 3.5 might be acceptable.

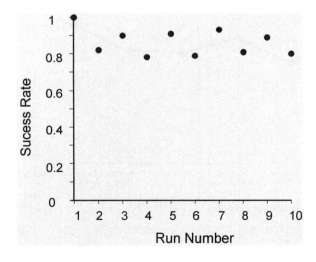

Figure 3.5

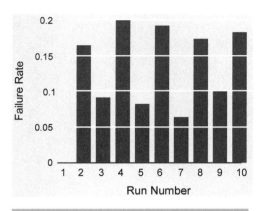

Figure 3.6

The little dots representing the data do look rather lost, however. An advantage of the bar graph representation is that what are essentially the same data are a lot more substantial in appearance! Is that alone sufficient justification? Surely not. One might say, however, that since each of the data represent multiple occurrences of the same process, we are entitled to use the bar chart approach, where each bar is independent of its neighbors, although in this example, each bar represents the same kind of thing.

Just as an aside, note that we could have chosen to present the same data in a way that makes the process look bad. Simply by showing the failure rate instead of the success rate, as in Figure 3.6, we exaggerate the failures and make the worst cases stand up, as they say, like tent poles. Now it is clear that there is a difference between the odd and even numbered runs. The failure rate is about twice as high on the even numbers. This version of the graph has made it very clear that there is likely some kind of systematic effect that is worth tracking down.

EXAMPLE 2A: DIFFERENT PARAMETERS

We have seen that when the data represent independent results, we should use a bar chart, or a scatter plot with no line. Here is another case. If the data represent different *things*, we should certainly use the bar chart and certainly *not* connect the dots. For example, suppose we had a table of data representing the electrical insulation capabilities of various materials: glass, rubber, wood, and so on. If the data are to be presented graphically, it should be obvious that there is no meaning to a line joining, say, glass and rubber. They do not even form a useful alloy! It therefore makes no sense to join the dots.

And yet how many times have we seen examples such as the creation shown in Figure 3.7? Here's an example adapted from a management tool used to help bosses be more effective. It shows how some employees viewed their boss (results shown are averaged).

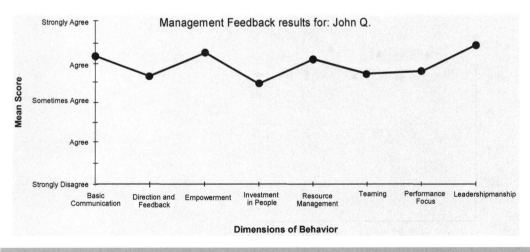

Figure 3.7

It is almost superfluous to examine the case for and the case against, but just for consistency let's do it.

Case Against Joining Dots	Case for Joining Dots
While the possibility exists to adopt a position between "Agree" and "Sometimes Agree," it is hard to see how there could be a position between "Investment in People" and "Resource Management."	It adds weight to the otherwise isolated dots.

Generalizing the "case against," we note that the layout of the horizontal axis is quite arbitrary, and that important fact makes nonsense of joining adjacent dots. We could have used a bar chart, or presented the results as a scatter plot with no line. Figures 3.8 and 3.9 show bar charts.

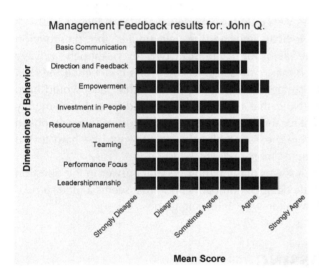

Figure 3.8

Figure 3.9

The horizontal axis in a situation like this is quite long: each "Dimension of Behavior" has to be separately identified. It is often more convenient to rotate the chart 90° clockwise to accommodate the labels. The result, as in Figure 3.8, is clearer and more compact.

Since the ordering of the vertical axis is just as arbitrary as it was when it was horizontal, we could reorder to show the individual's strong and weak points (which is presumably *why* we were making these measurements anyway). Figure 3.9 shows the result.

Showing a trend this way illustrates a reason to use a graphic instead of a table. In general, tables require fairly close scrutiny in order to extract information, and a complex table is often a deterrent for the reader. Graphics often give a sense of what the author intends with merely a glance.

Figure 3.10

EXAMPLE 2B: DIFFERENT PARAMETERS

Other *styles* of presentation are possible, too. Suppose we wanted to show how much money certain cities spent cleaning up chewing gum from the sidewalks every year. We could have made a graph as in Figure 3.10.

Here, the vertical line on the left unifies the presentation and establishes the zero for each city. But, even though a wavy line suggests itself, it makes no sense to add a line connecting the data points, because these are different things. A bar chart also would have been appropriate. (Note that the cities were listed in order of money spent. The argument for this is the same as for the cost-of-living graphs in the first chapter: were the listing alphabetical it would be hard to tell which city was spending more and which less.)

Thus, so far it seems that the most general answer to the question "Should I connect the dots?" is "probably not." That seems a little surprising, considering how often we see lines in graphs.

EXAMPLE 3: MISSING DATA

One possible kind of *correctness* of joining the dots occurs when a trend is being drawn and the sample rate changes, or when there are some data missing from a sequence. Here it is appropriate to join the next-nearest dots, as in this example of the use of a bar chart and a graph published in *The Economist*.[2] Note that the graph in Figure 3.11 differs from the "missing data" graph that we saw back in Figure 1.46 because in this example, there are dots that make it very clear that data are

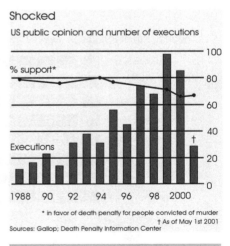

Figure 3.11

---

[2] *The Economist*, May 12, 2001, in the section on the United States, page 34.

missing. Figure 1.46 was a trend line, and the missing data should have been indicated by means of a scale break.

There are two things to note in Figure 3.11. First, the graph combines two essentially different sets of information: the number of executions in the United States by year, and the percentage of the population supporting capital punishment. It is, of course, mere coincidence that the numerical values of the percentage support for capital punishment and the number of executions in the United States have recently been between 50 and 100. Without this coincidence, the scales might have to be different, and the presentation would not have been so simple.

The chart correctly shows the execution data as a bar chart, and the support data as a line graph. The execution data are independent things, and therefore the dots are not joined. The percentage support data are the all same thing sampled in time. This chart should serve as a reminder (albeit a macabre one) that similar looking data sometimes require different treatment.

Second, because the Gallup organization does not give any data on the support for the death penalty between 1995 and 1999, the dots at the ends of the interval are joined. This may be taken as implying that for four years support for the death penalty was slowly and quite uniformly declining. Possible, but unlikely. No reasonable person would draw that inference.

So why is it OK to join the dots here? The purpose of the graph is to show the decline of support over the long run, rather than the details year by year. This purpose would not be served so well without the line. Had we wished to take additional data, we would have expected the results to be close to the straight line. The linear interpolation may not be particularly accurate, but at least there is no deception in showing one. One could have used a dashed line for this section of the graph, if it seemed necessary to further emphasize the lack of data for this period.

By the way, this graph provides us the opportunity to reverse the question: Is it OK to have the line and *not* show the dots? Here the result would look like the graph in Figure 3.12. Without the dots, it is no longer made clear that data were available for only half the years shown. That omission should be considered

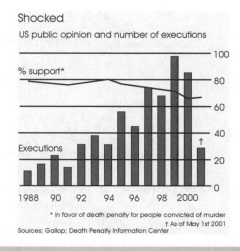

Shocked

US public opinion and number of executions

* in favor of death penalty for people convicted of murder
† As of May 1st 2001
Sources: Gallop; Death Penalty Information Center

Figure 3.12

misleading: therefore, the dots *are* needed. In general, it is best to show the dots if the series is missing some data points.

EXAMPLE 4: A TREND IN TIME

Now let us create a situation where a bar chart doesn't look right. Suppose we have a table of data representing the number of races won by athletes at a series of games. Suppose we awarded 3 points for first place, 2 points for second, and 1 for third place. We followed the performance of 3 athletes (the entire field consists of athletes A, B, and C) at a dozen meetings and obtained the following:

| | Meeting | | | | | | | | | | | |
Athlete	1	2	3	4	5	6	7	8	9	10	11	12
A	3	3	3	2	3	2	3	3	2	2	2	2
B	2	2	1	3	2	3	2	2	1	3	3	3
C	1	1	2	1	1	1	1	1	3	1	1	1

We would like to present the results as a graph. What to choose? We could present the data as a bar chart, with a key identifying the athletes, as in Figure 3.13.

Figure 3.13 is not very informative. It is very hard to read (though it might be better in color). What if we try a line-based graph, as in Figure 3.14?

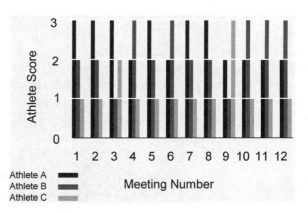

Figure 3.13

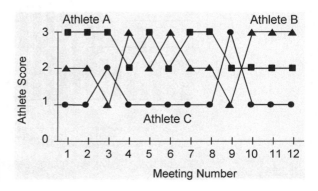

Figure 3.14

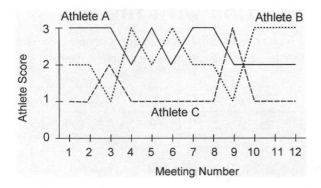

Figure 3.15

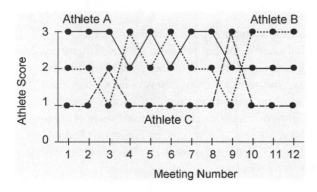

Figure 3.16

The dots represent the individual results. They would not copy well, and frankly they do make the graph look a bit cluttered. Let's leave the dots off, and use different lines for different athletes, as in Figure 3.15.

Now, the graph of Figure 3.15 is misleading because the impression is given that there was a continuous measurement process going on, rather than a sampled measurement. Really, the dots are needed, though they could all be the same shape (as in Fig. 3.16) if the line style were used to identify the athlete. Perhaps that would be a fair way to look at the data.

Case Against Joining Dots	Case for Joining Dots
The numbers are effectively independent measured results.	Each line represents the performance of an individual, and each line interpolates other measurements whose results can be inferred. One might argue that if one had sampled the performance between the meetings, one would have observed approximately the value shown by the straight-line interpolations.

Based on this graph, the reader might infer that athlete A started off as the best of the three, and athlete B emerges as best at the end. This notion is not evident in the bar chart version. There is still an implication that the even spacing of the dots (shown as "Meeting number") represents time and that the meetings were periodic. This kind of presentation is therefore not fair unless the spacing of the dots along the horizontal axis represents the spacing of the meetings with reasonable accuracy.

EXAMPLE 5: INTERACTION WITH THE AXIS

Now suppose we change the problem and look at the number of people who won Nobel prizes in the years 1989 to 2000. We can produce a table as before.

Subject	Year											
	1989	1990	1991	1992	1993	1994	1995	1996	1997	1998	1999	2000
Physics	3	3	1	1	2	2	2	3	3	3	2	3
Chemistry	2	1	1	1	2	1	3	3	3	2	1	3
Economics	1	3	1	1	2	3	1	2	2	1	1	2

Therefore, we use a bar chart showing the results, as in Figure 3.17.

As it happens, the bar chart of Nobel prizes in Physics, Chemistry, and Economics is similar in appearance to the one for our imaginary athletes. We chose a period of 12 years. The number of prizes was never less than one during the years shown, and never more than three. Of course, in some years three prizes were awarded in two categories, a feature that has no counterpart in our athletic example. Nevertheless, the results in Figure 3.17 do bear a strong resemblance to the runners graph in Figure 3.13. The bar chart with key is no more legible or attractive than before.

As before, we could improve the presentation by making a graph instead of a bar chart, as in Figure 3.18. But the question remains, does it make any sense to join the dots?

Case Against Joining Dots	Case For Joining Dots
There is no meaning to a point half-way between one data point and another. The numbers are independent of one another. (It is reasonably certain that none of the individuals who shared the Physics Prize in one year was among the winners in the following year. In this, the data represent a different *kind* of thing from the case of the athletes.)	For similarity to other graphs, or to make a difference from other bar charts.

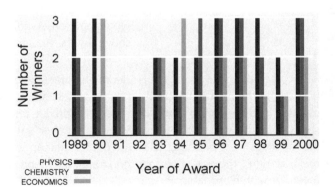

Figure 3.17

PHYSICS
CHEMISTRY
ECONOMICS

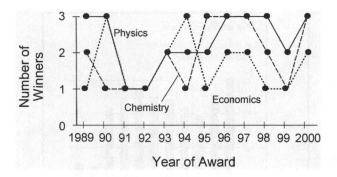

Figure 3.18

There is another issue here, too. We might say we are plotting the number of prizes as a function of time, but in reality, the number is not a nice *function* of time. The prizes are awarded once in the year, and the number of awards at all other times is zero. A sample for a given year would be zero until the awards were announced (in October) and nonzero afterwards.

In other words, there is an *interaction* between the data and the horizontal axis. In the case of the Nobel prizes, we are plotting prizes won *per year* against the *year*. If we plotted prizes won *per month* against the *month*, we would, as noted, get a lot of zeros. The *interaction* between the data and the horizontal axis can be viewed as the cause of the change needed to the vertical axis.

In fact, a plot of the number of Nobel prizes awarded per year constitutes a histogram. We will explore this more with the next examples.

But bear in mind that it is perfectly possible to have a set of numbers that can be presented clearly and unambiguously as a line graph, and yet careful consideration says that the line is inappropriate. The data "look" as if a graph would be appropriate, but appearances are deceiving. Although there is nothing special about the fact that these are Nobel prize data, we have given the effect the nickname "the Nobel prize problem," in the hopes of making it memorable—like winning the Nobel should be a problem!

EXAMPLE 6: HISTOGRAMS IN ALL BUT NAME

In the graph earlier showing the number of executions taking place, the case was made that the executions were independent things. The *interaction* argument could also have been made. The data could have been plotted on a monthly basis, with smaller numbers. The interaction case can be made for using a bar chart in the pair of examples in Figure 3.19, adapted from *The Economist* of February 17–23, 2007.

In the left graph of Figure 3.19, bars are used for the capital expenditure, an annual number. The years for which the numbers apply are shown underneath the bars. If the data were plotted monthly, the amount would have to be reduced by an average factor of 12. There is an interaction between the axis and the data.

Similar arguments apply in the example of the right graph. The percentage of the population who gave to charity is a number that depends on the period to which the data applied. In any given week, the number would be small, presumably a little over 1%. Although the label does not say so explicitly, the annual number is what is shown. Since the number interacts with the axis, a line presentation is inappropriate.

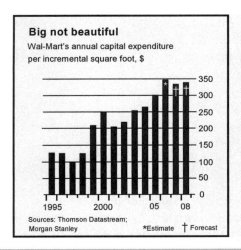

Big not beautiful

Wal-Mart's annual capital expenditure
per incremental square foot, $

Sources: Thomson Datastream;
Morgan Stanley   *Estimate † Forecast

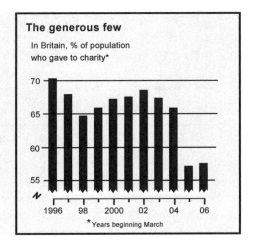

The generous few

In Britain, % of population
who gave to charity*

*Years beginning March

Figure 3.19

This interaction is what you would expect from a histogram, and in fact, both of these bar graphs are really histograms, although not quite of the usual type. In an ordinary histogram, used to show the distribution of a variable, the horizontal axis covers the *range* of the quantity being studied and is divided into intervals usually called *bins*. The height of a rectangular column represents the frequency with which the parameter value was within a particular bin. It is customary to have the rectangular bars touching, as the bin boundaries are adjacent to one another.

What is different here is that the bars do not touch. Is that the only difference? Yes. The graphs are, in other respects, histograms with calendar years as the bins. The height of the column represents the *count* of the parameter of interest (executions, number of $ per square foot, % of population) within the particular bin. Surely, each of these parameters represents the same notion as the *frequency* with which a parameter value was within a particular bin.

What we have here is a variation of a histogram, rather than a bar chart. It looks like a bar chart because the bars do not touch, but it qualifies as a histogram because it is displaying a frequency of occurrence within a bin. *The test is the interaction*: if you change the bin width, you change the frequency of occurrence. It would be *correct* (although not customary) to draw such a graph with the bars touching, since (as in the ordinary histogram) the upper boundary of one bin is the lower boundary of the next. Furthermore, it would be *incorrect* to draw a line joining the center of their top lines, as in Figure 3.20.

The dashed line in Figure 3.20 is quite wrong, because of its implied interpolation. This is exactly the kind of thing that Barker and Wheeler had in mind when they instructed their electronics students not to join successive data points with straight-line segments.

A value for the number of executions in each year can only be obtained at the end of the year. At all other times, the number must be smaller, as in the 2001 result for the first four months. If we multiply the number (which seems to be 28) by three (assuming

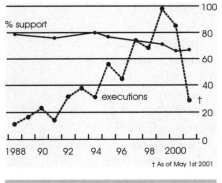

Figure 3.20

a constant monthly rate) we get 84, which seems to be the same as for 2000. The apparent dip at the end of the graph of Figure 3.20 is purely a result of the bin not containing a year's worth of data.

EXAMPLE 7: INCREASING RESOLUTION

Normally, if you increase the rate at which you sample a function, you get the possibility of a better frequency response. In the case of the execution numbers or the Nobel prizes, you clearly do *not* get a better resolution plot, and so you must use a bar-like presentation. If increasing the sample frequency does not change the data significantly, it may be permissible to join the dots.

Of course, this only helps if you know what "significantly" means.

If we plot the frequency response of a loudspeaker by testing with sine waves and taking data at some selected frequencies, we might get a curve showing a flat response from 100 Hz to 1 kHz, and a roll-off in sound output outside this range. Given the data shown, presumably obtained by sine-wave injection at 100 Hz, 1 kHz, and 10 kHz, we could be forgiven for drawing the line shown in Figure 3.21.

In Figure 3.21, we have used a Bezier curve as a modern version of a French curve to obtain a smooth line. (Bezier was French, as it happens.) It is well known that the frequency response will turn down at the high- and low-frequency ends. A few more data points would remove some of the smoothness, as in Figure 3.22.

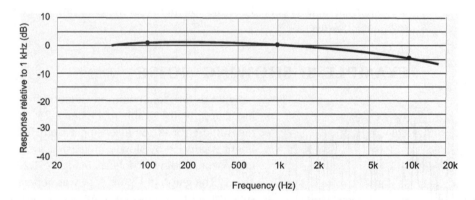

Figure 3.21

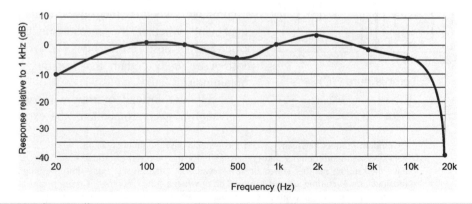

Figure 3.22

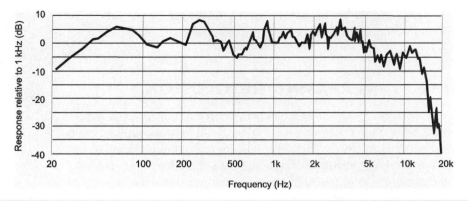

Figure 3.23

A much more thorough examination, using modern test equipment, would be quite revealing, as in Figure 3.23. Even though the loudspeaker no longer seems to have a smooth response, the actual sound level in the region of interest does not change much depending on how many data points are taken. The response may be a few decibels (dB) or so higher or lower than the value at 1 kHz, but such changes are not truly significant.

Thus, for a graph in which interpolation is fair, and would (for example) lead to increased resolution in the time or frequency domain, it may be acceptable to join the dots. A curved line may be more realistic than simple straight-line segments, however.

EXAMPLE 8: SHOWING NOISE

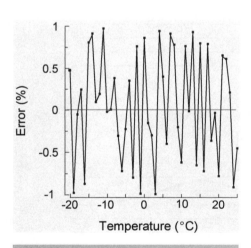

Figure 3.24

Now let us turn to a common situation in engineering. Suppose the graph in Figure 3.24 shows the measured error in some sensing system. The experiment was done by slowly increasing the temperature. We will join the dots, as in Figure 3.24.

The graph in Figure 3.24 was generated by a spreadsheet from random numbers.[3] However, had the data been real, the experimenter doing the work would not know, before the temperature test is complete, whether there was a temperature effect, and whether the measurements contain noise. Seeing these results, however, would make most of us pretty certain that the measurements contain noise.

Suppose instead of simply recording one observation at each temperature, the experimenter

[3]A random number between −1 and 1 is generated by the function @RAND * 2 − 1, since @RAND generates a random number between 0 and 1. For the purposes of ordinary use, it is convenient to have the random number generator in a spreadsheet "start afresh" every time the program does a recalculation, including when it is turned on or when a cell is modified. For our purposes, however, this starting afresh had the effect of changing the set of random numbers every time we accessed the spreadsheet, and became inconvenient in the extreme. We therefore "froze" the set of numbers by selecting the cells to freeze and doing a copy / paste special / values.

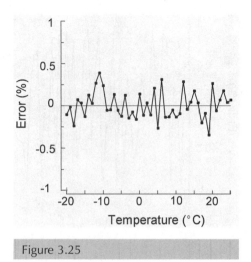

Figure 3.25

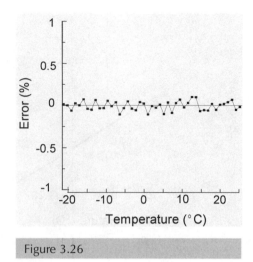

Figure 3.26

had recorded 10 and taken the average, plotting the result as in Figure 3.25. Clearly, there is a lot less noise in the data set represented by Figure 3.25.

Suppose 100 observations were averaged, as in Figure 3.26. Now the noise level is even lower. It is well known that the average of random noise is zero, so that if an infinite number of samples were taken, the effect of the noise would disappear completely. At some point between a small number of samples and an infinitely large number, does a straight line joining the dots represent the trend of the data, rather than the noise? We don't think so.

The average of the noise only tends to zero if the observations are *independent* and *uncorrelated*. Generally, whether real observations are in fact independent depends on how closely spaced (in time) they are. It is possible to design the experiment of measuring the temperature dependence to ensure that the measurements are independent.[4] Automatic data acquisition and processing are practically a prerequisite.

But if the observations *are* independent, does it really makes sense to connect the dots of the scatter plot with a line? Because the random noise in the situation described in the first temperature-effect graph is so large, the line does not help illustrate the trend. In some way, however, it does help underscore the randomness of the results. Figure 3.27 is the same graph as Figure 3.24, but without the line.

The problem is that the eye tends to see data trends that are not there. Probably the same mental processes are at work that enable us to see bears and hunters and tea trays in the night sky. For example, across the middle of the graph is what looks like a sequence of observations in a straight line. They are highlighted by adding a line, as in Figure 3.28.

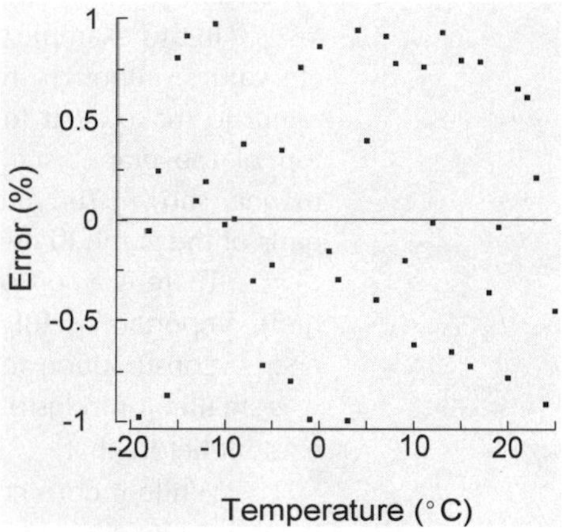

Figure 3.27

4 The ideas come from the world of time and frequency metrology, and use a parameter called the Allan variance. A fairly recent paper on the topic is by Thomas J. Witt, Using the Allan variance and power spectral density to characterize DC nanovoltmeters *IEEE Transactions on Instrumentation and Measurement* **50**(2), 445–448 (April 2, 2001).

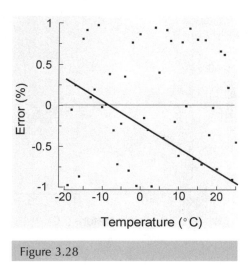

Figure 3.28

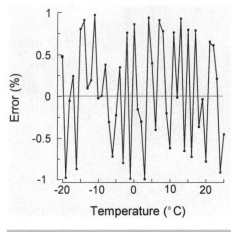

Figure 3.29

Is this evidence of a noisy situation with an underlying trend? In fact, relatively few of the points near the line in Figure 3.28 are adjacent in temperature, a fact that is clear with adjacent dots joined by a line. Figure 3.29 is a repeat of Figure 3.24, for easy comparison.

We must conclude, then, that the connecting line *should* be drawn, in this case, not to show the trend but to show the random nature of the noise.

EXAMPLE 9: LEADER LINES

Another example of lines that appear to interpolate between data points is shown in Figure 3.30, in a graph adapted from one first presented in *Scientific American.*[5]

In this example, the dotted lines are not really there to suggest intermediate values, although intermediate values presumably exist. They are intended rather to make clear to the reader that the various categories identified at the top of the graph, such as Home and Commerce, apply throughout in the manner shown. The lines serve the function of *leaders*, connecting the different parts of the graph to the labels at the top.

There is a sort of logarithmic time axis in the vertical direction, but the more important point, emphasized by the lines, is that (for example), the energy consumption for Transportation, presently a large fraction of the total, was smaller for Industrial Man (about a hundred years ago), and practically nonexistent before that.

While it does contain an inordinate amount of white space, the presentation is quite effective.

The question of joining the dots has no simple answer. There are cases, it seems, where it makes sense to join the dots only to avoid creating an erroneous impression, and others where joining must be avoided for exactly the same reason.

[5]Earl Cook, The flow of energy in an industrial society, *Scientific American* **224**, 134–144 (September 1971), reprinted in *Energy and Man: Technical and Social Aspects of Energy*, IEEE Press, New York, 1975.

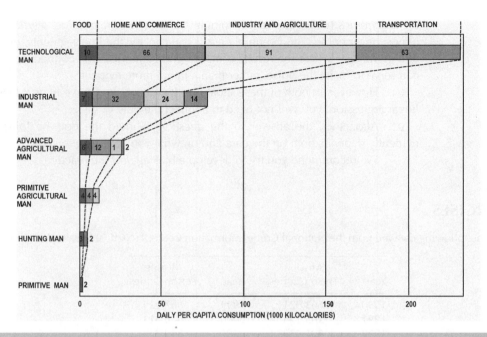

	FOOD	HOME AND COMMERCE	INDUSTRY AND AGRICULTURE	TRANSPORTATION
TECHNOLOGICAL MAN	10	66	91	63
INDUSTRIAL MAN	7	32	24	14
ADVANCED AGRICULTURAL MAN	6	12	1	2
PRIMITIVE AGRICULTURAL MAN	4	4	4	
HUNTING MAN	3	2		
PRIMITIVE MAN	2			

DAILY PER CAPITA CONSUMPTION (1000 KILOCALORIES)

Figure 3.30

The only advice must be this: think carefully before joining those dots. As Omar Khayyám[6] wrote, about 900 years ago:

> The moving finger writes; and, having writ,
> Moves on: nor all thy piety nor wit
> Shall lure it back to cancel half a line,
> Nor all thy tears wash out a word of it.

That line, once published, cannot be unpublished!

SUMMARY

The question of connecting the dots is far from simple to answer.

There are some situations where connecting the dots is definitely *wrong*. You need to know how to avoid this mistake. *In general, do not join the dots.* Find another way to show the data.

In particular, do not connect the dots with a line if the data (a) represent different *kinds* of things or (b) *interact* with the horizontal axis. This last is an example of the *Nobel prize* problem, and a bar-like histogram is appropriate.

There are some situations where connecting the dots is definitely *needed*. If you are trying to show a trend, but the data are rather sparse, or irregular, show and join the dots, provided the dots (a) represent the same *kinds* of thing (e.g., samples) or (b) are at least a little predictive or capable of extrapolation.

These are examples of the *support for the death penalty* situation.

There are also some situations where you have a free hand and may do as you please. If the data are a time series, where the same parameter is sampled regularly, the connecting lines may highlight the data. However, the old *Mathematics*

---

[6]The *Rubáiyát* (poetry) of Omar Khayyám (1048–1131) (Fitzgerald translation, 1859).

for Electronics textbook is right: joining the dots by straight-line segments is almost never correct, and certainly not if you can *see* the line segments, except that in the case where the results contain significant noise, the connecting lines can serve to make the noise (rather than a trend) more evident.

However, in both of these situations, if you can do a curve fit or at least a linear regression, you will not need to connect the dots.

Altogether, the answer to the question "Should the dots be joined?" evidently depends both on the data and on what you are trying to show.

We recommend you try to develop a bias *against* the practice.

EXERCISES

3.1. The following data are from the National Crime Information Center's (NCIC's) Missing Person File:

Year	Missing Person Entries	Year	Missing Person Entries
1990	663,921	1999	867,129
1991	705,175	2000	876,213
1992	801,358	2001	840,279
1993	868,345	2002	821,975
1994	954,896	2003	824,699
1995	969,264	2004	830,325
1996	955,252	2005	834,536
1997	980,712	2006	836,131
1998	932,190		

Show the data as a graph to be presented at a conference.

3.2. You wish to present data graphically. Should you join the dots in the following:

(a) A plot of impedance magnitude as a function of frequency?

(b) Percentage of U.S. teenagers using alcohol 1990–2006?

(c) Per capita energy consumption for a list of countries?

(d) Percentage change in annual average of house prices? Percentage change month by month or week by week compared to previous year?

(e) Annual trends in the population of birds in a particular study region?

(f) Number of extrasolar planets discovered each year from 1980 to the present?

3.3. The following data represent the calibration of a voltage measuring system, where the input quantity is accurately known compared to the device being calibrated. Plot the data. Would you regard the instrument as accurate? Do you think there are any problems?

Input	Reading	Input	Reading
0.1	0.100986	5	5.13043
0.2	0.201897	7	7.005818
0.3	0.307644	10	10.23438
0.5	0.520924	20	20.91677
0.7	0.7302	30	31.17337
1	1.014027	50	51.6171
2	2.03374	70	72.8402
3	3.117381	100	104.0637

The nondata parts of the graph

4

One of us (HK) has a memory from childhood of a book of beautiful letters. He does not now remember what it was, nor why it was printed, and there is no one left to ask. But the memory of page after page of letters of different design is with him still. Even as a child, he found the designs' balance and poise fascinating, their individuality striking, and their ability to have emotional context quite amazing. Who could deny the appeal of these

austere, yet classically beautiful letters?

Who would not be transported back to an era of art deco classiness by

these upright, yet not-quite-formal letters?

And who would not take seriously a

DANGER – KEEP OUT

warning that looked like this? You have been reading too many spy novels if you would not believe a carton that said this on the side

and you've been watching too many reruns if you imagine Pa Cartwright getting Hop Sing to burn this in the side of the barn:

PONDEROSA

Well, that was fun! Having got that out of our systems, let's move on. That childhood book must have been a challenge to print, for certainly it was published before the invention of the computer, and of software-defined fonts. Each letter that came from a different font had to be cast in metal. The different bits of metal had somehow to be fitted together to make lines and pages. It's easy on a computer. Very tough with hardware.

Used properly, the words tell you what a graphic is all about. They may help you identify some particular part of your graphic, or they may explain some difficult part of your presentation. The *font* is the set of letters of a particular size and design in which your words will be printed. There are thousands of designs (a handful of which appeared above) and dozens of sizes.

Choosing the letter design (called a typeface and often informally referred to as a font) and size is but one part of making a technical graphic, albeit an important one. This chapter is all about helping you to deal with the words on your graphics.

The words on a graphic may be

Axis labels

Captions

Callouts

However, before we start discussing the details of these parts of the graphic, we will offer some guidelines on the general aspects of words in technical graphics.

First, Choose the Typeface for the Graph. It is quite likely that the typeface used in your graph will be different from the body text of your document, and it makes an attractive presentation style to deliberately choose complementary fonts. Typefaces, like people, come in two basic types, with just a few individuals that are hard to classify because they seem to have some mixed-up characteristics.

Typefaces of the Roman kind, supposedly similar to those used by the ancient Romans, are characterized by having both thick and thin vertical lines. The manuscript for this book was set on our computers in a typeface called Times Roman, so named because it was developed (long ago) for a newspaper called *The Times* in London. Times Roman was designed so that it could be used for headlines as well as news articles. Here is a well-known name written large:

Harold

See the difference in weight (thickness) between the two vertical strokes and the horizontal one of the first letter? See how the curved lines vary in thickness?

Roman typefaces also (most often) have serifs, the little wedges at the end of the strokes that the Roman masons added to define the end of the line neatly when it was chiseled into stone. The Romans did not have moveable type, remember!

The "opposite" of Roman fonts are called sans serif fonts.[1] They tend to have even-weight lines throughout. Here, for example, is the same name in a font called Arial:

Harold

A pair of complementary fonts might be one from the Roman collection, and one from the sans serif. Typically, body copy (the bulk of the material) is set in a Roman type, and display text (section headings, figure captions, and the like) are in sans serif. There are thousands of fonts to choose from (and most are available free for desktop computers from Web-based sources) but most of the classic ones fall into one or other of these types, and most look well together.

Exceptions do exist. A group of fonts that you might think of as "modern" falls between these two categories. For example, a font called Ottawa has some variation in line weight, but no serifs:

Harold

EnviroD and Technical give the impression of having been written with a stencil and a round-tipped draughtsman's pen. You can use these to good effect on drawings for house plans:

Harold

For most technical documents, however, we will be satisfied with Roman and sans serif, with Roman in the body of the material, and a sans serif font in the headers, the graphics, and the captions.

An exception occurs in the case of equations. In books and papers produced using a program called TEX, a product (and trademark) of the American Mathematical Society, equations are set in a font called Computer Modern Math Italic, or CMMI. It is very reminiscent of the math books (and the exam papers) of my childhood. It is characterized by several of the letters, but in particular the letters (here in Times Roman italic) i, j, and x. In CMMI they look a lot like this: i, j, x. (Those letters are actually from a font called ItalicT that you might be able to find in a form downloadable for your computer.) TEX produces equations the way they *should* look. In particular, note that functions are shown in a Roman font, and variables in italic.

If you have equations in a graph (rare, but possible), or if you label some part of a graphic with variable names from an equation in the paper, you should endeavor to reproduce the equation font as closely as possible.

Adjust the Font Size in Graphs. Font size is measured in units called points. The *point* was originally one 72nd of a French inch, and the system was developed in an early attempt to standardize type sizes. The *point* in question was almost certainly the size of the period, or full-stop, in some typeface in

---

[1]Of course, the opposite of sans serif should really be serif, but for some reason that designation is not used.

pre-metric France. In the United States it was originally about 12/13 of that size. It is now generally taken to be 0.0138 inch. An uppercase 12-point font thus measures (from the top of the ascender to the bottom of the descender) about 0.0166 inch. This is the vertical space needed to set a line of general text, with uppercase and lowercase, and letters with ascenders and descenders. The blank space between lines is called *leading* (from the metal originally used to cast the type letters). A 12-point font would typically have 2 points of blank space between the lines: this would be described as being "set 12 on 14," and written "12/14." Smaller type may have less leading. Most desktop computers can create letters in font sizes from around 6 point (which is illegible to someone with poor eyes unless they have a magnifying glass), up to 72 point.

Small type (5 or 6 point) may be used for telephone books and classified ads. The body type in papers, books, and reports is usually between 8 and 14 point. Display type may go from 16 point to 72 point. Anything larger is called poster type. Presumably there is no upper limit.

Just as line thickness is important to a sense of balance and style in technical graphics, font size and weight are also important. For example, very often you can make the font a size or two smaller in a graph than in the body text. For example, if you are writing a two-column paper and supposed to be using 10-point Times Roman for the words, you could try 9-point (or maybe even 8-point) Arial or Univers for the figure. Even though you set the size a point or two smaller than the Times text, the words will look bigger because of the open design of sans serif fonts. In fact, if you are using a sans serif font in the graphic and you don't make it smaller than the body text, it will look clumsy.

Text in 10-point Times Roman	Text in 9-point Arial
By the end of 1346, therefore, it was widely known, at least in the major Euro ean seaports, that a plague of unparalleled fury was raging in the East. Fearful rumours were heard of the disease's progress: "India was depopulated, Tartary, Mesopotamia, Syria, Armenia were covered with dead bodies; the Kurds fled in vain to the mountains. In Caramania and Caesaria none were left alive . . . ." But still it does not seem to have occurred to anyone that the plague might one day strike at Europe.	By the end of 1346, therefore, it was widely known, at least in the major European seaports, that a plague of unparalleled fury was raging in the East. Fearful rumours were heard of the disease's progress: "India was depopulated, Tartary, Mesopotamia, Syria, Armenia were covered with dead bodies; the Kurds fled in vain to the mountains. In Caramania and Caesaria none were left alive . . . ." But still it does not seem to have occurred to anyone that the plague might one day strike at Europe.

You can see in the extracts above (from *The Black Death*, by Philip Ziegler, Penguin Books, 1982) that the sans serif font looks bigger than the Roman, although it is actually 1 point smaller. The illusion occurs because the ascenders and descenders are *relatively* smaller, so the type can be set tighter vertically.

The difference in the construction of the letters in this example should ring a warning bell: don't mix these typefaces arbitrarily! There is a book in my possession that is set mainly in a sans serif font, but when it mentions the name of a ship, it adopts the convention of using an italic font. That is quite common, but what this particular book does that is not common is *it switches to a Roman typeface as well!* The result is that when it names a ship, it looks odd:

For example, to look at these words you might get the impression that instead of being a very large ship, the *Titanic* was actually some kind of midget ship that sank when it hit an ice cube!

Honestly, all the type in the previous paragraph has the same point size! We selected two that are very different: the sans serif font was Albertus Medium, and the Roman was Baskerville Old Face.

One of the difficulties that the drag-and-drop era of software has bequeathed us is the nonfixed nature of the size of embedded graphics. It used to be, back before drag and drop, that you could get a graphics package to produce a graph that was, for example, 3.3 inches wide and however tall you wanted, so that you could make a figure that was exactly the right size for a two-column format. Then, if you selected a 9-point font in your graphics package, that's what you got in your paper.

Nowadays, drag-and-drop technology has made it easy to change the size of the figure after you have inserted it in the paper. You can enlarge or shrink the whole thing, or you can squeeze it a little in just one direction. In fact, things are so flexible now they are *limp*! (Are you listening, software designers? No, of course not!) With most (maybe all) graphics packages, you can use a built-in ruler to size the image. But when you insert it into the document, the size changes. And there seems to be no default way to ensure that an image is the same size in your final product as it was when you drew it. Figure 4.1 is an example. You have to instruct your software specifically what you want.

Figure 4.1 was drawn twice the size it is shown here. At the instant the figure was inserted into the computer file for the manuscript, its size was set by the software to a width of maybe an inch. So then the image had to be stretched until the appropriate final size was achieved.

Note that in this case, both a Roman and a sans serif font are used. The Roman font is used here because the quantities shown here are variables in equations in the paper from which this figure was borrowed. The paper in question used Times Roman italic (instead of CMMI) for the variables.

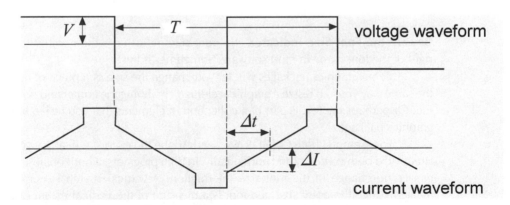

Figure 4.1

If we needed to reduce the size of the figure, we could just shrink it, as in Figure 4.2, but the type is now really too small to be practical. The proper solution is to rescale the words, even if that means redrawing the figure. The result is shown in Figure 4.3. Here the text size adjustment has meant that the arrows have had to be redrawn and relocated, and the two parts of the graphics had to be separated vertically by more space relative to the original.

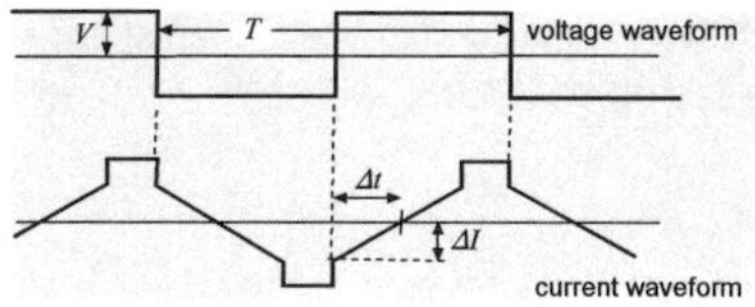

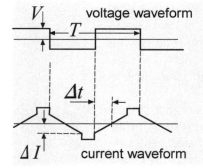

Figure 4.2

Figure 4.3

While there is almost certainly a way to fix the size of the graphic in your word processing program, or there are rulers as there were in your graphics software, it is tedious to have to adjust a figure to be the correct size. Again. But however tedious it is, you have to do it! Very often we review papers in which the graphics have been resized to fit a page limit, or a two-column format, simply by scaling the whole image. The result is often nonsense, because the text is illegible. And the paper is rejected, by us or by someone else!

Resizing graphics may not always involve such major rework as the previous example, but you should always check to make sure that, after the fonts have been reset to an appropriate size, the result is still clear and uncluttered and legible.

We suggest the following font sizes in your graphics:

	Two Column	Single Column	Viewgraph
Axis labels	8–10 pt	10–12 pt	14–18 pt
Tick mark labels	7–9 pt	9–11 pt	12–14 pt
Graph labels	7–10 pt	9–11 pt	12–14 pt
Captions	9–10 pt	10–12 pt	16–18 pt

Our preference is for a sans serif font in all graphics. Some publications have font requirements, however, and these may have to be adhered to. On the other hand, many publications will tolerate small changes from the "official" style, so it's often worth a try to submit something that looks good, even if it doesn't comply.

Beware the Font Size Indicator. The font size indicator is the number to the right of the font name in most software. Sometimes it lies!

Some graphics packages will let you change the size of a piece of text in the same way you can resize a graphical element: grabbing the corner and stretching. Or perhaps the text is part of a collection of elements that you resize in the graphics package.

For example, both CorelDRAW® and Adobe Illustrator® will allow you to stretch text both vertically and horizontally. In both programs, a horizontal stretch causes no change in the font size information. A vertical stretch is correctly tracked. Since the point size of a font is a measure of the vertical extent of the letters, this seems appropriate. Neither Visio® nor PowerPoint® will allow you to treat text like a graphic element. In both those programs, only the box containing the text changes size, the text itself does not. The indicated size therefore correctly does not change.

If you are using a graphics program other than these, it is worthwhile to check what the font size indicator in your graphics package does when you resize the text this way. It may give a misleading number.

For example, in Corel Presentations™, the font indicator does not appear unless the selected element is text. When text is selected, the font name and size appear in a pair of windows in the top left of the screen. The size indicator seems to show the size of the font when the text was created. In the previous example, the italic "V" identifying the amplitude of the voltage waveform, is shown as 72 point. The span from the bottom of the descender to top of the ascender should be exactly an inch high if this were true. Figure 4.4 is a copy of the screen obtained when the letter is selected. The ruler at the left shows that the height (at least of the uppercase "V") is in fact more like $\frac{1}{4}$ inch.

Strictly speaking, you should not resize letters simply by grabbing them and letting the software scale them. When a letter is printed smaller, it is common practice to have the strokes of the letter print slightly lighter (thinner) than simple scaling would dictate. However, it seems that software that lets you treat text like graphics just takes care of that little detail on your behalf. Unless

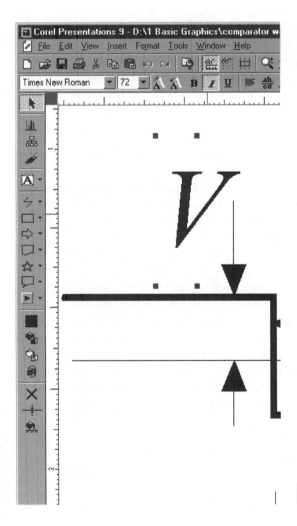

Figure 4.4

you are rescaling the text by some very large numbers, and unless you happen to notice the letters start to look odd, don't worry about it.

On the other hand, do be sure that the text in the final size graphic is legible without the use of a microscope, and of a size listed in the table.

Avoid the Temptation to Write Everything Uppercase. Books on typography and books on style will tell you that the reader will have a harder time with uppercase lettering than with lowercase. Apparently this is because the uppercase letters are all the same height, and words have no "shape" to them. Good readers recognize the shape of a word, without seeing the letters.

Well, your readers are bound to be an intelligent lot, and they will not have any trouble comprehending the words, whatever case you set them in. However, you should bear in mind that uppercase letters have a tendency to draw attention to themselves. That's why they are used to say important things, like "DEWEY WINS!" or "BIG REBATE ON OLDSMOBILES!" Since we are discussing graphics, we can presume that the message is in the graphic, and not in the words. Therefore, lowercase, perhaps with initial caps, will do nicely for all parts of the graph.

Copy and Edit Words to Save Time. Once you have selected the typeface you want, the size you want, and the appearance (bold, italic, etc.), it is almost always quicker to copy and edit words than it is to start from scratch. If there are two axis labels and five numbers on each, this process will save dozens of keystrokes and mouse-clicks compared to formatting each block of type from scratch.

AXES

When it comes to axes, or at least the lettering for them, the rule is pretty simple. The reader should be in no doubt about what is intended. Thus, each axis of a graph should be labeled, and the label should specify both the quantity being plotted, and the units in which it is presented, for example, "armature current (amps)." It is also a good idea to include the symbol used for the variable, if that is appropriate: "armature current, I_a (amps)."

When it comes to the horizontal axis, things are usually straightforward, but vertical axes offer some choices. Thus, for the horizontal axes, something like Figure 4.5 will do. You may decide not to capitalize the initial letters, that is really a matter of style. But things are easy because the words are all horizontal, and written left-to-right.

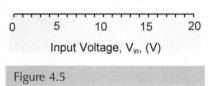

Figure 4.5

With most software packages, you will have to fiddle[2] with the numbers to get them to line up properly under the correct tick mark, because the horizontal size and location of the actual numbers seems to be a mystery to the software. At least you can use the software to line them up vertically. Figure 4.6 shows how the axis labels should look.

For the vertical axis, you can space things evenly in most graphics packages, and since the vertical extent of the numbers is simply the point size, the software gets it right. On the other hand, you may have to fiddle with the

[2]See Chapter 5 for how to deal with "snap to grid" problems.

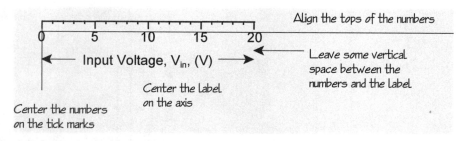

Figure 4.6

numbers to get them lined up along the right edge, which is usually what you want to do.

For the vertical axis, horizontal space may be at a premium, and even a rudimentary consideration of style would say the layout of Figure 4.7 is silly.

The layout of Figure 4.8 is a little better and, for labels with only short words, may be adequate.

For longer labels, the words really do have to be rotated through a right angle. The custom is to rotate counterclockwise, so that moving your head over to the right and twisting it to the left would let you read the words.

Simple as that seems, do not rearrange the text as in Figure 4.9 or, to avoid that approach, as in Figure 4.10! This last one is called *hotel style*

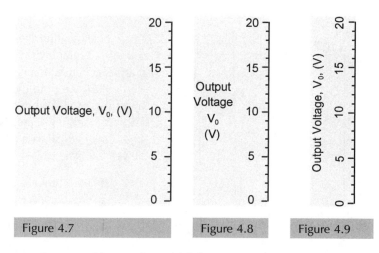

Figure 4.7   Figure 4.8   Figure 4.9

because it is reminiscent of neon signs outside certain establishments in the sleazier parts of town.

The vertical axis label should be rotated, the numbers should not, with the result looking as shown in Figure 4.11.

Figure 4.12 summarizes our advice.

Now, there may be times when the software that you are using to plot the graph will want to do something bizarre all on its own. Do not let it get away with any nonsense. The subject is covered in Chapters 7–10, where we discuss the use of spreadsheets.

Use conventional Scales and Avoid Multipliers. Technical people are so accustomed to graphs that their minds make assumptions about them, sometimes without basis. A small deviation from what is expected, sometimes even a large one, may go unnoticed. Getting it right is always worth the effort, even though it may involve some work.

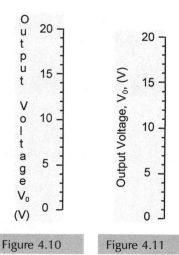

Figure 4.10   Figure 4.11

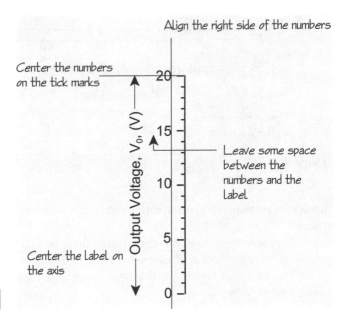

Figure 4.12

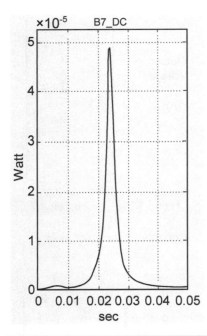

Figure 4.13

Let us illustrate the point with a couple of examples. A graph very much like Figure 4.13 appeared in an *IEEE Transactions*.

First, we note the unidentified characters at the top of the graph. Whatever it was for, the label should not have been left there. No matter—our concern here is with the vertical axis label. We have, associated with the vertical axis,

- The numbers 0 through 5
- The word "Watt"
- The multiplier 10^{-5}

However, we do not see what the quantity is that the axis covers,[3] and while the multiplier is not unclear in its meaning, it is expressed in a nonstandard way and located in an odd place.

Actually, multipliers *are* often expressed in an unclear way. The label might say, for example, "Output Power $\times 10^3$ W" or "Output Power W $\times 10^3$" or some similar and unconventional wording. The question always comes down to this: Do we multiply the quantity by 10^3 to get a watt, or are the units thousands of W? There is a difference of 10^6 between the answers—large, but not always large enough that the answer is clear because of context alone.

There is no excuse for creating this sort of confusion with your graph label. There exists a set of standard multipliers, in factors of 3 orders of magnitude, so

[3] A careful study of the paper reveals that the letters and numbers probably identify a signal measured at a location shown in another figure. The paper in question does not make this clear. This measured signal is probably the quantity being plotted, but the text does not explain that, either.

that the axis numbers can be kept reasonably sized and the quantity expressed unambiguously. In Figure 4.13, the peak value of the curve is about 4.9 in terms of the axis numbers, and $\times 10^{-5}$ in terms of the multiplier. In other words, the peak corresponds to about 4.9×10^{-5} W. Using the conventional multipliers, this is 49×10^{-6} W, or 49 μW.

Figure 4.14 shows what the graph should have looked like. (We also took the time to fix the size of the rectangle around the graph, so its upper bound was on an integer line.)

For our second example of axis multiplier problems, we will look again at the Horowitz and Hill graph introduced in the first chapter. Look at Figure 4.15.

Now, the reader sees the ordinate is labeled "Resistance" and sees that the units are given as μA/V. A reasonable assumption, given that μ in the label, is to think "Oh, they must mean megohms," and to spend any remaining thinking time trying to figure out what the expression "(Current $\times \frac{1}{2}$)" might actually mean.

This $\frac{1}{2}$ multiplier is here because the original graph included a second line showing how the thermistor curve was linearized by the addition of a series resistor. This doubled the total resistance in the circuit, and therefore halved the current for any given voltage. We have omitted that line here for clarity. Assuming that $\frac{1}{2}$ is the same as 0.5, one is still left to speculate. Is the current in the calculation multiplied by half? Is the resistance value multiplied by half? Of course, more typically, the multiplier is a factor of 1000, or maybe 10^6, but the point is the same. It is always hard to tell what is meant by a label such as "Resistance times 1000." Furthermore, it is always possible to make unambiguous labels.

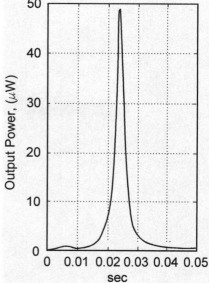

Figure 4.14

The thermistor shown has a nominal value of 10 kΩ at 25 °C, so one could scratch out the "50" on the vertical scale, and write in "10 kΩ." However, label notwithstanding, the vertical scale on the original graph is actually the *reciprocal* of resistance. Because of this, there is not much you can do with your graphics package to rectify the situation.

With the reference point "50" becoming 10 kΩ, you could grab a calculator and relabel the vertical axis, as in Figure 4.16. But that isn't too useful. The nonlinear nature of this ordinate should reinforce the idea that it would be better to use an ordinary, conventional axis. (Note that we took the opportunity to extend the axis to cover the entire graph area, too.)

The reader could certainly be forgiven for seeing the original axis, with its "Resistance" label and numbers going in even steps from 0 to 130 and assuming a scale linearity that does not exist!

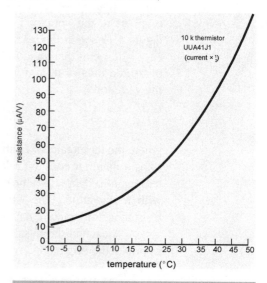

Figure 4.15

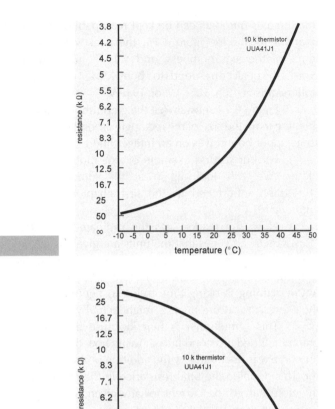

Figure 4.16

Figure 4.17

As this version of the graph shows, the resistance of a thermistor decreases with increasing temperature,[4] so the ordinate is now very odd, with the large numbers at the bottom. Even if you flip the axis around, it is still not linear. Figure 4.17 shows the result.

It seems that there isn't much choice but to redraw the curve, either from measured data or from a mathematical model. A convenient approximation to the variation is

$$R_{T_1} = R_{T_2} \exp \beta \left(\frac{1}{T_1} - \frac{1}{T_2} \right)$$

where the temperatures T_1 and T_2 are in Kelvin. The coefficient β can be thought of as a material constant related to the temperature coefficient. It is somewhat temperature dependent, however. Its value is typically around 4000, increasing with temperature. (The value for any particular thermistor can be got from resistance/temperature data for a couple of points.)

An actual thermistor characteristic is shown in Figure 4.18. This is about as unlike the original as it's possible to get!

---

[4]Not long after writing these words, the author went on a field trip to southern California's Mojave Desert, where the temperature was a sweltering 4800Ω!

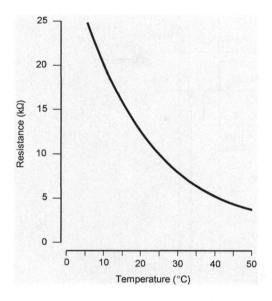

Figure 4.18

CAPTIONS

Use an Appropriate and Relevant Caption. There are three styles of caption: brief *descriptive* ones, used in most papers and reports; longer *explanatory* ones, used in *Scientific American* and the like; and *bad* ones. The bad ones can be bad because they are too long, too short, or simply irrelevant.

There is a story that the image shown in Figure 4.19 once appeared in a report. The caption was reputed to be "Horse." Well, er, yes. So it is. Not a very informative caption, however. A better caption would have perhaps told us why this particular horse is being shown. Was this the winner of the third race at Santa Anita? Is it an intelligent horse teaching its owner how to count? We may never know.

The name *horse caption* has stuck for understated captions. There seems to be no simple name for long overdone ones. We therefore offer the name *"buffalo,"* inspired by a line from a song by singer/writer Jake Thackray, "On Again!, On Again!" is about his wife, who apparently never stops talking.

Figure 4.19

On Again!, On Again!

JAKE THACKRAY

She'd just go on a-gain, on a-gain, on a-gain! e-ven more. The hind leg of a

don-key is no-thing for her she can bore the balls off a buf-fa-lo

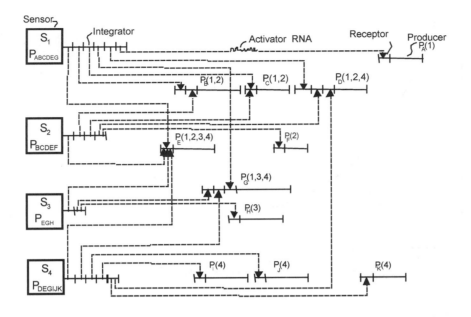

This diagram is intended to suggest the existence of overlapping batteries of genes and to show how, according to the model, control of their transcription might occur. The dotted lines symbolize the diffusion of activator RNA from its sites of synthesis, the integrator genes, to the receptor genes. The numbers in parentheses show which sensor genes control the transcription of the producer genes. At each sensor the battery of producer genes activated by that sensor is listed. In reality, many batteries will be much larger than those shown and some genes will be part of hundred of batteries.

Figure 4.20

There are a couple of examples of over-long and boring buffalo captions in Chapter 14, Case Studies. Figure 4.20 is an example from *Science*.[5]

The caption contains 98 words, which, if we may borrow from Shakespeare, are full of sound and fury and signify, well, almost nothing. *This diagram is intended to suggest the existence . . . ?* Good grief! Perhaps *Science* did not employ editors in those days.

If we take away the waffle words, and cut the material that should be in the body of the paper (the last sentence), we are left with:

Overlapping batteries of genes and how control of their transcription occurs. The dotted lines symbolize the diffusion of activator RNA from its sites of synthesis to the receptor genes. Numbers in parentheses show which sensor genes control the transcription of the producer genes. At each sensor the battery of producer genes activated by that sensor is listed.

This is still 57 words long, and much too long. The last three sentences should be incorporated into the diagram as annotations. This leaves:

Overlapping batteries of genes and how control of their transcription occurs.

[5] Roy J. Britten and Eric H. Davidson, Gene regulation for higher cells: a theory, *Science*, pp. 349–357 (July 29, 1969).

So it looks like only 11 words of the original 98 actually belong in a caption!

To generalize, for a descriptive caption:

1. Say what the graphic is, and leave it at that: not necessarily a proper sentence.
2. Don't start by describing what kind of graphic it is. ("A scatter plot showing ... ") Your reader can see that.
3. Don't include comments that are not relevant to the graphic. (See, for example, Case Study 5 in Chapter 14.)

A descriptive caption may have you merely summarizing the axis labels ("Measurement Error as a Function of Temperature"), or it may mean you can use jargon from the field ("Amplitude Probability Distribution of the Measured Noise")

For an explanatory caption:

1. Say what the graphic shows, rather than what it is. This will have to be a sentence.
2. You may describe what kind of graphic it is, but try not to start the caption by doing so.
3. Don't include comments that are not relevant to the graphic. (See, for example, Case Study 5 in Chapter 14.)

An explanatory caption will likely be longer than a descriptive one, as it must do more than merely summarize the axis labels. It must say, as succinctly as possible, what the point is that you are trying to get across. This does not mean you must write an arbitrarily long caption: one or two sentences should be enough.[6] (Continuing the sample captions above, we might write "Measurement Error Increases as a Function of Temperature" or "Amplitude Probability Distribution of the Measured Noise Corresponds to an Impulsive Source.")

Don't Explain Abbreviations or Acronyms in Captions. Thus, do not put "Amplitude Probability Distribution (APD) of the Measured Noise Corresponds to an Impulsive Source" in the caption even if you are going to use the abbreviation APD in the text. The same remark applies to headings and subheadings: these just are not the place for such explanatory devices.

---

[6]The newspaper *The Economist* always summarizes its articles in one or two sentences under the title. If you adopt the one-or-two sentence guideline, you will need to become very focused. Here are some examples from recent copies of *The Economist*:

The Bush administration is worried about an "energy crisis." That is the wrong way to frame policy

The new census figures show that America is an increasingly multiracial society. It is time to end affirmative action

America and Europe should not let missile defences come between them

The appeals court ruling has not brought the Microsoft antitrust case any closer to a conclusion. It could even turn the firm into a regulated monopoly of sorts

Note that, in each case, there is no period at the end of the last sentence. This is a matter of style, not rule. It does provide a "clean" look to the summary.

Put abbreviations and acronyms in the body text, after the first mention. If you are not going to make frequent reference, do not introduce the abbreviation. (On the other hand, if the abbreviation is more widely known than the full name—particularly for an organization—there is no need to use the full name. In context, this probably applies to NASA, the CIA, the FBI, and several other government agencies or departments. If you are going to make frequent reference, try to use some device other than the abbreviation once in a while. For example, the European Space Agency (ESA) can be referred to as "the agency" once in a while.)

CALLOUTS

Sometimes separate elements of a graphic have to be identified. The tradition is to use a label, and sometimes a line or an arrow. The combination is called a *callout*. More or less the same guidelines apply whether there is an arrow or a leader, or not. The identification has to be clear and unambiguous. A fair example is given in the diagram of a dc/dc converter, Figure 4.21.

The expert reader may sense that much of the labeling is redundant, since the symbols inside the blocks carry the same information as the words. This is explained by the fact that the diagram came from a report to management. (They need a little more by way of explanation.)

The parts of an electric field mill are shown a cutaway view in Figure 4.22. Here the intent was to explain (in a final report) how the various aspects of it worked, including its mechanical design, its air drive, and its sensing electronics with optical output. The explanations are all in the text. What is in the diagram is only what is needed to clarify the text.

In many instances, there are many choices for the location of the leader lines. Provided the intent is clear, you can use any option that seems to make sense. Our practice for this kind of image is to imagine the diagram as its

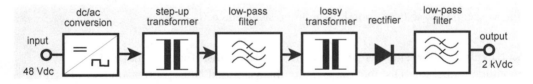

Figure 4.21

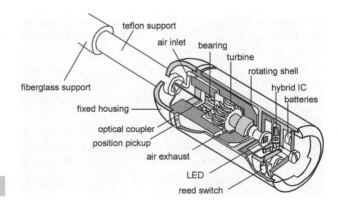

Figure 4.22

three-dimensional real self and pretend that the leader lines are bits of thread. Provided that it is easy for the reader to identify the location of both ends of the bits of thread, they have been attached well enough. In this example, the bits of thread are mostly parallel to the cut surfaces of the section.

With a complicated diagram such as this, the leader lines go all the way to the objects they identify. With a simpler problem to solve, they can stop short of the object, provided there is no ambiguity if the reader mentally extends them, as in Figure 4.23, which shows an electricity substation.

It is common practice to have leader lines not parallel to anything in this kind of drawing, or even to the edge of the page. Angles of about 30° or 60° degrees seem better. Just why this should be we do not know, but it does "look" odd when this rule is broken, as in Figure 4.24.

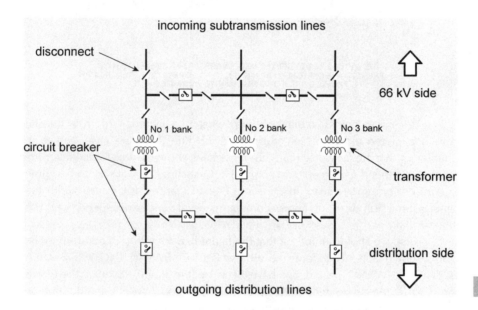

Figure 4.23

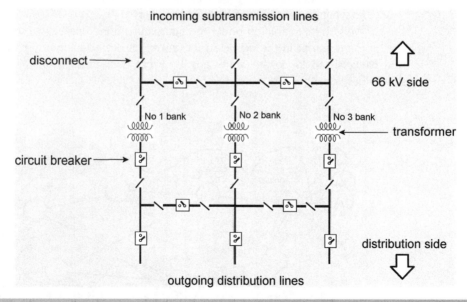

Figure 4.24

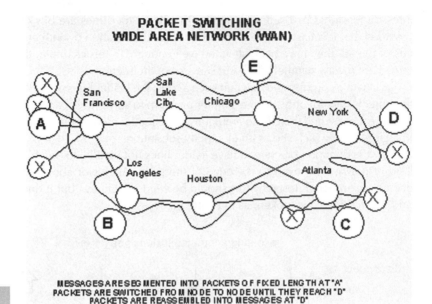

Figure 4.25

We came across a diagram somewhat like Figure 4.25 in a tutorial from one of the technical societies of IEEE. While there are case studies in Chapter 14 that are far worse than this, it contains enough errors to make it interesting. Let's spend a moment and analyze it. One should always try to find time to consider one's own work in the same critical manner. (Of course, with the paper submission deadline approaching at its usual supersonic speed, who has time for *that*?)

First, we should point out that the figure had a caption in addition to the words shown here. Therefore, the words "PACKET SWITCHING WIDE AREA NETWORK (WAN)" should not have been on the figure. (Even if there was some valid reason to retain them, the abbreviation "WAN" should not be explained on the diagram, and the entire text should not have been set uppercase.)

So what else could have been done better?

Well, the explanation under the figure (also done uppercase) should be part of the text, not in the figure, so let's take that out. (And in any case, who said packets had to be the same length?) What does this leave? Figure 4.26 shows these changes.

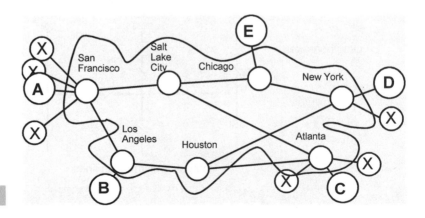

Figure 4.26

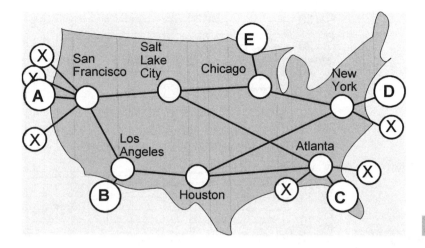

Figure 4.27

This is pretty bad. Let's make the labels a bit bigger, and see if that helps. And, although it adds nothing to the information content of the graphic, it would be better style if the map were a little closer to reality, and were shaded. (It was quite a while before we figured out that it *was* a map, and not one of those cloud-objects that are used to indicate an Internet-like switching fabric.) The outline could be lighter (instead of being the same thickness as all the other lines in the figure), too. Figure 4.27 shows these changes.

It may be presumed that the possible routes from "A" to "D" are meant as examples only, and thus there is no need to add a connection from Atlanta to New York, or Los Angeles to Salt Lake. Presumably, the meaning of the "X" circles could be explained in the text.

After review and examination of the context, you might decide to remove the figure. That's OK too: never include a figure merely because you don't want to waste the time you spent drawing it.

Sometimes, you must label the lines in a graph. You can do this in several ways, and you might choose a different solution on a different occasion. (Except for the need to develop your own style, which requires consistency, see Chapter 13.)

There were some examples of this need in Chapter 1. You could use a key (which we also touched upon in Chapter 1), or you could use labels with leaders, or you could add labels right next to the graph lines. If you assume that at some time the graph will be copied in monochrome, identifying the curves in one of these ways will seem like a good idea. Figure 4.28 is an example.

This was adapted from a paper in the *IEEE Transactions on Instrumentation and Measurements*. The work had to do with a new kind of device for converting the energy of an ac parameter into dc. The transfer ratio for such a device is customarily expressed in the dimensionless unit $\mu V/V$ (rather than 10^{-6}) because of the application in transferring ac calibrations to dc. It is the error in this ratio that is shown here as a "difference."

The little explanation of the various curves is clear enough, and it will copy, but the drop-shadow effect is unnecessary.

A simpler way might have been as shown in Figure 4.29.

The labels in Figure 4.29 are a bit reminiscent of the pasted-paper effect shown in Figure 1.27. If the graph area had not been covered in lines, we might have seen something more like Figure 4.30.

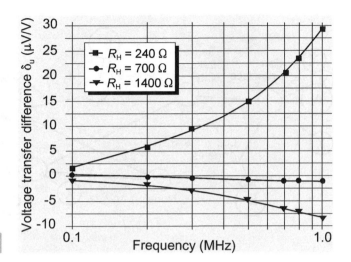

Figure 4.28

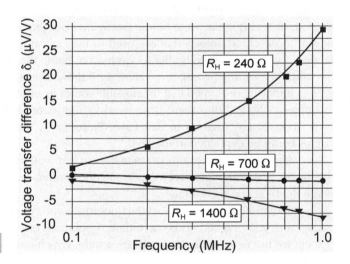

Figure 4.29

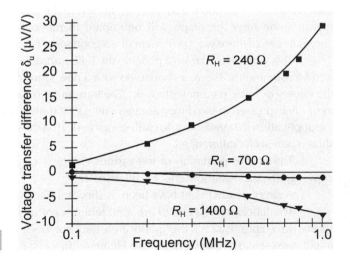

Figure 4.30

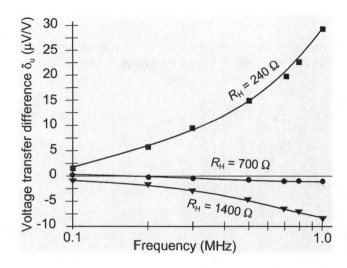

Figure 4.31

Finally, Figure 4.31 shows an alternative that we do not recommend. Not only does it take longer to angle the text, it really doesn't look "right," somehow.

As for shaping the text to follow the curve of the lines, save that for when you're doing greeting cards!

SUMMARY

The basic rule is strive for clarity and avoid ambiguity.

- Choose a font for your graphics that complements the one used in the body text—but not necessarily the same font.
- Make sure the type is the right size in the final graphic, about one point size less than the body copy, if you are using a sans serif font.
- Don't write everything in the graphic in uppercase.
- Label graph lines in the simplest way possible.
- Label axes with conventional scales, and don't use multipliers.
- Rotate the label for the vertical axis, but not the axis numbers.
- Keep captions appropriate and relevant.
- Don't explain abbreviations or acronyms in the graphic.
- Don't put leader lines parallel to lines in the graphic.
- As always with graphics, once you have a set of words formatted correctly, copy and edit them, rather than typing from scratch and reformatting.

EXERCISES

4.1. What is wrong with the words in Figure 4.32? Rewrite them. By the way, the original caption for Figure 4.32 was "Spectrum measurement $S_1(f)$ with L_1." Does that strike you as a good caption?

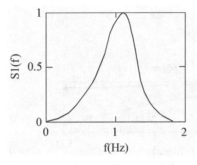

Figure 4.32

4.2. What is wrong with the words in Figure 4.33? Rewrite them.

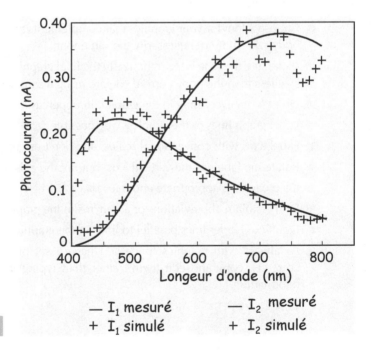

Figure 4.33

4.3. What is wrong with the words in Figure 4.34? Rewrite them.

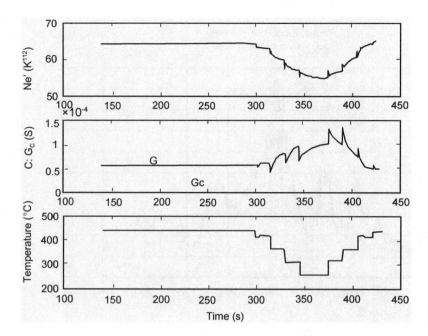

Figure 4.34

4.4. What is wrong with the words in Figure 4.35? Rewrite them.

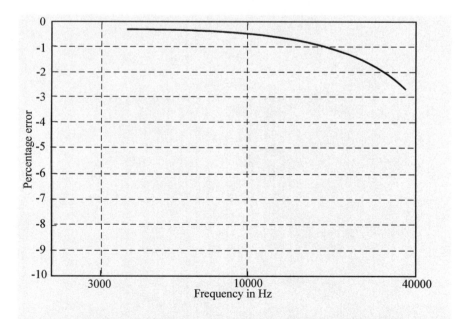

Variation of percentages of error between the nonideal and ideal frequencies of oscillation

Figure 4.35

4.5. What improvements should be made to the horizontal axis of Figure 4.36?

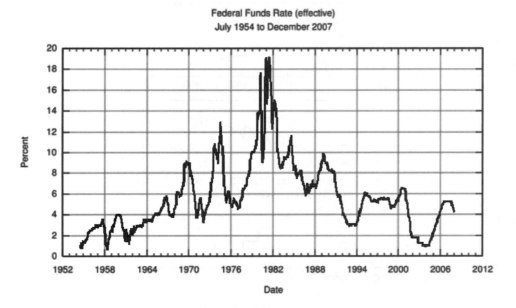

Figure 4.36

Getting the most out of your software

5

So you sent in the Abstract, and they accepted your paper. All you have to do now is write it. If you're an engineer, you likely want to write a two-column paper, set in 10-point Times Roman, with embedded graphics. Camera-ready copy, it's called. Engineers like camera-ready output: the production process is part of the creative urge. If you're a scientist, maybe you're using something like TEX, or the paper is submitted in double-spaced Courier, with the figures stuck together at the end, and the captions collected somewhere else, probably in a separate file.

Either way, you need to produce the words that say what you want to say, and you need to produce the graphics that will help you make your point. W. Somerset Maugham used to try to produce 2000 words every day (that's about 3 pages of this type, with no figures), so at that rate your paper should take less than a week, assuming you know what you want to say! And assuming you have nothing else on your calendar.

You are writing the paper in a word processing program of some kind (Microsoft Word® and Corel WordPerfect® are the two leading contenders, but there are others), and you will be doing your graphics in a graphics package. That's where this book comes in.

Back in the good old days B.P.P. (before PowerPoint®), you might have to wait a week while the graphics department produced the figures for your paper. This was always a testing time: while the end result was always very professional, you often had to spend a good deal of time telling the artist what you wanted. And the process was often iterative, occupying more precious days. Nowadays, you do your own graphics, possibly in PowerPoint or some such program, and unless you are careful, you get what the software people in Outer Mongolia—or wherever they actually *code* that stuff—deem you want. No matter; you can in fact get the output *you* really want from your software, provided you are willing to invest the effort to persuade the program who is in charge!

The first thing to remember is this. An important part of producing graphic images is to *find ways to avoid wasting time*. If you approach the task of making a graphic the same way you would a page of text, you are going to waste time—in *large* quantities. The difference in the time it takes to produce a graphic efficiently or not might easily be a factor of 5, maybe a factor of 10. (We are not talking about spending extra time to improve the appearance or style. *The same graphic* that might take an hour to produce might be produced in 10 minutes, if it was done efficiently.)

So this chapter has two purposes: to help you get what you want, and to show you how to get it efficiently.

KEYBOARD SHORTCUTS

Let's begin with some basics. Doing graphics on a computer is a two-handed job.[1] Sure, there's a mouse that can only be managed with one hand, but the simple fact is that the other hand can play an important role. Avoiding the keyboard and just using the mouse for *all* actions is going to maybe triple the time it takes to produce a graphic, and give you a sore wrist as well. Let's have a look at some of the things the left hand can be doing while the right hand is on the mouse.

First, there are the well-known shortcuts[2]:

- **Ctrl-c** copies the current selection into the clipboard.
- **Ctrl-x** copies the current selection into the clipboard and deletes the original.
- **Ctrl-v** pastes the contents of the clipboard.
- **Ctrl-z** undoes the last editing operation. (It does not undo the last save to disk or clipboard.)
- **Ctrl-s** saves the file.

You probably know these keystrokes from whatever word processor you customarily use. The shortcut Ctrl-a acts so as to *select everything* in most graphics programs. In CorelDRAW® before Version 10 it brought up the *alignment* dialog box. Corel must have reckoned this a more useful shortcut—a reasonable thing, considering that the number of times you need to align things is probably much greater than the number of times you need to select everything. In later releases, they fell into line with the more usual function of selecting everything.

Suggestion: *Learn all the shortcuts you can for your program.*

Here is a handy shortcut that works in PowerPoint and CorelDRAW. Ctrl-d works to make duplicates of your selection; that is, it is the same as Ctrl-c followed by Ctrl-v. But there's an additional feature: if you use Ctrl-d then *move* the object while it is still selected, then use Ctrl-d again, the command not only duplicates the object, it also duplicates the move. We will use this later in this chapter.

---

[1] In making this observation we do not wish to appear insensitive to those who have difficulty with one hand, or indeed those who (like the daughter of one of the authors) have only one hand. Such people will have no choice but to switch back and forth between mouse and keyboard, and for them the optimal solution may be different. There are also one-hand keyboards on the market that may speed things up. But for the sake of simplicity, the remainder of the chapter assumes the reader is a common-or-garden right-handed person.

[2] In these shortcuts, it does not matter whether the letter is entered uppercase or lowercase.

DRAWING SHORTCUTS

Here's another shortcut for PowerPoint. Suppose you want to draw a box and put some text in it. Draw the box, and while it is still selected, start typing! There is no need to click on the text icon, or to insert a text box. What you will get will be the default font, centered on the box. The text is not constrained to fit inside the box, however, so unless you have a wide box you may need some hard returns.

This was a PowerPoint shortcut, and it works in SmartDraw® and Visio®, but it won't work elsewhere. There are many other differences between the various programs. Here's a handy trick that works in Corel Presentations™. Suppose you are drawing a line drawing that calls for part of an ellipse (e.g., it could be a view of something cylindrical). First, you draw the ellipse by selecting the appropriate tool. Then, you deselect it, and reselect by double clicking. This brings you to the Edit mode, shown on your screen by a little square on the ellipse. If you now click on the little square, you can drag it around the ellipse. If you have your cursor *inside* the ellipse, you get two straight lines that go from the ends of the line segment that will remain to the center of the ellipse, as in the left side of Figure 5.1. If you have the cursor *outside* the ellipse, these lines disappear, as in the right side of the figure.

This same feature exists in CorelDRAW®. We are not aware of it anywhere else.

The various other programs also have some handy features not shared by the competition. To some extent, the differences arise because the various software designers just have different ideas about what works best. To some extent, they also reflect the perceived need to avoid a "look-and-feel" lawsuit. It is worthwhile to examine some of these differences before we proceed.

We will look at several programs that you might be using to produce your graphics. However, bear in mind that programs keep evolving—as soon as one bug is fixed, the software geniuses have to create new ones to guarantee their employment. Therefore, some of what follows may not apply by the time you read this!

To make it a solvable problem, we will not be able to explore all possible programs. We have looked at (listed alphabetically) AutoCAD®, CorelDRAW®, Adobe Illustrator®, PowerPoint®, Presentations™, SigmaPlot®, SmartDraw® and Visio®. We will not have space to study all these closely—so let us apologize if your favorite feature or your favorite program is not mentioned.

The problems of exchanging data between these programs is addressed in Chapter 12; for now we are concerned only with *operating* them. Spreadsheets, also capable of creating graphics, are covered separately, in Chapters 7–10.

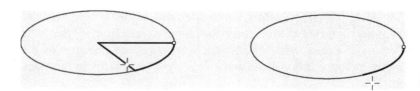

Figure 5.1

LINE SHORTCUTS

To begin with, let us turn our attention to the matter of drawing a straight line, and getting it exactly where you want it, perhaps the most important function of a graphics program. In all the programs we looked at, there is a `line` icon on the screen somewhere, so you can quickly grab it and start to draw.

CorelDRAW®. Let us start with CorelDRAW. The `line` icon is one of several in a flyout menu (that has different kinds of lines for you to choose from) on the left of the screen. You can click on the icon, or you can hit `F5`, if you are into the shortcuts. The cursor changes to four lines in the shape of a little cross, with a wiggly line under and to the right, as in Figure 5.2.

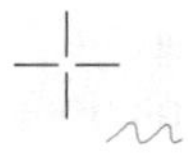

Figure 5.2

If you click (down-and-let-go) on the screen, it has the effect of anchoring the end of the line right where the center of the little cross was when you clicked. If you now move the cursor, a straight line will be drawn, dragged from the anchored end to the middle of the cursor. When you click again, the line is released from the cursor, and anchored on the screen. The little wiggle turns into a bent arrow, as in Figure 5.3.

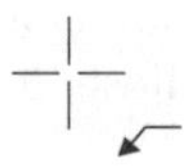

Figure 5.3

This change indicates that the system is ready to continue the line with another segment. If you click again, the process of dragging a line from the anchor point repeats. If you first move the cursor, the arrow turns back into a wiggle, and the system is ready to draw you a new, independent line.

Now, if you leave your finger down when you click the first time, CorelDRAW will draw you a line wherever you move the cursor, essentially treating the cursor like the end of a pencil. CorelDRAW will also turn the ragged line you draw by hand into a beautiful curve, using something they call a Smart Drawing Tool, but let's not go into that here.

Suppose you want your line to be a horizontal line. It's easy to see in most graphics programs: they show you on the screen that a line isn't vertical or horizontal by putting a little sideways step in it, as shown in Figure 5.4.

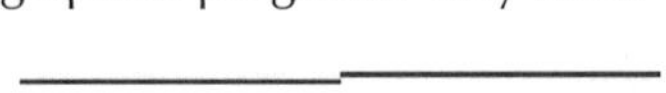

Figure 5.4

There is another way you can get a line that is horizontal: press the `Ctrl` key as you drag the line. In this program, `Ctrl` forces the line to be horizontal or at increments of 15° to the horizontal.

Once you have a horizontal line, you can stretch it or compress it by clicking on the *middle* square of the outer squares that indicate the line is selected. That is, either of the ones indicated in Figure 5.5.

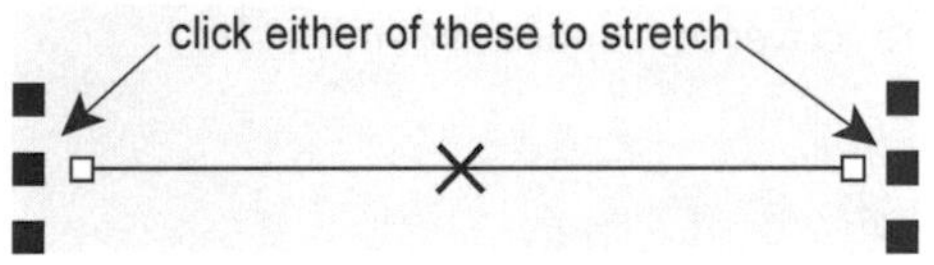

click either of these to stretch

Figure 5.5

Thus, if you want to get two horizontal lines, exactly lined up but with a gap in the middle, one way to do it is to draw one line, duplicate it (`Ctrl-c`, `Ctrl-v`), and then stretch the ends of the two items until they are where you want them. (You can also "nudge" a selection, by using the up/down/left/right arrows, with `Ctrl` as a way to reduce the size of the nudge. But this approach is quantized and may not get you exactly to the place you wanted.)

Figure 5.6

Adobe Illustrator®. Now let's go through the same thing in Adobe Illustrator. As before, the tool is available as an icon, here called a `Line Segment Tool`. When you move the cursor back onto the screen, it will appear as a little cross, with a dot in the middle, as in Figure 5.6.

If you click with the cursor in the screen, and let go, a dialog box appears asking how long you want the line, and at what angle. You could, of course, use the dialog box to tell the program how long you want the line (in points or inches or whatever) and at an angle of zero (for horizontal). But since that would involve taking your hands off the mouse, or since you might not know the length of line you want, you might also hold the left mouse button down and draw the line yourself. In this program, holding down the `Shift` key constrains the possible angles, to 0°, or 45°, or 90°.

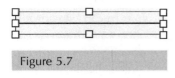

Figure 5.7

Now you can trim the length of the line by stretching or squeezing from either of the middle squares at the ends of line, as in Figure 5.7, in the same way as you could for CorelDRAW. If you do, a convenient label `anchor` will appear next to the cursor, to indicate you have got the anchor, not the line segment, which would pop up and say `path`. As before, you can get two short lines lined up by copying and pasting, except that the shortcut `Ctrl-v` puts the copy in the middle of the page, no matter where you have the original. However, if you use `Ctrl-f` (Paste in Front) or `Ctrl-b` (Paste in Back) you will get what you were after. You can also "nudge" a selection with the arrows. In Adobe Illustrator, the `Shift` key gets you a bigger nudge.

Corel Presentations™. This program is very similar in operation to Adobe Illustrator. The `line` icon is on the left of the screen, one of several options in a flyout box. The screen cursor is a cross with a dot in the middle, and the constraints on the angle are obtained by holding down the `Shift` key. They are at 45° increments. You hold down the mouse key to drag the line away from its

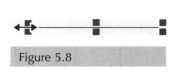

Figure 5.8

anchor. Stretching the line is a little different, however, as there are only two boxes at either end of the line. When you put the cursor in the right place, it turns into a double-headed arrow as in Figure 5.8, and you can stretch or squeeze the line.

While the rest of the operation for a double line is the same as Adobe Illustrator, there is no "nudge" feature. And when you have finished drawing the line, the cursor turns back into the selection tool, so to draw another line, you have to go back and rechoose the line tool. It may be quicker to `Ctrl-c` and `Ctrl-v` to get a copy of the line, instead. This is true for any program where the tool disappears when you have used it once. PowerPoint is another example of this.

PowerPoint®. In fact, let's turn now to consider PowerPoint. It may be the most used of the programs we will look at here, in spite of the fact that is not the best in some ways. It has many of the features we have seen so far, but it implements some of them badly.

First, the basics. The line icon is one of many at the bottom of the screen, although—if you like to work hard—you can also find it from Autoshapes/ Lines/Line. Having clicked on the icon, you can now take the cursor over the area where you want the line, and click again. This has the effect of anchoring the end of the line right where the center of the little cross was when you clicked. However, unlike in Presentations, if you fail to hold the mouse button down, not only will there be no line when you move the cursor, the cursor will have changed back to the selection tool.

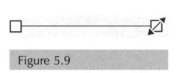

Figure 5.9

The Shift key constrains the angle to increments of 15°. Once you have drawn the line, it will appear with just two little squares (total) for anchors. When you put the cursor over the end, you get the arrow shown in Figure 5.9.

Now, when you try to stretch the line, you are allowed to change its angle, too. This turns out not to be convenient. Sometimes you get a new angle when you don't want one. If you press Shift while trying to adjust a line length, instead of constraining the angle, it just might make the line blow up out beyond the boundaries of the slide. If this happens, all you can do to rescue the situation is to right click and choose Edit Points.

Visio®. Originated as an independent program and was bought up by Microsoft, Visio has a different look and feel from the other programs. The shapes that you might want to draw are presented to you in a window to the left of the drawing window. The selection can include ready-made symbols for whatever you may be working on—electrical, mechanical, civil engineering, for example. Sets of "stencils" organize the symbols. To add a "shape" to your drawing, you simply click on it and drag it onto your drawing.

Each shape comes with a set of "connectors" that allow you to join the symbols. By default, the connecting line will bend to create right-angle segments. You add the connectors by selecting a connector tool. When you hover over a connection point, it lights up red. Click on it, and you can drag the line to another connection point. If the line is not drawn where you want it, you can drag each segment. Sometimes they stay where you put them—we have never managed to sort out when changes are stable.

Still and all, the program will allow you to assemble quite complex drawings quite quickly, though the drawings tend to have something about their style that makes you know they came from Visio.

The idea of having collections of ready-made drawings is a good one, but the ones that come with the program are a bit devoid of character. Whatever program you are using, you can create partial drawings as building blocks for your drawings, in your own style. It won't be as easy to use your drawings as simply dragging them into the drawing window, but you may still find it worthwhile. For drawing circuit diagrams, for example, we have a set of the common symbols in several file types. And they are made to be easy to use: the transistors, for example, can be placed *on top of a line* and will appear to connect to it, and will obscure the part that is not needed. Such a part can be slid up or down a line very easily.

OBJECT SHORTCUTS

Another thing graphics programs do differently is they use different ways to *select* an object and different ways to indicate that it has been selected. Suppose we have the collection of objects shown in Figure 5.10 on the screen. We can select the bottom three objects by clicking on the first object we wanted, and then hold down the Shift key while clicking on the others in turn. Alternatively we could click some empty space with the mouse (using the left button) and dragging (with the mouse button down) around them all. This is sometimes called rubber-banding or *lassoing*. Let's look at lassoing.

In **PowerPoint**, if we start at the top left of the three, and drag to the bottom right, the screen looks like Figure 5.11 (apart from the fat dotted line we added to illustrate the path of the cursor) just before we unclick. Right after we unclick, the screen would look like Figure 5.12. The little squares are the way the program tells us we have selected the bottom three objects.

In **Presentations**, the rubber band looks like Figure 5.13 and the selection of the three objects would look like Figure 5.14 after you unclick.

The **CorelDRAW** display would look fairly similar, as seen in Figure 5.15. A difference here is that the set of three objects is indicated by the black squares, but the individual parts contributing to the whole are indicated by little open squares. There are three of them, barely visible: one at the top left of each square, one at the top of the circle.

Adobe Illustrator adds some lines (and turns the outline of everything selected into an interesting shade of purple). We show a monochrome version in Figure 5.16.

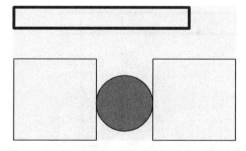

Figure 5.10

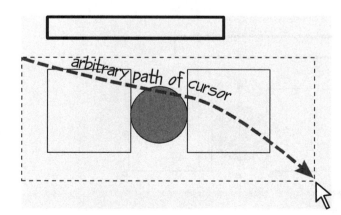

Figure 5.11

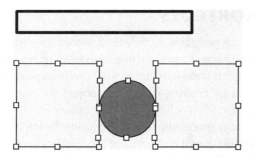

Figure 5.12

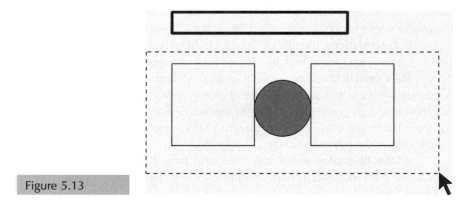

Figure 5.13

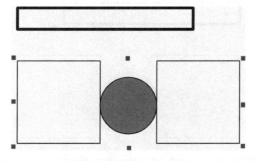

Figure 5.14

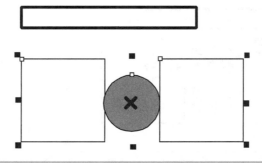

Figure 5.15

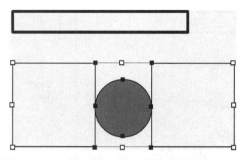

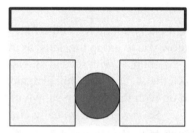

Figure 5.16

Figure 5.17

Now suppose we wanted to resize the bottom three objects until they were about the same size as the top rectangle. In other words, we wanted the result shown in Figure 5.17. In anything except PowerPoint, you could adjust the size of the bottom three objects together simply by selecting them, and dragging a corner. In this case, the lower right corner would be the obvious choice, since the top left corner is already in the right place. In **PowerPoint**, the same action would adjust the size of each object, but as if it was being altered on its own, with the result shown in Figure 5.18.

It would be interesting to see if anyone on this entire planet thought this kind of edit was useful. We have not met such a person. To achieve the desired result in PowerPoint, you must first *group* the bottom three objects together by selecting them, and then clicking on Draw and then on Group.

That's too much mouse activity. The shortcut is a bit better, but takes two hands: Alt-r, Enter. Enter works only because the Group option is at the top of the menu, where the cursor is. It is probably safer, in case Microsoft reorders things to do Alt-r g. It makes no difference if you hit the g while still holding down the Alt key from the Alt-r or whether you let go the Alt.

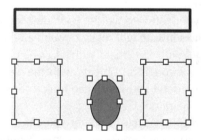

Figure 5.18

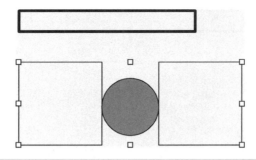

Figure 5.19

Alternatively, you can right-click, but there's still a little questionnaire to fill out before PowerPoint will allow you to group the parts, as in Figure 5.19. So it goes.

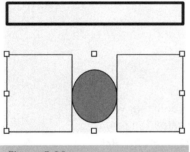

Figure 5.20

Then, as with several other programs, you must hold down the Shift key while dragging the corner, or you can end up with the situation shown in Figure 5.20.

So PowerPoint is really clumsy at doing this rather simple edit. It gets worse. Or better, depending on how you look at things. If there is any text in the group, it will not be resized in proportion. You will have to select the group (thereby including the text) and manually set a new point size. When you do that, you may find that the boxes that contain the text now have to be resized individually. Oh, and if you try to copy and paste text from one slide to another, you may find that the text has reverted to the default font, even if you were not using the default.

PowerPoint also has the added attraction of being not well integrated with any other software (e.g., not even Microsoft Word), so we can't really recommend its use for making document-quality graphics. However, since you probably own a copy, we realize you'll end up using it. Probably all of us have been part of research collaborations that decided to use it, as a sort of lowest common denominator.

Grouping is simpler in most programs. Some have an icon on the screen that you can click on, some allow Ctrl-g. Find out what the technique is for your software.

We mentioned clicking and lassoing earlier. There is a useful difference between these two methods. Sometimes you want to select one element that is right in the middle of a lot of others, and you cannot seem to select that particular element by clicking on it.

You can see if your software will let you "dig" for hidden objects at one location. This means the software will rotate through all the objects in the neighborhood. CorelDRAW will do that with Alt-click. PowerPoint and Presentations will cycle through all objects with the Tab key, once you have selected an object. If you don't know, you can look for something like "selecting a hidden object" in the help index.

If there is no "dig" or "cycle" feature, you can try zooming in, to get a better look at the scene. But some software has limited zoom capability—PowerPoint, for example, limits you to a gain of 4.

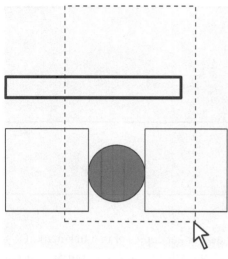

Figure 5.21

If these attempts don't work, you could try lassoing instead of clicking. Thus, if you lasso from the top left to the bottom right as shown in Figure 5.21, only the circle will be selected. This is because only the circle is *completely* surrounded.

This trick of not completely surrounding the objects you *don't* want works for groups, too. If you take the lasso across only part of a group, it will *not* be selected. This is a useful thing to remember.

You would think *selecting* would be a simple matter, but there is more. In Adobe Illustrator you can select an object and then select objects that are *like* that one in terms of color or some other attribute. Illustrator calls this the Magic Wand— very useful feature.[3] Adobe Illustrator will also allow you to lasso an object and *a node* in an object (which it calls *an anchor*), and you can then move either the object or the node and maintain the relationship between the elements selected, thus changing the shape or slope. This is very useful if you have a diagram with an object that has a callout, for example, and you want to move the object while taking the end of the leader line along, too.

Suggestion: *Learn all the little quirks of* your *software.*

LINING THINGS UP

Whatever software you are using, you need to know how to use it to get your graphic *organized.* Many of the things you need to do to make a respectable looking, publication-quality graphic require arranging things neatly.

For example, the tick marks along an axis can in most programs be made evenly spaced and lined up by using something in the software. This will save you a lot of time fiddling around placing the lines yourself. For example, suppose you want the axis of a graph with seven tick marks on it, labeled 0 through 6, as shown in Figure 5.22.

You might start by drawing the horizontal line, at a suitable thickness, and a tick mark, a little thinner, as in Figure 5.23. (Well, at this point, you don't

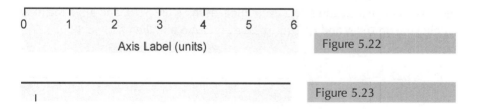

Figure 5.22

Figure 5.23

---

[3] The only other program we are aware of with this capability is the (now extinct) AccuDraw. The AccuDraw referred to here was a program that produced files that were compatible with WordPerfect at around revision 4 or 5. There is now a CAD program with the same name. We have no idea if they are related.

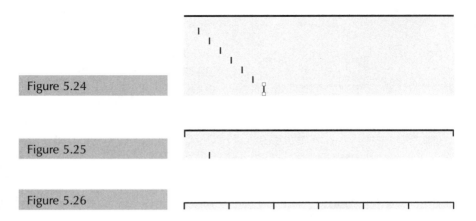

Figure 5.24

Figure 5.25

Figure 5.26

need to line things up.) Next, make six additional copies of that tick mark. Do it with `Ctrl-c`, `Ctrl-v`. Just what the result will look like on your screen will depend on what program you are using. In Presentations, the copies appear in exactly the same place as the original, so the graph will appear the same. In PowerPoint, each copy is placed down and to the right, so you get something like Figure 5.24. Here the last tick mark is shown selected.

Next, drag a tick mark to each end of the line. In Presentations, it will appear as shown in Figure 5.25. All the "unused" tick marks are still stacked up under the one on the bottom. Select everything by lassoing it, deselect the long line by clicking on it, and then finish by clicking on `Align Top` under the symbol, and `Space Left-Right` under the symbol. The result is the finished axis shown in Figure 5.26.

All that remains is to add the text. There are a couple of things worth mentioning. First, if you don't deselect the long line, the process of spacing left–right gives you the result shown in Figure 5.27. Maybe this is a reasonable thing for the software to do in response to the instruction to space left–right, because it puts the two ends of the long line at an even space from the short lines, but it isn't quite what we wanted.

Second, if you instruct the program to space *text* left–right, you may not get what you want right away. Odd things can happen if the text (such as axis numbers) consists of different lengths, where "length" is indicated by the size of the selection marks on the screen. This is a minor detail, just be on the lookout for it.

In PowerPoint, the process is similar, but there are a few more mouse clicks involved because of the need to click `Draw` before you can do anything else. (Use `Alt-r` to save one mouse click every time.) We can start with the long line and the seven tick marks, and select everything, as in Figure 5.28, and align the tops by clicking on `Draw` and `Align or Distribute` and then `Align Top`.

Figure 5.27

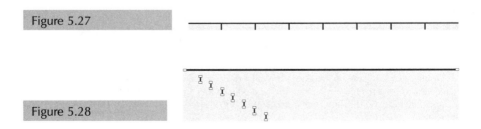

Figure 5.28

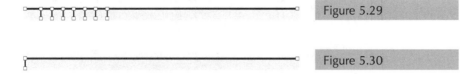
Figure 5.29

Figure 5.30

Figure 5.29 shows the result.

However, the instruction to Distribute Horizontally produces the perplexing result seen in Figure 5.30.

If you do what you did in Presentations, place a tick mark at each end of the long line, deselect the long line after aligning the elements vertically, and then Distribute Horizontally, you get the desired result, as in Figure 5.31 (unless you have inadvertently clicked on Relative to Slide at the bottom of the Align or Distribute menu. This is unlikely, however, as this particular feature is normally hidden and has to be found by clicking on the little chevron mark at the bottom of the dialog box.)

Now, if you want to use the shortcut mentioned earlier, start the same with the pieces shown in Figure 5.32. Select the vertical line, but then use Ctrl-d instead of Ctrl-c to duplicate the line once. The result is the familiar offset copy shown in Figure 5.33.

While the line is still selected, use the arrow keys to move it up and to the right. Note that it snaps to the grid. Once you have it aligned vertically and positioned to the right horizontally, duplicate it (with Ctrl-d) five more time and you get the result seen in Figure 5.34.

Now you can use the Distribute Horizontally as described earlier to line your tick marks up properly.

Another way you can distribute the lines is to group them, align them with one end of the horizontal line, then drag the end of the group to align with the other end. The first step looks as shown in Figure 5.35.

The finished lines look as shown in Figure 5.36.

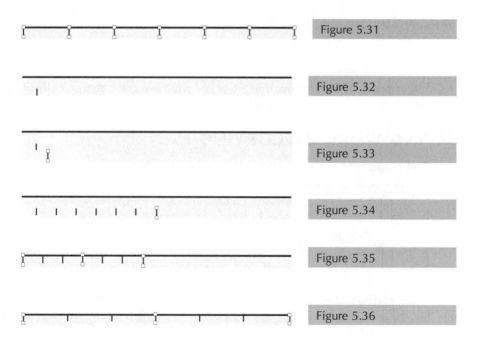

Figure 5.31

Figure 5.32

Figure 5.33

Figure 5.34

Figure 5.35

Figure 5.36

Figure 5.37

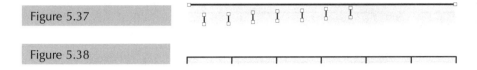

Figure 5.38

Figure 5.39

If you don't happen to have the snap-to-grid option turned on, your efforts with Ctrl-d may end up giving you the sort of result shown in Figure 5.37. The northward drift of the lines can be fixed, but the effect makes the whole Ctrl-d business a bit less useful.

If you can draw a horizontal axis, you can draw a vertical one. Or you can simply rotate a horizontal one to the vertical position. Beware odd changes, however. Visio Technical had the odd habit of turning an axis such as the one in Figure 5.38 into the slightly modified form of Figure 5.39, when rotated. Go figure! (This feature has evidently been removed from Visio since the product was acquired by Microsoft.)

The idea is this. Get to know *your* program, and all its little quirks.

Suggestion: *Learn to avoid precision control and adjustment in your work*—let the computer do the work for you.

SNAP-TO-GRID IS YOUR FRIEND (OR NOT)

It is good to let the applications do the work for you, but a feature to be cautious about is snap-to-grid. Sometimes you should use the snap-to-grid feature, and sometimes you need to circumvent it. In general, if you are creating axes for charts and graphs, it is easier to use the snap-to-grid feature. (By default most applications have this turned on.) So snap-to-grid is your friend most of the time.

But let's say we want to get finer resolution in our alignment on a drawing that is not an axis for a chart or graph. How do we get around the snap-to-grid? In most applications you can either use Alt or Ctrl while moving a selected item and it will temporarily (i.e., only while you are pressing the key) turn off the snap-to-grid feature and allow the object to be moved or sized any tiny amount. In PowerPoint, use the Alt key. In other applications, you may need to turn the snap-to-grid feature on and off. For example, in Presentations you can go through a sequence of View then Grid/Guides/Snap to turn the feature on or off, or you can press Alt-F8.

By the way, a way to move things that you want to move only vertically or horizontally is to press the shift key while you drag.

SAVING WORK BY COPYING

Here's an example of a method for producing complex diagrams.

Let's assume the graphic you want to produce is not numerical. That means it won't be a graph or a histogram. Let's assume, in fact, that it's a diagram showing the structure of a chemical that goes by the catchy name of 4'-methoxy-2-methyl-1,2'-azoxynaphthalene. It should look as shown in Figure 5.40.

Now, there are lots of ways to put this together, and ours may not be the most efficient. But it is *reasonably* efficient, as you will see. The process of creating this graphic is quite unlike the process of writing its name.

We start by looking to see how we can break the diagram down into pieces that we can build and then duplicate. Obviously, there are a lot of hexagons in

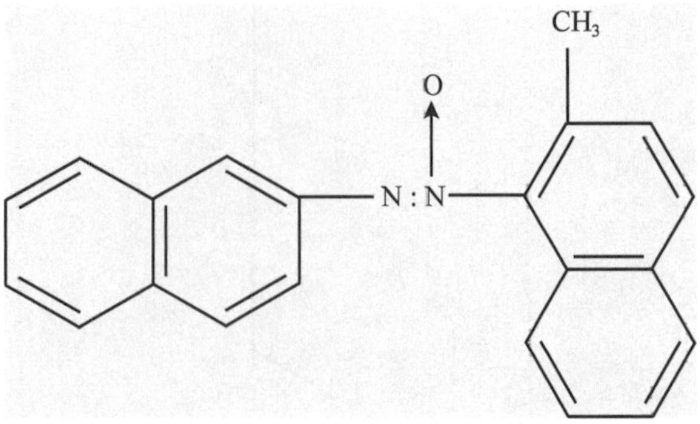

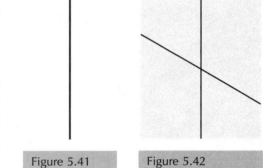

Figure 5.40

this figure—we'll start by building one of them. (Maybe your program can draw a hexagon straight off. If so, fine, why not!)

We note that the left edge of the leftmost hexagon is a vertical line. This must be *exactly* vertical, so it could be drawn pretty long to start with, so that any discrepancy will be obvious. But remember, in PowerPoint and Presentations, holding down the Shift key while drawing will constrain the line to be vertical or horizontal, or to be at one of a small number of angles. (Other software may use a different key for the same purpose, remember.) This is a useful feature at a time like this.

Anyway, you make a long vertical line, to be the left edge of the left hexagon, as shown on Figure 5.41. Now because the angle in a hexagon is 60°, you copy it, and rotate the copy 60°. With most graphics packages, you can rotate by some specific number of degrees. With some, "handy" amounts like 30°, 45°, 60°, and 90° are available on a pull-down menu. With others, you hold down the Shift or Alt keys when you rotate or draw the line. The result of copying and rotating a line by 30° is shown in Figure 5.42.

In this example, the rotation occurs about the center of the thing being rotated. Now shift the leaning line to touch the vertical one, as in Figure 5.43. Select both lines, duplicate them, and flip them left-to-right, as in Figure 5.44. With just the same two lines selected, move them across to touch the first two, as in Figure 5.45.

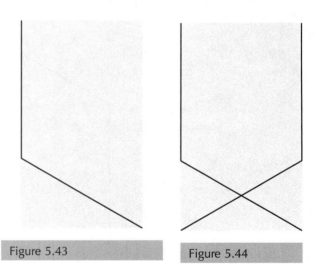

Figure 5.41

Figure 5.42

Figure 5.43

Figure 5.44

Only the top two lines of the hexagon are missing. There are several ways to make them. We suggest you select the bottom two lines, copy them, and flip them top-to-bottom. Figure 5.46 shows the result.

If you slide them up, you complete the hexagon, as in Figure 5.47.

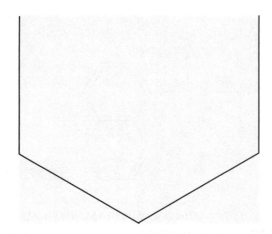

Figure 5.45

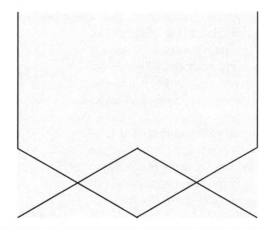

Figure 5.46

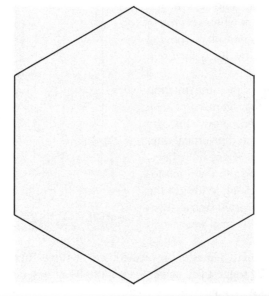

Figure 5.47

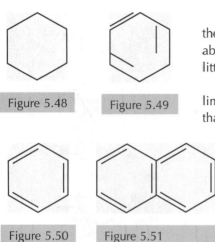

Figure 5.48   Figure 5.49

This is, of course, too large to be used for the rest of the diagram, so you can shrink it suitably. At the same time, you can make the lines a little fatter, as in Figure 5.48.

Now for the double bonds, the extra lines inside the hexagon. Select the three lines that need the extra line. Make a copy of these lines and make them into a group. Shrink the group somewhat. In Figure 5.49, it is clear that we shrank from the bottom right.

Make a group of the lines of the hexagon, and center the two groups in the vertical and horizontal directions, as in Figure 5.50. Now we have the basic building block of the molecule, the hexagon with the double bonds. Select everything, make a copy, and move the copy across to the right, as in Figure 5.51.

Note that in Figure 5.51 the double bonds are not all in the right place: flip the right side left-to-right, as in Figure 5.52. Ungroup the three inside lines on the left, and delete the extra vertical line, leaving two lines in the left hexagon, as in Figure 5.53.

Things will start to go fast for a while now. Select everything, make a copy, and move the copy to the right, as in Figure 5.54. Horizontal alignment of the two sides is not important.

Rotate the copied stuff 90°, as in Figure 5.55. As before, the rotation is about the center of the thing being rotated, so the right group is not in the proper place. We'll fix that later. Now we need a vertical lines and two horizontal lines. We have four vertical ones already. Grab one, copy it, make it shorter, and stick it where you see it in Figure 5.56.

Make a copy of the line, and rotate it 90°. This step makes sure the horizontal line is the same length as the vertical one. In Figure 5.57, the rotation is around the center of the rotating object.

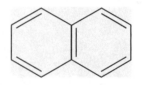

Figure 5.52

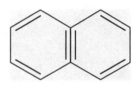

Figure 5.53

Figure 5.50   Figure 5.51

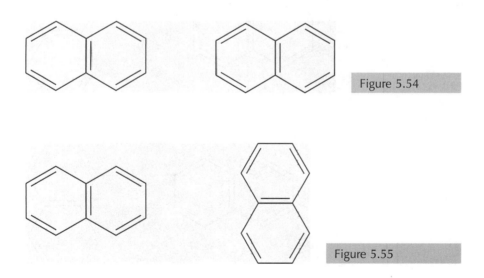

Figure 5.54

Figure 5.55

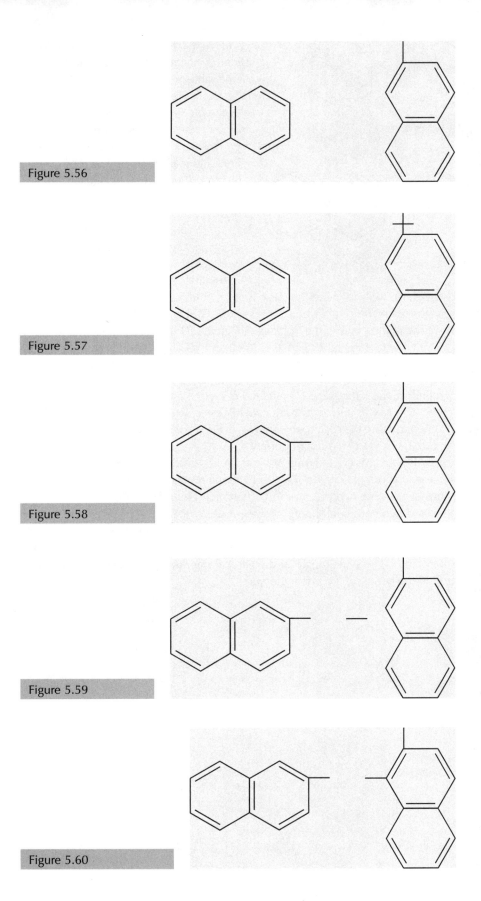

Figure 5.56

Figure 5.57

Figure 5.58

Figure 5.59

Figure 5.60

Move the new horizontal line into place, as in Figure 5.58, and make a copy over to the right, as in Figure 5.59.

In Figure 5.60, we line up the right side of the diagram.

That's about it for the lines; now we can add some text. We decided to start with the text in the center of the diagram, as in Figure 5.61.

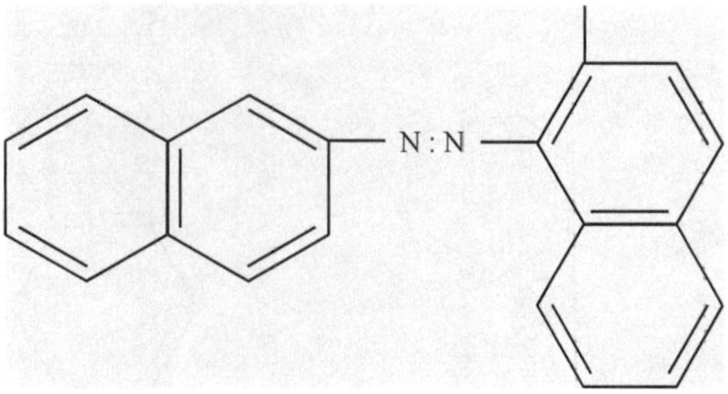

Figure 5.61

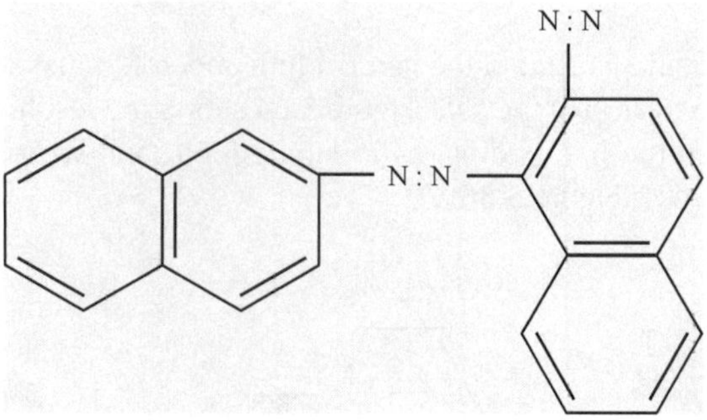

Figure 5.62

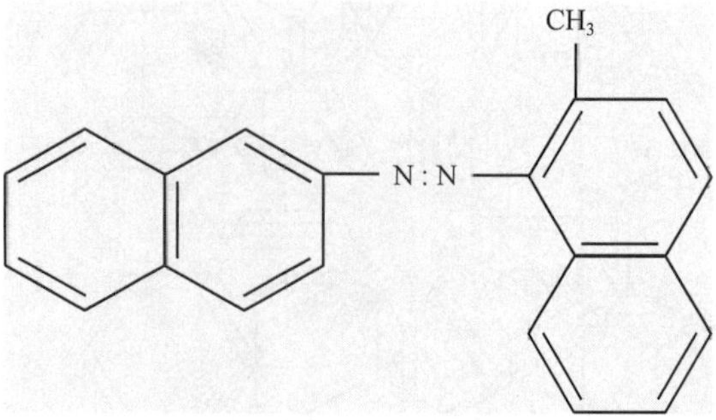

Figure 5.63

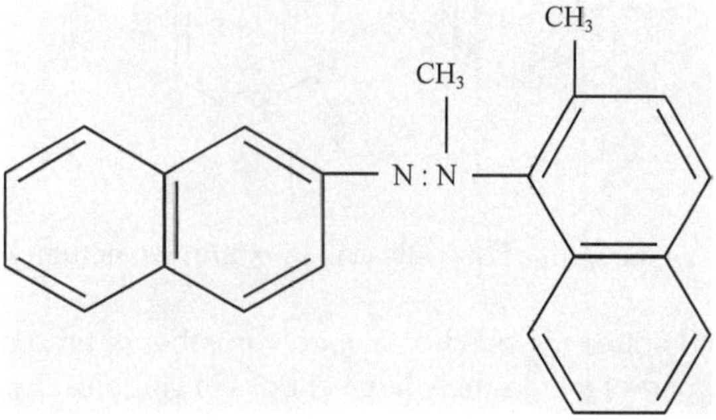

Figure 5.64

We guessed right about how much space to allow the text, and we chose a font and size to match our target diagram. If we had made a poor estimate, it would be a simple matter to lasso a double hexagon and the line or lines attached to it, and move the whole set appropriately. The rest of the text is going to be the same font, size, and weight, so rather than add new text, we copied the existing text, as in Figure 5.62, and edited it, as in Figure 5.63.

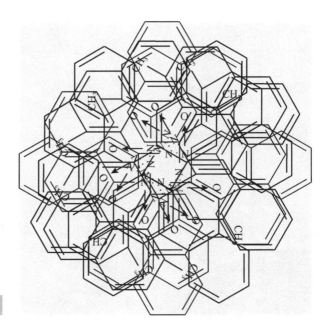

Figure 5.65

Nearly finished! Lasso the new text and the vertical line under it, make a copy, and move the copy over to the top of the "N." Figure 5.64 shows the result.

Edit the text, and change the line ending, as in Figure 5.65, and we're finished. Time for some exuberance (Figure 5.66) . . . .

Oh, please excuse us!

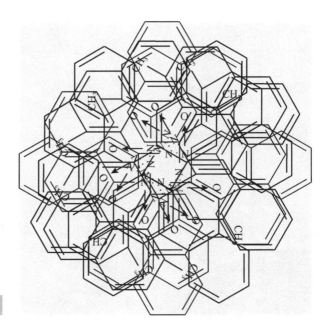

Figure 5.66

Notice that in all of this, we actually *drew* only one line, and we actually *wrote* only one piece of text.

Of course, we copied and edited the one line a goodly number of times, but we did it as the line became part of increasingly large chunks of graphics. By copying and editing, we not only made sure the various lines were the appropriate relative length (mostly identical), but we also saved considerable time. Drawing a line is relatively tedious and painful—you have to worry about where the starting point is, whether the line is at the proper angle, and whether it's the right thickness. By copying and editing the line we started with, we answered most of these questions automatically. By grouping our work in larger and larger modules, we reduced both the accuracy with which we had to align things, and the number of keystrokes.

The number of lines in the diagram increased fast. The sequence of the number of lines was 1–2–4–6–9–18–17–34–35–36–37–38. A couple of these lines are unnecessary (where two hexagons overlap), but provided the alignment is good, they won't show so there's no need to delete them. There are only 12 major steps in the process of generating 38 lines.

If you happen to have software that will let you produce a hexagon from a menu, even fewer total steps are needed!

The key in this example was to recognize the biggest grouping of graphical elements that was repeated (or nearly repeated—some editing is always a possibility). Here, it was the double hexagon.

The same principles go for the text in graphics, too as shown in Figure 5.67. We wrote (from scratch) only one piece of text, then we copied it and edited it as needed. This meant that the font characteristics were bound to be consistent all over the diagram, and that saves time compared to writing original text every time.

Suggestion: *For both graphics and text, get in the habit of copying and editing what you have already done.* You can save hours[4] this way!

Mind you, be sure you actually *do* the editing on the text. Figure 5.67 is a gray-scale version of a graphic that appeared in an article[5] on the challenges of photovoltaic power in fossil-fuel based power systems. Obviously, the professional graphic artist that put this together used the copy-and-edit shortcut. Obviously, too, that artist forgot the second part of the shortcut: *the edit.* Most likely the numbers were supposed to be 7, 8, 9, and 10, but it is a bit much to expect the reader to feel confident about that! They might have been 4, 5, 6, and 7.

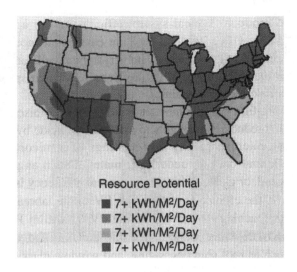

Resource Potential
- 7+ kWh/M²/Day
- 7+ kWh/M²/Day
- 7+ kWh/M²/Day
- 7+ kWh/M²/Day

Figure 5.67

[4]In fact, evidently in recognition of this principle, the people at Corel have added a slick feature to the Presentations software that is part of the WordPerfect suite. On the assumption that you have some data and you want a fairly standard kind of a graph, there is an icon in the top left of the screen that will bring up a ready-made spreadsheet and your choice of graph. The idea is that you type your data over the data they give you. You edit the labels; you reformat the graphic. Some of the graphs in this book were done that way. The approach can save a lot of time, depending on what you're trying to do.
[5]In *IEEE Power and Energy Magazine* for May/June 2004, pages 40–48.

At this point, let us assume that you have created the perfect graphic to demonstrate your point, and that the graphic is in your graphics program. That's all well and good, but you need it in your document in order to be any use. How to get it there? (We will examine how to get spreadsheet files into your document later.)

There are two parts to the question of getting the information from a graphics program and putting it into a word processor: the aesthetics and the mechanics. Not only do we have to insert the graphic information into the word processor, we have to organize the whole document. We are concerned with the topic of *layout*.

LAYOUT BASICS

For any document, there are some basic rules of layout. The rules are there to help the reader to follow your writing, and connect the words to your graphs. Following these rules will also have the effect of making your paper *look* better.

Layout is the art and science of combining the visual elements (text and graphics) of your paper into a *pleasing* and *readable* arrangement of pages that can be printed and published. Note the italicized words. We want to create pages that are pleasing to the eye, but are also readable–this means that your readers should be able to follow the flow of your paper as it rambles through long and complex sentences and in and out of graphs.

Here are some of the rules that we think should be followed for all layouts. We'll list all the rules first, and then discuss each one in turn.

Rule 1. All figures and tables that are to be used in the paper must be *referred to* (or "called out") in the text.

Rule 2. Figures and tables should be placed in line in the text *as soon as possible after* they are called out in the text.

Rule 3. Figures and tables should be called out *in order* and inserted into the text in the order they are called out.

Rule 4. Figures and tables must be combined with text in such a way that the text is still *easily followed* by the reader.

Rule 5. The *font size* in the graphic must be adjusted so that when the graphic is inserted at its final location and size, it "matches" the local environment.

Here are some more general guidelines:

- Layout is intended to help clarify and support the content. Don't let bad layout deflect the reader's attention from your content.
- Don't delude yourself. Great style and layout can help bad content—but only so far. Logic, clarity, and meaning should drive the layout.
- Remember to allow enough *time* for layout. It is much quicker to solve layout problems with a WYSIWYG program than with the old cut-and-paste paper approach. But it still takes time to get it right.
- Be consistent in your style and layout. It will help the reader recognize, identify, and comprehend different types of information, which in turn will help them understand your big picture concept.
- If you *really* want to be different, do it right.

We are not trying to teach you how to write a paper, but there is an obvious interaction between the process of writing and the process of layout. As you write, keep in mind the *message* you are trying to communicate. When you are writing your text and creating your tables and graphics, make sure that everything helps tell your story. Don't spend time on material of dubious relevance.

You also need to keep in mind the page limits for your publication. Some publications have hard limits (some as low as one or two pages), while others suggest a page limit or set a minimum and maximum page limit. Some publications will allow additional pages, but at additional cost to the author. Make sure you know before you start if there is a page limit and what it is. As you are writing your text and creating your tables and graphics, keep that page limit in mind.

Keep in mind also the layout restrictions that the publication imposes. Do they specify one column or two? Do they specify column and gutter widths? Is there a font size? All of these things will affect your layout decisions, so you need to know the answers before you begin the layout process.

Layout is concerned with the *flow* of the paper. One of the reasons we wrote this book is that we dislike publications where all the graphics are collected at the end, and the reader has to keep turning pages between the text and the graphics to follow what is being said. Embedding the graphics is more difficult than collecting them at the end, but it will make it easier on your reader to follow the flow of your story. Respect your readers!

Now, let's look at those rules a little more closely.

Rule 1. All figures and tables that are to be used in the paper must be referred to (called out) in the text.

This rule is pretty simple. It means that you need to *mention* each table and figure somewhere in your text. This mentioning may be more than a simple reference such as a parenthetical "(See Figure 5.1)." On the other hand, simple references do come in handy sometimes to solve layout problems. (More on that later.) This rule also means that you shouldn't use figures or tables that you haven't talked about in the text. *If you haven't talked about it, leave it out.* If you want to include a table or figure that you haven't mentioned, you must edit your text to include a reference to it.

Rule 2. Figures and tables should be placed in line in the text as soon as possible after they are called out in the text.

This rule sounds simpler than it really is. When you lay out your tables and graphics with your text, you need to place the tables and graphs as close to where they are called out as possible, but you still need to maintain a flow to your text and readability for your readers. Ideally, for a single-column paper, you should place a table or graph at the bottom of the page if it is called out in the first half of the page of text, or at the top of the next page if it is called out in the bottom half of the page of text. Figure 5.68 is an example.

The graph in Figure 5.68 is a kind of electronic strip chart and needs the width of most of the page. In a lot of situations, however, you will end up with white space at the sides of the graphic. So it goes. The callout A for the figure is too near the bottom of the page to be followed by the figure. The arrow shows the "connection" between the callout and the figure. The figure is almost page width and has gone to the top of the next page.

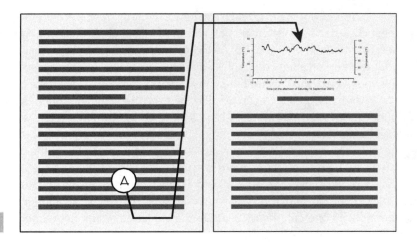

Figure 5.68

Note that the text just flows around the figure. In a single-column layout it is possible, but not mandatory, to put a wide graphic at the end of a paragraph. With a two-column layout, and a smaller figure, it is customary to put the graphics at the end of the paragraph in which it is called out.

Things get tricky when you refer to several tables and graphics either together or in quick succession. You may have several tables and graphics together on one page—and that may not be the most aesthetically pleasing layout. If you get into this situation, you may really need to think about what would be involved to change it.

It can sometimes be simpler to do the layout for a two-column paper, because you have more flexibility. A table or a graphic may often be adjusted to fit in one column or two. Sometimes you don't have to put a figure on the next page, you can put in it the next column. However, mixing one-column tables and graphs with two-column tables and graphs on the same page can cause readability issues, so you need to be careful. The example in Figure 5.69 looks OK.

Rule 3. Figures and tables should be called out in order and inserted into the text in the order they are called out.

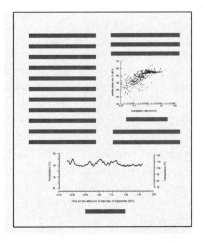

Figure 5.69

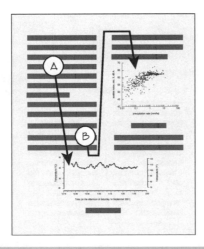

Figure 5.70

Rule 4. Figures and tables must be combined with text in such a way that the text is still easily followed by the reader.

These rules sound straightforward, but may be difficult to follow. Complying with both could cause a lot of work and rewriting once you actually start layout. Keeping the figures and tables in the order that makes sense to the overall concept is important, and generally you should make a serious effort to keep them in that order[6]. The previous figure, for example, could have been a way of dealing with callouts as shown in Figure 5.70.

The callout A for the wide figure is in the left column, first paragraph. The callout B for the smaller figure is in the left column, in the next paragraph. The arrows show the "connections." The first figure is two columns wide and has gone to the bottom of the page, and the text just has to flow around it.

Since the layout had the paragraph with callout A finishing in the middle of the page, the graphic logically goes to the bottom of the page. Suppose there was insufficient space. One might be tempted to put it at the top, instead. Since it is a two-column figure, it would have then ended up appearing *before* it had been called out. In other words, it would have been the situation shown in Figure 5.71.

Do not do this! The top of the following page is a location that solves the problem.

What sometimes happens, however, is that as you are laying out your paper, tables and figures do not occur in an order that facilitates layout. What if the callouts had been as shown in Figure 5.72?

Here, the first graphic has been moved to the end of the first paragraph after the callout, but there wasn't room for the second figure to follow the end of the paragraph containing its callout. Therefore, the figure had to go below the first figure, as shown.

This is acceptable, but not a *good* solution, for two reasons. First, the text of the paragraph following the B callout is now interrupted by the graphic, and second, the two graphics run into one another.

---

[6]The layout rules for a book are broadly similar because of the same need for clarity and straightforward reading. However, it is sometimes necessary to put a figure before its callout, on a facing page, in order to keep it from moving into a subsequent subsection.

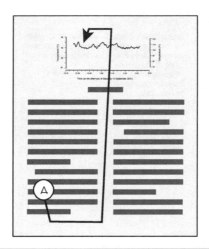

Figure 5.71

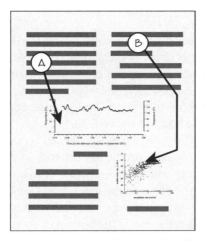

Figure 5.72

There are several solutions. One solution is to move the two-column figure to the bottom of the page, as before, in Figure 5.70.

This solution relaxes the rule about putting the graphic *as soon as possible* after the callout. Here, we allowed another paragraph to start before we inserted the graphic. Sometimes you just have to be bold enough to break one of the rules of layout, and this is not a terribly bad one to break. (It would be much more of a no-no to insert a figure *before* its callout, for example, or to have the figures out of order.)

Alternatively, if there is space in the paper, another solution is to put the second figure on the following page.

An alternative that is often overlooked (at least by infrequent writers) is to change the text to call out one or more of the graphics earlier or later. We realize that every word you have written is a pearl, and that this kind of forced change would be a travesty, an insult to the beauty of your prose. However, it is often easier to make a simple wording change than to change the rest of the layout.

This is where a simple callout ("See Figure *x*") comes in very handy! It takes up little space and therefore has minimal impact on your text structure.

If you do make such a change, however, you should spend some time rethinking where you've called out figures, and (if necessary) you should rewrite more of your text to facilitate better flow.

Rule 5. The font in the graphic must be adjusted so that when the graphic is inserted at its final location and size, it "matches" the local environment.

We have offered guidance on font sizes in Chapter 4, but it won't hurt to discuss the topic further. A typical technical paper is set in 10-point Times Roman. The "ideal" graphic in this environment would use 8-point Arial. All too often, we see graphics that have been created in some software at some convenient magnification, and when scaled to fit the paper the font size has changed enough that it is illegible. Or maybe it got seriously larger.

The problem is that the process of putting graphics into word processing programs has become so flexible it is limp! Once you have dropped the figure in, you can stretch it and resize it to your heart's content. *Resist the urge!* Try instead to make the figure "work" in whatever software you are creating it, and at a size that it will need to be in the final paper. With practice, this approach will actually make it simpler for you to produce graphics whose font size in the document is consistent.

If you are working in PowerPoint, for example, and your graphic is going to go into a paper as a single-column figure, try drawing a box of the appropriate width (typically about 3.3 inches, or 8 cm) and keeping your graphic inside the box. There are rulers at the top and side of the screen for the purpose. You can set the font size, and then when you copy the image into the word processing program, it will not have to change.

Now that you know the rules, you're ready for the initial layout. Start by setting up your one-column or two-column pages of text, then insert your tables and graphics into the text (more on the mechanics of how to do this later). Insert all of your tables and graphs in what might be the "obvious" places, and see how it looks. If you're lucky, you won't have too many problems to fix. If you're not so lucky, be prepared to spend some time finessing your layout.

In the end, you may just have to *remove* some of your graphs or you may have to *rewrite* some of your text. Each time you make a change, be sure that the flow of the figures still works and that you haven't created any new problems. Be prepared to makes lots of revisions, continuing to work it through until you've eliminated all the layout mistakes.

Learning to do document layout is an evolutionary process. You can learn from mistakes, and even turn accidents into opportunities.

Now we need to look at the mechanics of getting your graphics into your word processor. In particular, we are concerned with Microsoft Word and Corel WordPerfect as the recipients of our efforts.

First, let us deal with Microsoft Word.

INSERTING GRAPHICS INTO MICROSOFT WORD

The easiest way to insert a graphic into your document is to use the clipboard, by selecting all the parts of the graphic, and copying everything to the clipboard

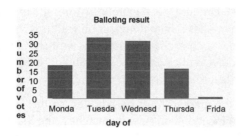

Figure 5.73

with `Ctrl-c`. Once you have the image in the clipboard, it matters little what the originating software was. The procedure we will describe works for PowerPoint and Excel, and even WordPerfectGraphics files.

You *can* simply paste the clipboard into your Word document. This is not recommended in the least. The labels are quite likely to get very discombobulated, and the text will pass under the image, rather than wrapping around it. Figure 5.73 is an example of something from PowerPoint that was just pasted into Word.

There's lots to fix here! Look, for a start, at the horizontal letters in the vertical ordinate label! "`n-u-m-b-er-of-v-ot-es`" indeed! Look, too, at the truncated words all over the place.

There is an improvement to be had by doing an `Edit`/`Paste Special`/ `Picture`, but even this is not recommended. There are a number of things about Word that do not work well and, sad to say, embedded graphics is one of them.

You can also do an `Insert` / `Picture`, though Word will not import many kinds of graphic files. It will not import PowerPoint files, for example. This is a considerable (and surprising) shame, as PowerPoint is the one graphics package a Word user is likely to have!

THE TRICK FOR WORD

The trick that most people use to reduce the problem of getting a graphic into Word to a manageable level is to `Insert` a `Text Box`, and paste the graphic into it. Once you have done this, the box (and therefore the graphic) can be anchored to the text in a way that you select. More about this in a moment.

It may be that when you create the text box, the graphic is bigger than the box you created. No matter, the box can be stretched and the graphic compressed. You can also specify the size of the text box and define how text is to wrap around it.

There are two ways you can paste the graph into the text box. If you have the file open in an application, you can lasso the graph, copy it, and paste the contents of the clipboard into the text box. Alternatively, if you have Windows Explorer open, you can highlight the file name, copy with a `Ctrl-c`, and paste the file into the text box without ever opening it! This method works only for some file types. It does not work for PNG files or JPG files, but it does work for TIFF files and WPG files. It is not recommended for PPT files, as it is very slow, and it includes all the white space around the graph for this type.

But once the box is in the Word document, you need to be careful with what you select. If you click on the pasted image, you get eight little black squares and some lines to tell you that's what is selected, as in Figure 5.74.

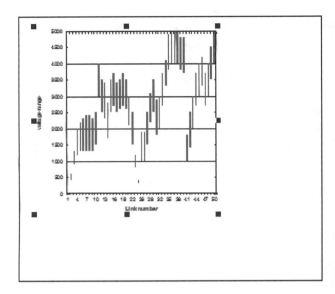

Figure 5.74

This will allow you to change the size of the image inside the box. If you stretch the (inside) box from the bottom right, it will maintain the original proportions, and will stretch the outer box to fit. If you compress the outer box until it is smaller than the inner box, the image on the page will be clipped. Making a further adjustment to the size of the inner box will cause the outer box to be resized.

An alternative approach is to right-click on the image and select `Format Object`. There is a tab for `Size` that will allow you to specify exactly the size of the finished drawing in your document.

If you need to make a minor change to a graphic that you already have inside a text box, you can return to the source software and make the change, and then swap it for the image embedded in Word. Thus, if you `Ctrl-c` the image onto the clipboard, and then select the inside box in Word, you can do a `Ctrl-v` to replace the image without going through the work of creating a new text box. However, if you delete the inside box (the image) it will automatically delete the text box too, and you will have to create a new text box to insert the new image.

If you click in the white space between the picture and the box, you get the effect seen on Figure 5.75. Now you have eight little white squares and a sort of diagonal line pattern to indicate that something else is selected. So far, we have not found a use for this selection. We think it may have value when there is text in the text box, but it seems to have no function when there is only a pasted graphic.

However, if you click on the diagonal line pattern, it changes into a sort of shading pattern, as in Figure 5.76, and now it *is* useful.

Now, if you right-click on the shading pattern (or on the black line itself), you can remove the black line around the text box. Books do not typically have black lines around their figures, and it is not in the style of some journals. It takes some work: it begins with `Format Text Box`, then you have to select the `Colors and Lines tab`, and then select `Color` and then `No Line`, before you click `OK`. This might be tedious, but if you then right-click on the graphic

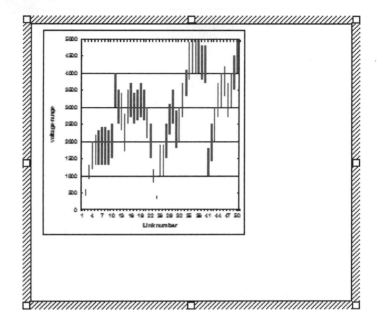

Figure 5.75

and `Set Autoshape Defaults`, then all subsequent text boxes inserted into the document will be similar.

If you look over in the left margin when you have the shading pattern around the text box, you might see that the anchor symbol ⚓ appears. (This feature seems to vary with which version of Word you are using.) If it is visible, you can move this symbol around without moving the graphic, so as to anchor the graphic to whatever piece of text you want.

In any event, you can right-click on the text box, select `Format Text Box`. Then, on the `Layout tab`, select `Advanced`. On this there is a

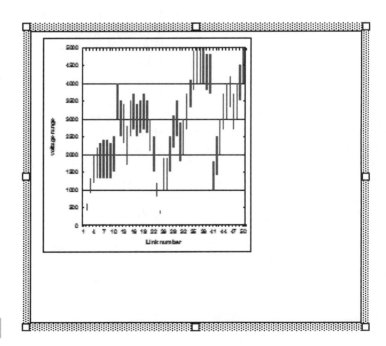

Figure 5.76

`Picture Position` tab that lets you lock the anchor to the text or to the page. If you are locked to the page, you can move the text box, and the anchor stays put. If you are locked to the text, the anchor moves when you drag the box. It should be that when you insert text, in one case the text box will move, and in the other case it will stay still. It seems not to work always, however.

INSERTING GRAPHICS INTO WORDPERFECT

The text box approach is available in WordPerfect, too, but it is *not* the recommended way to do things. No trick is needed. Instead, use WordPerfect the way it was designed to be used. Do `Insert / Graphics / From File`, and insert a WPG file from Presentations. If the original graphic was not done in Presentations, paste it into Presentations from the clipboard, possibly using a `Paste Special`. You should have no problem pasting into Presentations. (You may also observe that Presentations will open most file formats.)

Once you have inserted the Presentations graphic into WordPerfect, right-click on it and choose `Position`. (This allows you to change the kind of anchor used.) Change the anchor to `Paragraph` (typically) or `Character` (rarely). The default is `Page`, which is hardly ever useful. If you click on the graphic and drag it, you will see where it is anchored. Figure 5.77 shows a section of a page, where the figure is the anchor symbol and the push-pin is the symbol used in WordPerfect to indicate the location that the figure is attached to the text.

Here the little push-pin image in the margin is the anchor, and the line from it shows which graphic is being anchored. Once you let go of the graphic, the push-pin goes away. If the push-pin does not show when you move the graphic, you do not have the paragraph anchor.

You will notice in WordPerfect that, when you right-click on the graphic, you have a large number of options in addition to changing the type of anchor. You can change (or set) the size of the graphic, change the way the text wraps, and so on. If you want a border (the WordPerfect default is not to have a border around graphics), you can choose from a number of different styles.

There is one little bug that we can tell you about, that was particularly frustrating in the production of this book until we tracked it down.

Here's the deal. While most of the graphics in this book do not have a line around them, practically all of them have a little white space at the top and the bottom. That white space is inserted to give the right "feel" to the graphic. Therefore, it was necessary to control how much of it there is. We wanted to able to pack the text right up to the graphic like Figure 5.78 if we felt so inclined. We did not want to be obliged to put some other amount of white space. Sometimes, we found we were getting additional white space. More importantly,

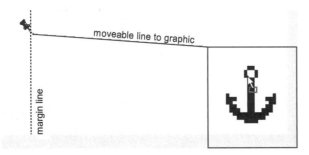

moveable line to graphic

margin line

Figure 5.77

we noticed that when we deliberately set the size of a graphic (in inches, as above), the size we set was attached to the overall image *including the added white space.*

What was frustrating was that it did not happen consistently. Some graphics would behave perfectly well, and others (seemingly created the same way) would come with attached white space. The white space could be moderate, like the version in Figure 5.79. Here we have added a border to show the outline of the graphic. Sometimes the amount of white space would be enormous, like the version in Figure 5.80.

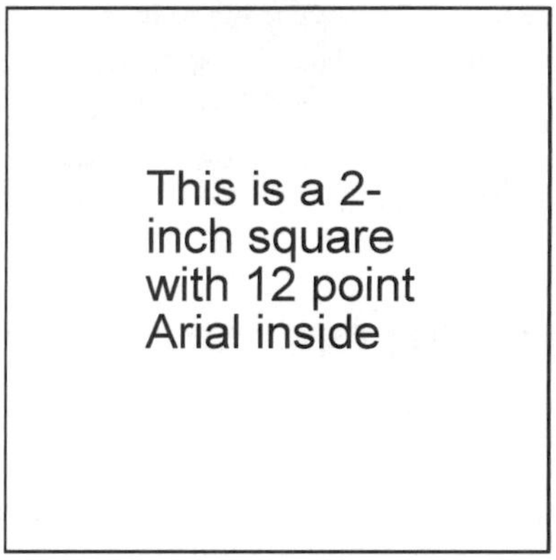

This is a 2-
inch square
with 12 point
Arial inside

Figure 5.78

In all the cases above, the graphic was set to be 2 inches wide. In a way (but not a useful way), it was. Naturally, we were curious about what was going on! Our curiosity was further aroused when we noticed that the attribute of having or not having a border would stay with the graph after extensive editing, including deletion of all the contents!

Evidently, there is some parameter in the WPG file that tells WordPerfect which of these options to use when embedding the file into the document. We

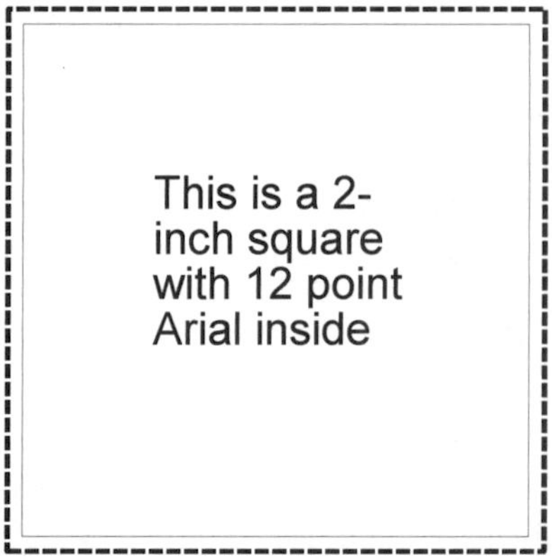

This is a 2-
inch square
with 12 point
Arial inside

Figure 5.79

This is a 2-
inch square
with 12 point
Arial inside

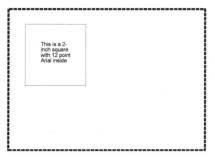

Figure 5.80

were unable to locate any help on the topic. It took us a while to find a way to control the effect once the WPG file had been created. Here is the trick.

If you just draw your graphic in Presentations, and save it in wpg format, when WordPerfect embeds it in your text file you get the entire page from Presentations inserted. If what you had was a viewgraph that more or less filled the screen, and you were setting the size by eye, you probably would not notice the white space around the image. On the other hand, if you had an image that was small, and you wanted the size reproduced accurately, you were in for a surprise!

On the other hand, if you had *selected* the contents of your graphic (all of them), when you did your save in Presentations you would be asked whether you wanted to save the whole file or just the selected items. Picking the option to save just the selected items would get you the version with the small border.

How about the version with no border? As far as we can tell, that came about by saving just the selected parts of the image into a format other than wpg first. This version of the image could be consistently created by selecting the entire image, saving the selection as a Windows Metafile (wmf) or an Enhanced Windows Metafile (emf), and opening it up as a new file in Presentations. Thereafter, the thing could be saved as a wpg without acquiring a border.

Now, you might be concerned that the process of taking your complex drawing and converting it to and from some other format might change something. After all, that is often the case (see Chapter 12). But here we can be of good cheer, as there is a trick! Draw something simple (e.g., a rectangle), select it, and save the selection as a wmf file. Open the wmf file, delete the rectangle, and now draw (or copy) your drawing. The neat thing is, you only have to go through the saving and conversion process once! Somehow, the file created this way is "conditioned" to ever after save without adding a border.

It seems to us most likely that this behavior is the result of failure to initialize something in the opening sequence and is an accident rather than a deliberate feature of the program. Let's hope the people at Corel don't fix it, because only this way can you control precisely the final size of your embedded graphic.

SUMMARY

Getting your software to do what you want, and do it efficiently, is essential if producing graphics is not to become burdensome. It is worth learning your software well—you might even benefit from cracking the binding on the manual.

- Learn as many two-hand shortcuts as you can. Using the mouse to select options from a menu is slooooooooooow!
- Whenever possible, copy and edit graphic elements, instead of drawing them from scratch.
- PowerPoint and Presentations will cycle through all objects with the tab key, once you have selected an object.
- Lassoing only works when an entire object or group is lassoed.
- Use the Control key shortcuts when possible.

  Ctrl-c copies the current selection into the clipboard.

  Ctrl-x copies the current selection into the clipboard and deletes the original.

> **Ctrl-v** pastes the contents of the clipboard.
>
> **Ctrl-z** undoes the last editing operation.
>
> **Ctrl-s** saves the file.
>
> **Ctrl-d** works to make duplicates of your selection and move. This does not work in all applications.

- Draw a box, and while it is still selected, start typing. What you will get will be the default font, centered on the box. This works in PowerPoint and Visio.

- Holding the Shift or Ctrl key while you draw a line constrains the angle in most programs.

- Holding the Shift key while you stretch a line or drag an object or group keeps the direction constant (some programs).

- Suggestion: *Learn to avoid precision control and adjustment in your work*—let the computer do the work for you.

- Remember the rules of layout:

  > Graphics and tables *must* be called out
  >
  > They should be inserted *in order as soon as possible* after the callout.
  >
  > The arrangement should not impede the *flow* of the paper.
  >
  > The *font size* should match the environment.

- Insert graphics into MS Word by using a text box.

- Insert graphics into WordPerfect by drawing (or pasting) the graphic into Presentations and using the Insert / Graphics / From File sequence.

EXERCISES

5.1. Find an efficient way to draw the reliability block diagram shown in Figure 5.81. You may improve the drawing, if you wish. Say which software you used, and describe the sequence.

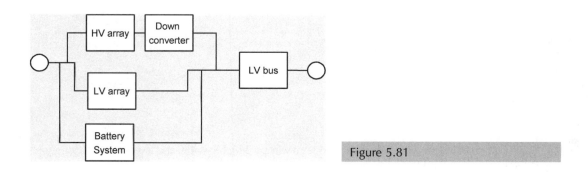

Figure 5.81

5.2. Find an efficient way to draw the proposed data structure shown in Figure 5.82 (without using a spreadsheet). Say which software you used, and describe the sequence.

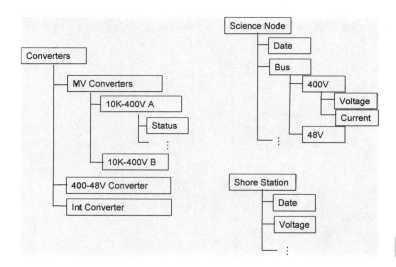

Figure 5.82

5.3. Scan (or download) a graph file in bitmap format. Trace just the axes and the labels. Apply your own made up data line.

5.4. For the graph in the previous exercise, produce a version 3 inches wide, and a version 6 inches wide. Adjust the line weights and font sizes to suit.

5.5. For the graph in the previous exercise, change the aspect ratio to make the graph taller and narrower.

Presentations or how to succeed in business

6

T est and Measurement World (August 2003) surveyed what it takes to get ahead in "this tough job climate." They found computer skills ranked number one, followed by communication/presentation skills, and then project management know-how. Maybe the same is not true in *your* field; maybe they never put anything about communication skills in the job ads where *you* work, but they sure do most places. Apparently, nearly 85% of the respondents to the *T&MW* survey thought communication and presentation skills important in their work. If you want to get ahead, get an overhead!

A remarkably similar result was obtained by the annual survey done by *Design News*. The *Electronics Industry Yearbook/2004* presented those findings as shown in Figure 6.1.

Of course, graphic communications—even overheads—have been used for a very long time indeed. They were once a matter of life and death (Figure 6.2).

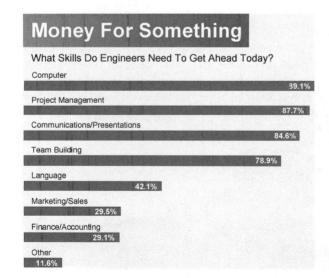

Money For Something

What Skills Do Engineers Need To Get Ahead Today?

Computer — 39.1%
Project Management — 87.7%
Communications/Presentations — 84.6%
Team Building — 78.9%
Language — 42.1%
Marketing/Sales — 29.5%
Finance/Accounting — 29.1%
Other — 11.6%

Figure 6.1

Figure 6.2

The technology has advanced, as in Figure 6.3, allowing modern presentations to be kept much more up to date: until these days, all you needed to take to the meeting was a memory stick with your presentation on it, ready to go. Our aim, in this chapter, is to help you improve what it is you have on that memory stick. We assume that you've never done a presentation before, and

Figure 6.3

we are going to go through all the steps along the way. If you really are a beginner, we suggest you study this stuff closely. Even if you have lots of experience as a presenter, you may find this process instructive on some point or other, so we suggest you skim through the chapter, at least.

PURPOSE OR OBJECTIVE

The first thing you have to do is get *organized*. You are making a presentation for a specific purpose. Almost always a presentation is aimed at persuading someone of something. Are you clear on what the objective of your presentation is? You must be, before you go any further.

In particular, the objective should be written down on a bit of paper that you can keep referring to, and should be in the form of what you expect out of your audience. This is an easy thing to do if you are "selling" something, such as the need for a new piece of equipment. You could have "Objective: to get management agreement to purchase a new XYZ machine." It is not so easy to define the objective if you are presenting the results of some research at a conference, but you have to try. Thus, a statement like "Objective: to explain how power lines make noise in rain" is not much use, because while it gives you a chance to present your results, it does not take audience reaction into account. "Objective: to enable the audience to see how the various parameters interact in power line noise" would be better. You could gauge audience understanding by the questions they ask. (If you make this assessment as you give the presentation, you should be able to make adjustments to what you say in real time, to help you achieve your objective—even though you have a fixed set of slides.)

If you are making the presentation to go with a paper that has been accepted for a journal and will be given at a conference, you have a little latitude. It may be that rather than present a short version of the paper, you would like to give emphasis to the science behind what you did, or perhaps to present some results obtained since the paper was submitted. Usually, such a variation from the material in the paper will be acceptable to the conference organizers, and may be appreciated by the audience—who can, after all, read the paper in the conference proceedings. But be sure you have an objective statement that matches what you will say. For example, you might write "Objective: to enable the audience to appreciate the science behind our decision to do *x-y-z* that we described in the paper," or "Objective: to have the audience understand the very latest findings in the *abc* field."

After you have fixed your objective or purpose, you must tailor the presentation to that purpose, *and* to the audience. Tailoring to the purpose means deciding *what* it is you say. Tailoring to the audience means deciding *how* and at what level of *detail* you say it. This means you have to know something about your audience. Are they your peers, your managers, or the general public? If you know little about them—perhaps because you are giving a presentation to a new customer and her associates—it is a good idea to think about adding a question or two in the early part of your talk to give you some guidance. Mind you, depending on what you learn, it can be difficult to adjust while you are already talking, and you have the slides prepared. So find out what you can ahead of time!

Will it be a large audience, as in a paper presentation at a society meeting? The bigger the audience, the more polished your presentation should be. Why that should be is not obvious, but it is a good general rule. Is the presentation a rapidly prepared impromptu one that you were invited to give after a meeting got under way? Unless you want to prove that you are an old-fashioned scientist, do *not* generate your slides in real time on transparencies while you are at the podium. Make them ahead of time during the meeting, and try to keep the presentation short, as you are probably going to create a schedule problem for the meeting organizers. (And you should hope that you have an earlier, and similar, presentation all ready to go, as your starting point for editing.) Is the presentation one of a series, for example, in a project review? Find out what material the previous speakers will have covered, so you can (a) use it without repeating it and (b) avoid boring your audience with a duplication. You need to adjust your presentation accordingly.

STRUCTURE AND OUTLINE

Depending on your objective, your outline may take on a different shape. There are a few common structures you can use for your presentation outline, just as there are in outlining a report. (Remember the old adage "First, you tell 'em what you're gonna tell 'em, then you tell 'em, then you tell 'em what you told 'em.") We suggest you follow the advice given to writers:

- For a *narrative*, where the purpose is to have the audience understand what you did or what happened in some particular experiment, use a three-part approach. Start with an overview and some background, then spend about two-thirds of the total time discussing the main topic, and finally summarize the whole thing.

- For a *proposal* for something new or to make a persuasive case for something, use a two-part structure. First, show that there is a problem; second, show that you have the solution. (Along the way, you can suggest that your competition has only an inferior solution, but it's best not to overdo negative things during a presentation.) Spend roughly one-third of the time on the first part, and two-thirds on the second.

In either situation, the exact time breakdown will depend on the audience, but the idea that the main topic occupies two-thirds of the total is a reasonable start.

For a routine report, such as a monthly progress report, the overall structure may be predetermined and may not seem to be either of the ones shown here. Normally, such a presentation would have a simple structure, and you can readily fit your material into it. However, it may be that you have encountered some problem during the reporting period. In this case, it is usually acceptable to modify the standard format and insert something along the lines of a two-part proposal structure within the overall outline.

Once you have your outline, you can allocate a number of viewgraphs to each of the main points, based on how long your presentation is allowed. Suppose you have been allocated 20 minutes for a talk, and asked to add 10 minutes more for questions. You have a three-part talk, so a split of 6 minutes for the intro and background, 13 minutes for your main point, and 1 for the summary would be a good initial allocation. In most circumstances, the rule of thumb is 1 slide per minute. Regard this as a rough preliminary allocation of time.

USING A SUMMARY

Some journals require a summary of the main points of your paper, in not more than five (short) sentences. If you are required to do this, you can use the sentences to start your presentation outline. Write five (or fewer, but not more) sentences that capture your argument, and map the sentences into the overall two- or three-part structure that the presentation demands. You might regard this as an intermediate step between the structure of the talk and the allocation of slides to each part.

But if the organization you are writing for does not require a summary such as this, you might as well skip the chore and go directly from an overall structure to an allocation of time. The details will work out later.

STORYBOARD

Of course, the division of time between elements of the talk is adjustable, and that is the next step. Some of your points may be easy to make; others may be more contentious, and therefore require more support. Starting with the outline and the (roughly) allocated number of slides, you prepare a storyboard. Just like in the movies.

Has it ever occurred to you what excellent quality the ordinary, average movie is, from the storytelling point of view? This excellence does not happen by accident: storytelling is a well-honed art, and many highly skilled people contribute. Always, the process of designing the movie includes a storyboard.

Figure 6.4 is a sample from a storyboard from `http://www.exposure.co.uk/eejit/storybd/ts7.gif`.

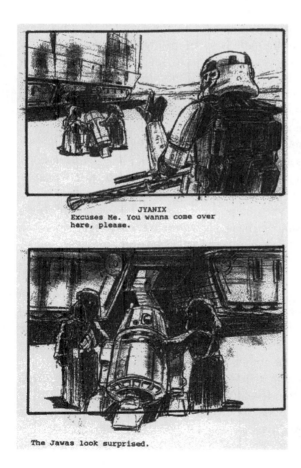

JYANIX
Excuses Me. You wanna come over here, please.

The Javas look surprised.

Figure 6.4

Well, as a matter of fact, the example of Figure 6.4 isn't quite the level of detail you need. Fortunate, really, in view of the significant level of resources required to storyboard a movie! Several people might take a year or more to storyboard a movie. Compared to that, your effort is very small.

What you need to do is expand on your outline, and state all the points you will make. You should not be producing complete viewgraphs yet, but the elements of your storyboard should be little thumbnail sketches of what you expect the slide to contain. You do not need many words, or any detail in the pictures (*pace* Star Wars), but enough to remind you of what the content was supposed to be when you get to actually preparing the slide.

In fact, it is a good idea at this early stage to use paper and pencil, and avoid the computer. You're still in the phase of building up the way you will tell your story, you're not actually telling it yet. Get several sheets of paper, and be prepared to throw some away.

One way to do the storyboard stage is to use a piece of paper with one side marked off into as many rectangles as you will have slides. At this point, you're still using the rule of thumb of allowing one minute per slide. (You can adjust that with experience. It's a fair starting place.) Often, restricting yourself to one side of paper won't leave you very much room to say what is in each slide, but that's fine at this point.

On the thumbnails, we suggest you draw a horizontal line to indicate a line of text, and a few lines to suggest a bulleted list. Add maybe one or two words in the thumbnail if you feel you will need a reminder. Draw a rough imitation of any graphs or pictures (or movies) you may have, adding words only if needed to remind you what you had in mind. The process of making the storyboard might take a while, but the time is spent mainly thinking about telling the story, not making the little wiggles on the paper. A good structure makes for a good presentation, and this is where you define the structure.

Since it's harder to do inserts on paper, you may end up linking a couple of pages together, as you go through the process, and find it necessary to add material. If so, it is worth taking the time to redraw the storyboard on a single side afterwards, so you can review it more easily and check the flow.

You *can* use your presentation software in the normal way to do the storyboard, but we don't recommend it, as you will be too tempted to get into making the actual slides. (Note that while none of the graphics software normally available has any storyboard capability, there is freeware and shareware to do the job. However, that is aimed at movie making and is more complex than we need.)

To some extent, you may be able to use some often overlooked capabilities of presentation software. PowerPoint has an Outline view at the left of the screen. You can make that window large, and you can type into the words and have the changes reflected in your viewgraphs. Corel Presentations has an Outline View of slides, that is roughly equivalent (it has a somewhat clearer organization, as the main titles and subtitles are identified in a separate column, but you don't see the slide at the same time). These views allow you to see all the surrounding content, and edit the words (but not the graphics) at will. The reason we don't particularly recommend this approach is that you are likely to get too involved in the words, and not get the top-level structure ironed out first.

Both programs have a Slide Sorter view that will let you shuffle slides but will not let you edit them. If you have an existing presentation that you are reorganizing for a new audience, the Slide Sorter is a powerful tool.

We urge you to get used to using these editing capabilities, but first and foremost to get used to starting with paper and pencil if the presentation is entirely new.

The following pages give an example of a storyboard for a viewgraph presentation. This presentation was aimed at explaining (to management) why the prototype of a new design of dc/dc converter had experienced unanticipated problems when first energized. The answer was that the controls had seen a kind of instability that had not been analyzed in the design phase: chaos. We start with the basic three-part outline, with some subtopics added, shown in Figure 6.5.

That might not look like a three-part story, but Figure 6.6 shows that it really is.

Here we have allowed 8 viewgraphs for background material, split about evenly between reviewing the general technical background and reviewing the particular case. There follow 11 viewgraphs examining the results of the converter in detail, in the light of the background just given. Two slides were allowed for the wrap-up. So Part One gets 8 slides, Part Two gets 11, and Part Three gets 2. A little light in the middle, but that gets fixed later. We were aiming at a 20-minute talk, and expected an additional 10 minutes for questions.

The amount of material available was large, as considerable time had been spent exploring the situation in the laboratory. There was, consequently, an urge to include a lot of experimental results in the presentation. However, many of the results could be viewed as similar, and adding more would be a diversion from the main point: chaos. The story could be told with relatively few results, taken as representative, and the words of the presenter could at least touch on this wealth of additional material. Some additional results could be added at the end of the official presentation as "backup."

When the storyboard was put together, some slides were added to help with navigation through the presentation. An outline of the presentation was

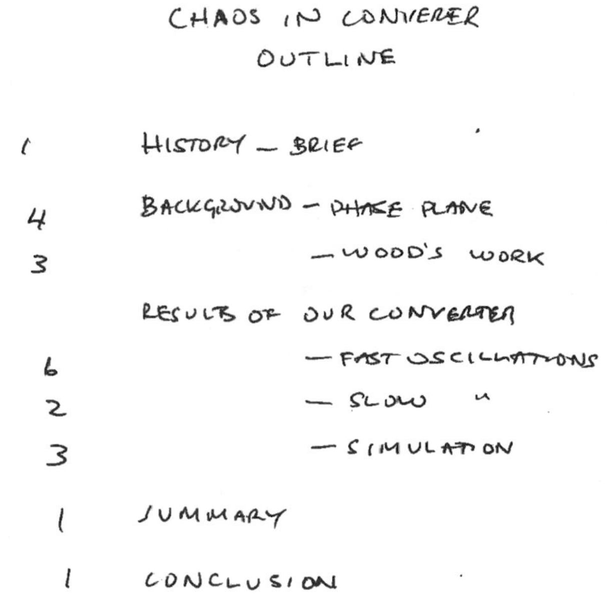

CHAOS IN CONVERER
OUTLINE

1	HISTORY — BRIEF
4	BACKGROUND — PHASE PLANE
3	— WOOD'S WORK
	RESULTS OF OUR CONVERTER
6	— FAST OSCILLATIONS
2	— SLOW "
3	— SIMULATION
1	SUMMARY
1	CONCLUSION

Figure 6.5

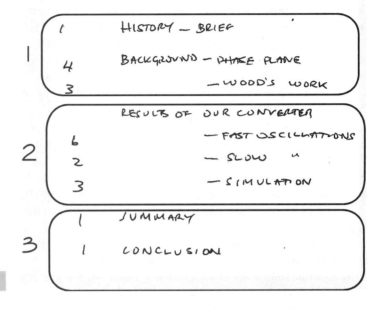

Figure 6.6

added up front and repeated every time the presentation moved from one major section to another. At such a juncture, an arrow was added to show the progress. Figure 6.7 is an example. You may consider this a useful trick. We use it often, especially for long presentations.

In the end, we realized that the argument for the converter having experienced chaos rested on the similarity of the results of a simulation to the actual

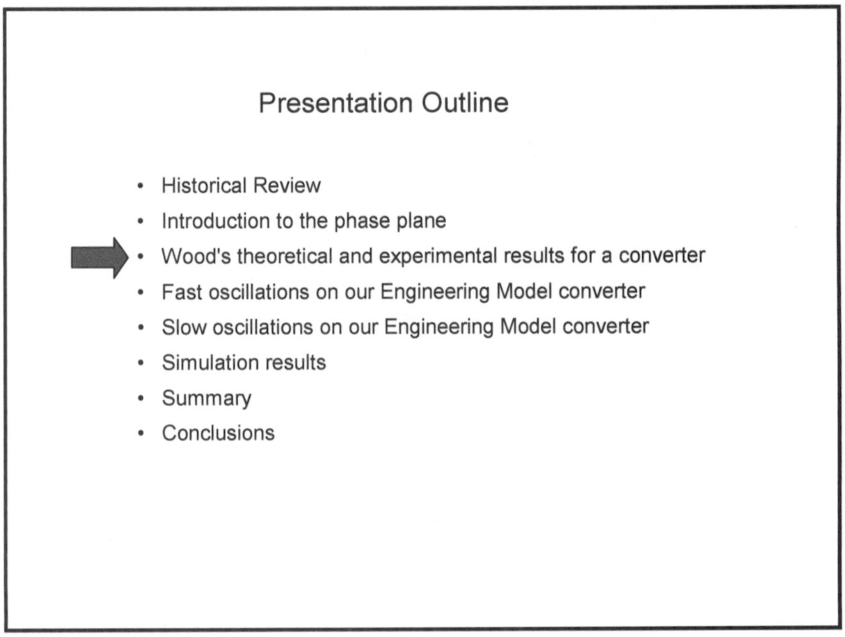

Presentation Outline

- Historical Review
- Introduction to the phase plane
→ • Wood's theoretical and experimental results for a converter
- Fast oscillations on our Engineering Model converter
- Slow oscillations on our Engineering Model converter
- Simulation results
- Summary
- Conclusions

Figure 6.7

results, and it would be worthwhile looking at the similarities in some detail. One of the slides was therefore repeated a number of times, with the points of similarity highlighted, and accumulating as the presentation proceeded. We ended up with about 40 slides altogether, but one was a title slide, seven were copies of the outline slide, and another seven were copies of the same slide comparing simulation with real results. So it was really only like there were 25 different slides with any real content.

Figure 6.8 shows the storyboard. It reads across then down.

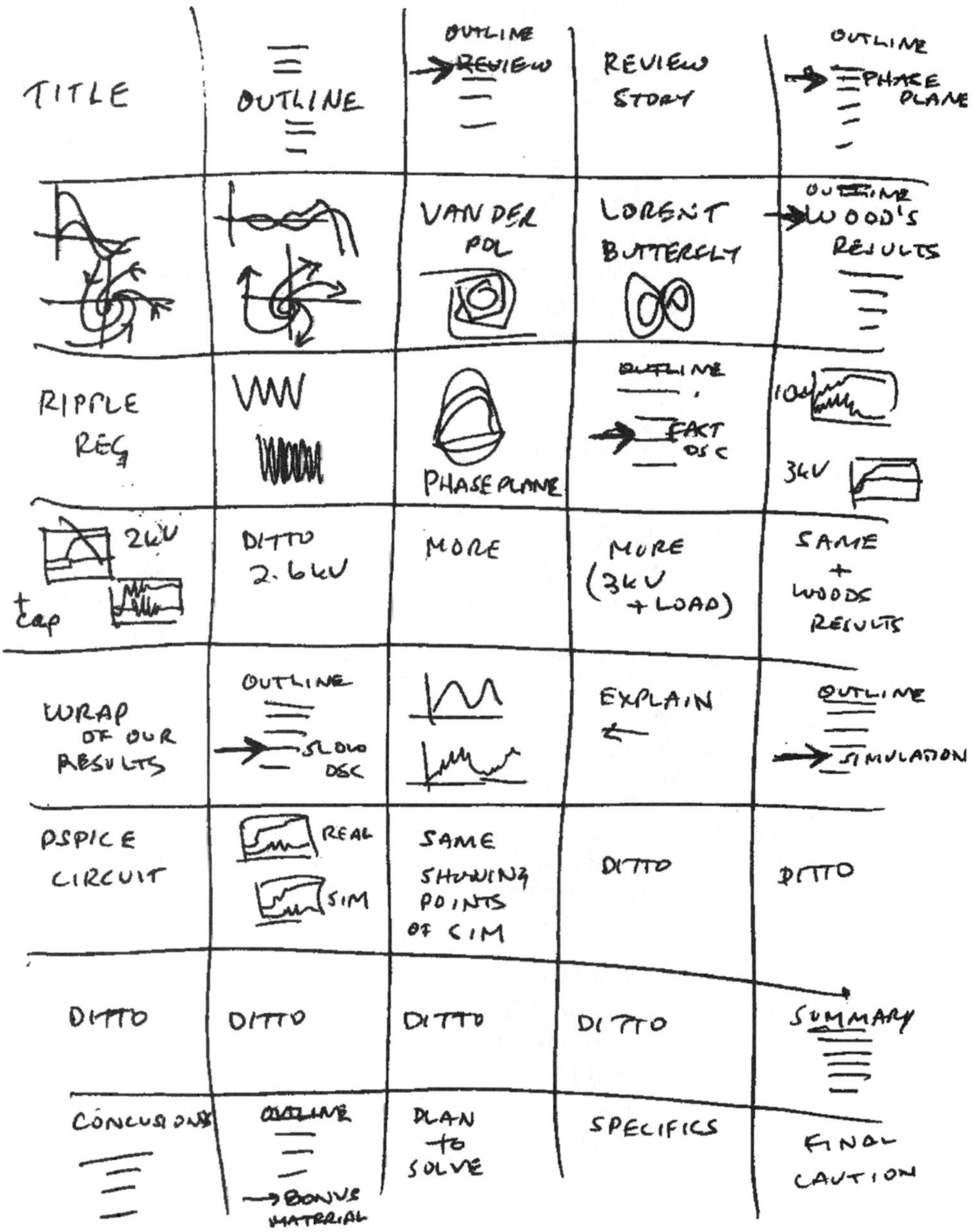

Figure 6.8

Once you have filled out the storyboard, you should consider taking a break. Give your mind some time to reestablish some sort of perspective. Create the opportunity to come back to your presentation with a fresh look.

When you do return to the task at hand, you should look for balance. You are not worrying about whether the point is well made, rather have you allotted each aspect enough space? You are looking at the whole presentation as a storyteller. Are you spending too much time on this detail, at the expense of time to clarify a difficult issue someplace else? Is there any place where there seems to be a gap?

You cannot do this assessment without knowing your audience. If you are giving a presentation to your management, the chances are you will have to include a lot of introductory material in order for the main points to make sense. This is because when you become a manager, they take out some of the bones in your head that let you think like a technical person. Or perhaps you will have to make your argument in teeny-tiny steps, leaving out none of the steps that you may feel are obvious.

Let's dwell on that for a moment. Whatever you are presenting, the chances are good that you are presenting it because you know more about the topic than anyone else. Fine, you should be able to do a great job. However, because you do know more about the topic, there are things that you will take for granted that others may not see unless you explain. Consider the following argument[1]:

> Expert testimony reveals that since the scientific revolution of the 16th and 17th centuries, science has been limited to the search for natural causes to explain natural phenomena. This revolution entailed the rejection of the appeal to authority, and by extension, revelation, in favor of empirical evidence. Since that time period, science has been a discipline in which testability, rather than any ecclesiastical authority of philosophical coherence, has been the measure of a scientific idea's worth. In deliberately omitting theological or "ultimate" explanations for the existence or characteristics of the natural world, science does not consider the issues of "meaning" and "purpose" in the world. While supernatural explanations may be important and have merit, they are not part of science. This self-imposed convention of science, which limits science to testable, natural explanations about the natural world, is referred to by philosophers as "methodological naturalism" and is sometimes known as the scientific method. Methodological naturalism is a "ground rule" of science today which requires scientists to seek explanations in the world around us based on what we can observe, test, replicate and verify.

Note the way that each sentence adds only an increment to the sum of what has gone before. Starting with "expert testimony" as his authority, the judge argues very carefully and in small steps about something most of us (as scientists and technologists) would take for granted. He sets the stage for saying, later in his Opinion, that Intelligent Design does not qualify as science, and that the School Board (the defendant in this case) was wrong to allow it to be taught "as

---

[1] The paragraph is taken from the Memorandum Opinion of the United States District Court for the Middle District of Pennsylvania, December 20, 2005, *Tammy Kitzmiller et al., Plaintiff, v. Dover Area School District et al., Defendants,* Judge John E. Jones III presiding.

science." This statement on its own, without all the closely argued background, may not be quite so obvious.

And so it must be with your presentation. Are you taking for granted that your audience will see—as you do—the *implications* of your statements? They may not.

If you are not sure, try your ideas out on a single member of your target audience. There is no point in trying things out on your peers—they will surely "get it" as you do.

Which brings us to the other extreme. Don't underestimate your audience. Too much background or explanation will bore them, or possibly give the impression you are looking down on your audience. Your peers will not expect the same kind of presentation as your managers.

Getting the right amount of background, or deciding how finely to break the topic down, is always a judgment call, but your presentation will benefit from your thinking through this aspect as part of the overall storytelling. Is the audience technically sharp, but not on your particular subject? You may decide that you can omit a slide or two that you used to fill in a missing step. Or are they more generalists? In which case maybe you even need to add a slide or two of explanation. (We will see later that there may be ways you can make this decision during the presentation, and do it without it being apparent to the audience.)

PUTTING THE PRESENTATION TOGETHER

Now you have your presentation storyboarded, it is time to start making the slides. If you are starting a presentation from scratch, this phase of making the presentation begins after you have all your ideas organized. If you are editing an earlier presentation, having your ideas organized will help you select slides from your earlier work, and decide what new material is needed.

Now is the time to think about how your ideas are going to look. At this point, you start to be concerned with the appearance of the slides. First, there are a few basic rules. For the most part, your presentation software will help you to obey them. Here's our list.

RULES FOR SLIDES

- Use no more than 18 lines of text on a slide (this is the same as selecting a certain minimum font size). You want the folks at the back of the room to be able to read your stuff.[2]

- Use no more than three typefaces on a slide, and no more than four in a presentation—that means if you use bold and italic, you must keep to the same font family. By creating a uniform style throughout you avoid the appearance that you have just thrown things together.

---

[2] Or maybe you don't. We would not recommend it, but we have heard it said that putting a whole lot of information on the screen, perhaps in the form of a really complex table, gives you "defensibility." That is, you can bury any embarrassing thing you wish in there, confident in the fact that your audience will not notice it, and you can "forget" to mention it. If questioned later, you can always point out that it "was on the viewgraph." Certainly, the viewgraph defends itself against being read!

- Unless there is some compelling reason not to, use a sans serif font for your presentation, and do not use large amounts of uppercase.

- Do not use justified text. Usually, there are so few words on a line that the additional space needed to stretch the text between the margins creates a poor impression. Stick with ragged-right.

- Assume that the top of the slide will be occupied by something other than your material. The top is often used to contain information about what the presentation is about, or where (in a long presentation) the slide belongs. Sometimes a Master Slide can take care of the top of the slide for you, sometimes is better to do it yourself.

Shopping List

- Cat food
- Vacuum cleaner bags
- Kitty-litter

Figure 6.9

You will see, in some books purporting to help you with putting together a presentation, an example like Figure 6.9 to illustrate some point.

We will not do that, as it is self-defeating. What we have in that shopping list example is a very legible picture of a slide that has been reduced to 1-inch height. The fact that you can still read it reflects the use of *enormous* type to start with. In fact, the heading was 96-point Arial, a size that means the uppercase letters are 1.3 inches tall! We don't think that an example like that is particularly illustrative of anything, so instead we shall go the other way and shrink ordinary slides down to the point where they are only just legible. But remember, a slide usually has a font size around 18 pt as its smallest size, so if we take a normal 7-inch height slide area and shrink to 2/3 normal size, we end up with a reasonably respectable 12-point type. We can put two of those on a page, and still leave room for some words. Figure 6.7 is an example.

Your software will probably make some initial selections of fonts for you, and you may want to set up the defaults in advance. Rarely is the software the way you want it right out of the shrink-wrap.

By the way, in general, do *not* use the Wizard feature when putting your presentation together. If the defaults need changing, that's nothing compared to letting the Wizard take over. It turns out there is a whole community of people out there who really do not like the Wizard. Some are quite passionate about it. There is a wonderful example of *why* you should avoid it on the Web site of Peter Norvig, `http://norvig.com/Gettysburg/`. We include one slide of the presentation (Fig. 6.10) just to show what can happen if you allow automation to take over.

Somehow, it doesn't have the same *ring* as "Four score and seven years ago our fathers brought forth on this continent a new nation, conceived in liberty and dedicated to the proposition that all men are created equal."

Which observation reminds us that we are *speaking* while our presentation is going on. The content of the slides is supposed to be a support for that speech, not a replacement for it.

Under no circumstances are you going to read from the slides. For you, the speaker, they may serve as a *reminder* of what you were going to say, but they are not to be your script. For your audience, they may contain *supporting detail* for what you say, but they should not hold the audience's attention away from you and your words for too long. We will have more on your role as speaker later on.

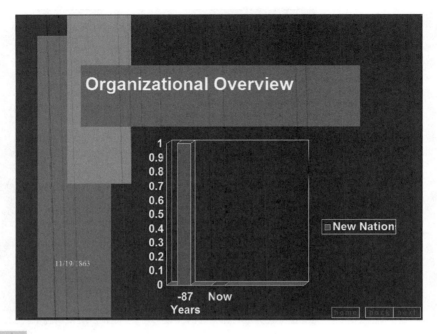

Figure 6-10 See insert for color representation of this figure.

There is a knack to putting the content on a viewgraph. With practice, you may establish your own style. Until then, here are some guidelines on content:

- Use bulleted (or numbered) lists with abbreviated statements of your point.
- At the top of such a list, add some words to set the stage. Use more complete sentences here than in the list.
- Plain text, with no bullets, should be used sparingly. You do not want your audience dwelling on your slides while they read the wisdom contained there. You want them listening to you.
- Use graphics as frequently as you wish for experimental results or analyses.
- Use contrived graphics (such as pie charts showing where the money went) less often.
- Use movies sparingly, unless there is some overwhelming reason to create a "movie-rich" presentation.
- Use detailed charts (such as Gantt charts) if necessary, but make them readable either by zooming on part of the chart or by adding larger-font text.
- Use equations sparingly or not at all, unless they are truly indispensable. If you must present an equation, by all means discuss its salient points, but do not just read it out. Audiences will thank you for that.

Next, we present some sample viewgraphs, with comments. We start with something fairly simple, in Figure 6.11. The slide captures all the important historical points at the start of the presentation. It is unusual in that all the words are complete sentences, but they are all short enough (even in the bulleted

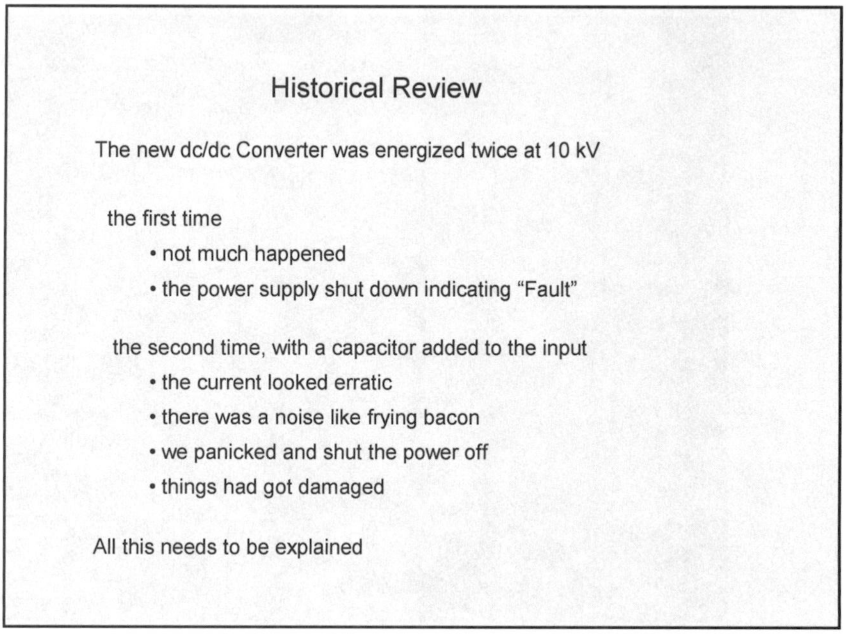
Historical Review
The new dc/dc Converter was energized twice at 10 kV
the first time
• not much happened
• the power supply shut down indicating "Fault"
the second time, with a capacitor added to the input
• the current looked erratic
• there was a noise like frying bacon
• we panicked and shut the power off
• things had got damaged
All this needs to be explained

Figure 6.11

lists) to work. The hard part of dealing with this slide is talking about the topic without being dragged into reading it. (That's one reason why we don't usually use complete sentences. Since we talk in sentences, a bulleted list that is only keywords won't drag you in, as speaker, quite so effectively.)

Figure 6.12 is what you might consider an "extreme" slide. This slide has only two bullets, yet the words in the bulleted list are so dense that you can pretty well guarantee that your audience is not paying attention to you while this slide is on the screen. They are trying to figure out what it says! Figure 6.13 does something that is usually avoided: it includes equations.

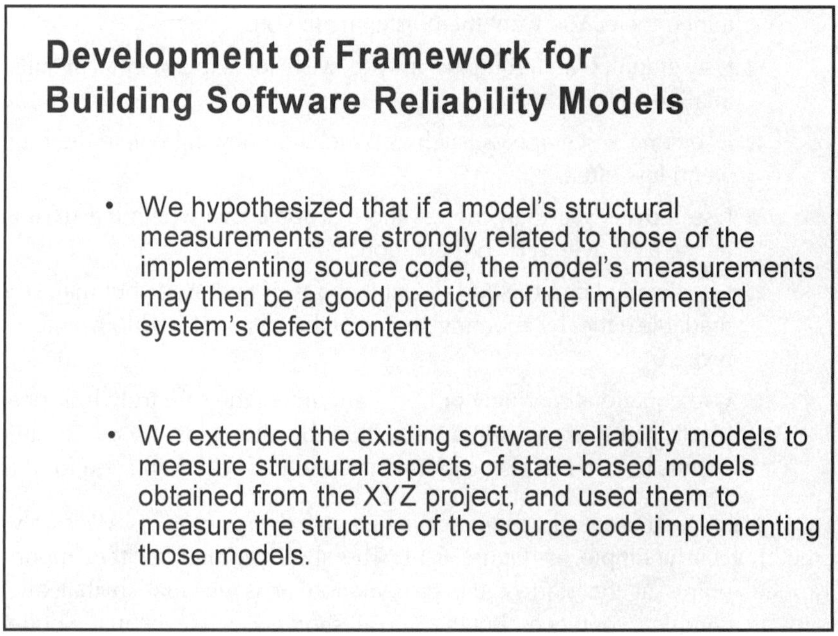
Development of Framework for Building Software Reliability Models
• We hypothesized that if a model's structural measurements are strongly related to those of the implementing source code, the model's measurements may then be a good predictor of the implemented system's defect content
• We extended the existing software reliability models to measure structural aspects of state-based models obtained from the XYZ project, and used them to measure the structure of the source code implementing those models.

Figure 6.12

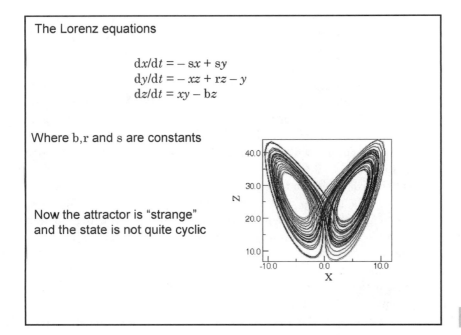

The point of this slide is to demonstrate the complexity of the motion compared to the simplicity of the equations that generate it. Therefore, it is acceptable to include the equations (which are simply stated on this slide, and are not derived in its predecessors) so that their simple form can be commented on. The detail of the motion is relatively unimportant to the argument, so here it is OK to include a rather small image of the well-known butterfly.

Figure 6.14 is one of the slides from the section just labeled with "Ditto" toward the bottom of the storyboard we saw earlier.

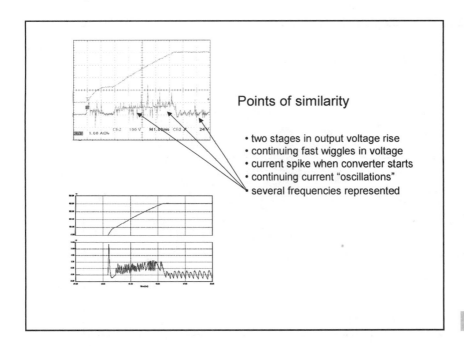

In Figure 6.14 the bulleted list is built up line by line, with arrows pointing to the relevant item at each step. (It was because of the arrows that a simple slide-in buildup of the bulleted text would not work. There was no way to slide the arrows *out*!)

In this slide, the top image is real data, the lower one is from a simulation. These images are small, but each had been introduced in larger form earlier in the presentation, so all the detail need not be perfectly legible here. By building up the case for similarity in great detail, step by step, the audience is gradually persuaded. Each step of a sequence like this requires less than the typical one minute per slide.

In all the examples so far, the graphics have been of a predetermined type (well, apart from Gettysburg). That is, they are screen images or something like that, and therefore not easy to edit. Normally, if you are producing graphics specifically for a presentation, you have somewhat more control. For example, you may want to present a sort of cost breakdown by means of a pie chart, such as the one in Figure 6.15.

Figure 6.15 is a very complex slide and will occupy more than a minute. It will be worthwhile, because, on a single slide we have identified all the major elements of cost, shown what percentage of the total they are estimated to be, and started a dialog on how costs might be reduced.

The image is very colorful, and that has come to be a hallmark of many presenters. The choice of color or not is yours, as presenter. All the presentation software will deal with color (though some colors seem to look different on the big screen). Unlike the situation of preparing a graphic image for a technical paper, there seems to be no particular reason *not* to use color.

However, it is good practice to first make the presentation work in monochrome, if you can. That way, the color enhances the presentation, but the presentation does not rely on there being color. Naturally, if you are borrowing graphics from your own paper, they probably start life in black and white, anyway.

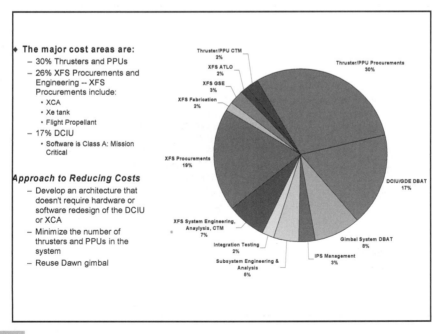

Figure 6-15   See insert for color representation of this figure.

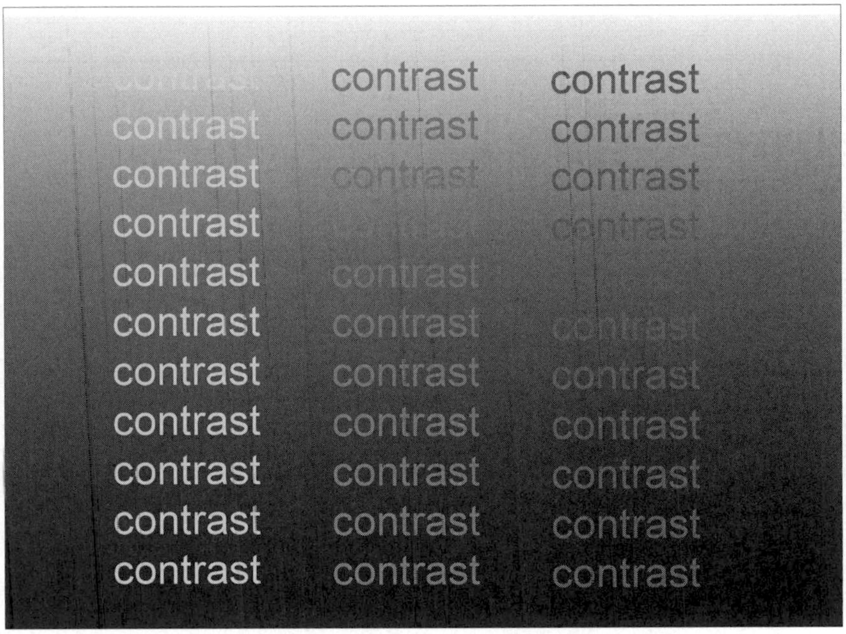

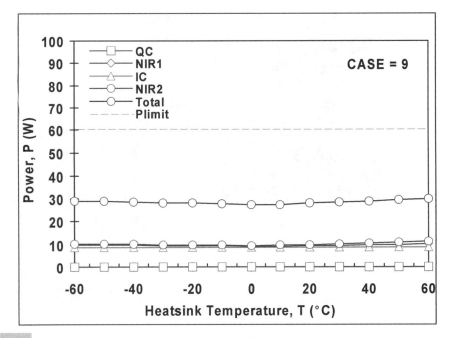

Figure 6-16 See insert for color representation of this figure.

However, there are some issues with color that don't happen on a white page. Consider the example of Figure 6.16.

In Figure 6.16, each column of words is made up of identical text. Because of the background shading, the words seem to disappear at some point that depends on what color the word is. The use of color on color is often problematical for the same reason. Therefore, be sure to check for legibility.

The same legibility problems may appear if you use color as a code in a graph, to separate variables. You need to check, and preferably on the big screen. Consider Figure 6.17.

Figure 6-17 See insert for color representation of this figure.

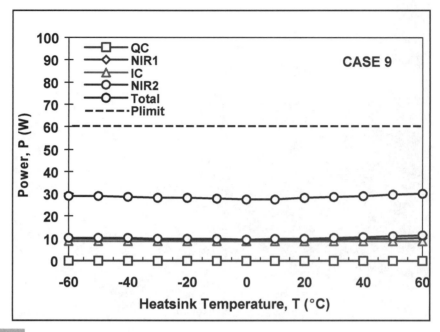

Figure 6-18   See insert for color representation of this figure.

The colors and line weights of Figure 6.17 might work well enough on the printed page. But for a presentation, you would do better to make all the graph lines thicker, so the colors can be more easily distinguished. Believe it or not, while the line weights in Figure 6.18 look "over the top" on the small screen, they would not be overkill. (Try an experiment: compare your computer screen with what you see projected on the big screen. It is quite revealing.)

When you use spreadsheet-generated graphs such a this, be consistent. If you change the *order* or *number* of curves, your spreadsheet will likely assign a different color and marker. Look out for this, and if necessary change them so

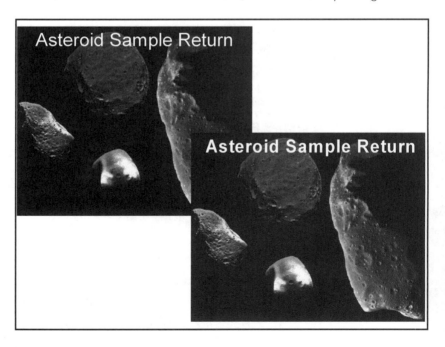

Figure 6.19

that the same parameter is always shown the same way. Consciously or not, your audience will associate these cues with the variable they first see them assigned to.

You may decide that the way to get the contrast you need against a dark background is to use a light or white color for the font. (Indeed, some software will make the change for you automatically.) In the event you use a light font on a dark background, consider the use of boldface at the same time. Figure 6.19 shows an example of the difference. Many viewers would regard the regular font as too thin against the dark background.

BACKGROUNDS

Presentation software offers you a choice of ready-made backgrounds. Should you avail yourself of them?

It's hard to offer a general rule. Some of the backgrounds are really attractive and will not be a distraction for your audience. But some are going to dominate your contribution, and obviously these should be avoided. Back in the time BPP,[3] a good slide for a formal presentation in front of a large audience had a 35-mm format and a medium-dark blue background. Dark blue was used because black on white was considered to have too much glare, and the blue softened it a little without reducing the overall contrast too much. Of course, the magic lanterns of those days didn't have the contrast of today's equipment, and you were expected to turn out the light and draw the shades. It was an environment that favored those of us able to listen with our eyes closed.

The spiffy backgrounds you can choose from now are the descendants of those simple blue slides.

The problem is there has not been time for natural selection to do its thing. Some of the descendants are mutants that will not survive long, but you need to be vigilant just now, as they are still around. Figures 6.20 and 6.21 are a couple of ready-made backgrounds from familiar software.

We reckon you could hardly do better than avoid such frivolities. If you can find a pleasant background that appeals to you, you might try it, but you might also consider creating your own. Remember, a simple blue background still works, but these days it should be pale. Consider, too, how your slides will reproduce if you are required to have black-and-white handouts.

Mind you, it is fair to say that the "formality rules" that applied (or that you *sensed* applied) to the preparation of the paper are relaxed a little when it comes to the presentation. For example, in the rare case that the thing you want to show actually *is* an area, there is surely nothing wrong with using some tricky presentation as an area chart, instead of a pie chart. For example, suppose for some reason you need to compare the area of Alaska and Texas. You could:

1. State the numbers: Alaska covers 656,425 square miles; Texas covers 268,601 square miles. This is what you might do in the paper. Two pieces of information scarcely justify the space of a graph in a paper. The bare information isn't very eye appealing, however, on a slide, even in a table.

2. Draw squares or circles to scale to show the relative sizes. For this, note that the ratio of the areas is 2.44, so that the linear dimensions of the

---

[3] Before PowerPoint.

graphic object should be in the ratio of 1.56, as in Figure 6.22. If you want to show that the size ratio is large, you could have one object only partly cover the other, as shown. Whether you chose squares or circles, this would make for a boring slide.

3. Search the Web for scale maps to copy and paste, and arrange them artistically together, as in Figure 6.23.

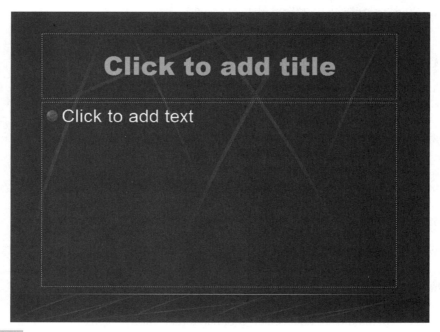

Figure 6-20   See insert for color representation of this figure.

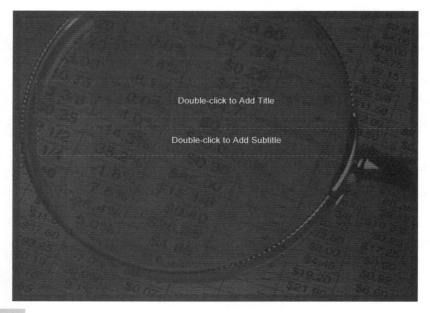

Figure 6-21   See insert for color representation of this figure.

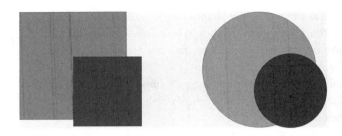

Figure 6.22

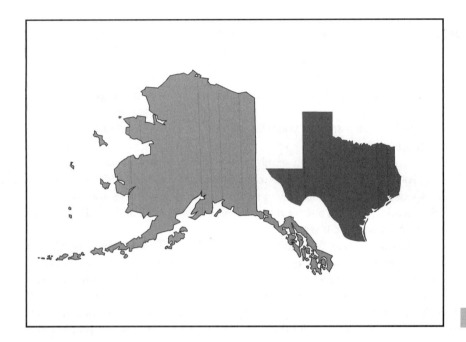

Figure 6.23

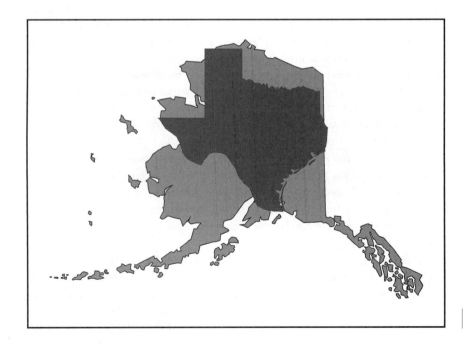

Figure 6.24

For these particular examples, the shapes of the outlines are sufficiently recognizable that no further identification should be needed, given the context provided by the caption or by your words. That may not always be the case, so you may need to add some identification.

The first approach, merely stating the numbers, has much less impact than either of the graphical approaches, but surely the last one is likely to create a better impression in the minds of your audience.

Of course, superposing the two as in Figure 6.24 to show just *how much bigger* one state is than the other is something you might only pull off in Alaska.

PUTTING THE SOFTWARE TO USE

Your software probably has several useful features that you should consider. Can you build up a list, by adding a line at a time? If so, do so when you don't want the audience reading ahead, but otherwise, try to resist the urge. Audiences seem rapidly to tire of "gimmicks" like slide-ins and fade-ins. Besides, it is more effective if almost all of your slides appear complete, and only one or two are built up as you talk.

If you are presenting a graph that you have created in a handy vector format, you can use your software to build it up, too. You could start with the graph space, axes, labels and everything, but only one curve. You talk about that for a while, then add a second curve. This kind of thing can be quite effective as an attention grabber, if you don't overdo it.

You may notice your software automatically increase the font size of your text if there is room. Override this feature. You may fix the font size based on the level of indent in a list, or based on having one fixed size for display and another for body copy. But you want consistency from one slide to the next. It doesn't matter if one slide has more white space than another because the font sizes are the same. Do not let the software appear to stress some points more than others by changing font size based on the amount of free space available! (Remember, your presentation software is not written by people who actually *give* technical presentations; it is written by the people who brought you Y2K.)

Can you set the software to do auto-timing? Sometimes this feature can be a way to ensure that a presentation stays within a time limit, but in general it is hard to adjust to the machine pacing. Usually, you can make each slide stay on the screen for its own particular amount of time, and that may be a good way to pace the presentation. If you do that, you will have time to consult your own set of notes, since you will not be worrying about moving the talk along. Practice beforehand, if you do this!

Presentation software usually has the ability to include "actions." These actions are instructions that are executed when you click on some particular part of the screen. They can move you to the first slide, the last, the next, and so on. They can crank the volume up or down, start a movie, or run some predetermined software. (Movies can be inserted in such a way that they run only after you click on them, for example.)

Here is a trick. If you put an action object on a slide, you can change its colors to make it invisible. You can also resize it, so you can make the entire slide into an invisible action object. Then, if you click on the screen during your presentation, you will start the predetermined action, for example, starting a spreadsheet program to give you a particular detail that you think will help.

Or, if you determine that such detail is not needed for this audience, or that time will not allow the diversion, you avoid the link simply by not clicking on the screen. And the audience is completely unaware that you made this decision.

The actions can be quite powerful, but since they are all predetermined by you, be sure that nothing can happen to invalidate them during your talk. For example, if the spreadsheet file you want is on your desktop, and you are giving the talk from the computer in the conference room, the link won't work. Avoid this mistake by putting everything on your memory stick. But be sure to check that it has the appropriate drive letter assigned in the hyperlink before you get started!

STARTING THE TALK

We have not discussed the title slide yet. It is a very important slide, as it sets the stage for what follows. The title should (usually) include your name and the date, and a few well-chosen words aimed at arousing interest in the audience. It will suggest the tenor of the talk to come.

Often, it sits there on the screen while the audience straggles in, stops talking, and gets comfortable. Often, it is on the screen while you are being introduced. So what should it say?

Should it be a boring summary of what you will say? Should it be a snappy phrase to catch the audience's attention? Whatever you choose, do it carefully and thoughtfully. You can leave the composing of the title slide till you have finished all the other slides, if it would help you. (The working title of this book was "Insert Picture." The actual title was not fixed until the material was in the hands of the publisher.)

In the case of the earlier converter story, the title was "What is going on with the MARS converter?" A plain and direct title, and a little bit of a deliberate challenge, as perhaps some of what was going on was the responsibility of some of the audience.

We once gave a talk to a group of physicists interested in using the ocean as a giant particle detector. The title was "Lessons learned from the NEPTUNE power system," which hints that all had not gone smoothly with this difficult endeavor, and we added a subtitle "and other deep-sea adventures," suggesting that there were also lessons to be learned from the mishaps others had enjoyed in their attempts to do science under the ocean.

Of course, you can even change the title slide to suit the audience!

Your actual presentation is done *ad lib*, since you are familiar with your subject. So you don't need a script. However, your opening remarks, made when the title slide is on the screen, should not be left to the last minute. A good way to get your presentation started is to launch into a more or less memorized opening.

Your opening is aimed at getting the attention of the audience, and at establishing a rapport with them. Since you have rehearsed this moment, it is a good way to get yourself heading off into the talk with confidence. You have a lot of choices about how you go about this. Since it pays to be prepared, you decide in advance how you are going to start.

You could use the direct approach. "This morning, ladies and gentlemen, I am going to discuss the problems we have seen on the MARS converter." *Wrong!* Remember back to your objective statement? You wrote that with a desired response from the audience at the forefront of your mind. So a start that says what *you* are about to do does not move you in that direction. How about "This morning, ladies and gentlemen, we are going to try to understand the problems we have seen on the MARS converter." Much better. That "we" will draw the audience in.

Or you could look at the audience and pose the question: "Do any of you know precisely what is going on with this converter? No? Well, this morning, we're going to review what we know, and see if we can figure it out."

Or you could wake them up with something along the lines of "The last time we turned on the converter, we thought it was frying!"

Or you could try to find a quotation. This is a risky approach, because if a quotation is going to work, it has to be *relevant*. Preferably relevant to the audience *and* to the topic. We looked in *Bartlett's* and there is nothing under "converter." In general, you'd be lucky to find the *perfect* quote. You might think of an anecdote instead. This is a good plan only if you have storytelling skills of a Studs Terkel or a Garrison Keillor.

THE BODY OF THE TALK

You know your stuff, you have a well-prepared presentation, a good set of slides, and a set of opening remarks that will set the stage for a good start. What could go wrong?

Well, plenty, if you let it! Mainly, things depend on your self-confidence in addition to all the things listed above.

One of the authors (HK) remembers the very first big conference he attended. He was not presenting a paper, but a colleague of his was, and he went to lend support. He didn't need it! Bill had presented only a few papers before, but he had a very natural, relaxed way of talking during his presentation. The audience sensed that here was a guy who knew his topic and was enjoying talking about it. He sounded, talking to an audience of a hundred or so, the same as when talked to our small group back at the lab. Only he had a microphone. HK determined then and there that stage fright was not going to help, and when he had a paper to present, he was going to emulate Bill's calm and friendly demeanor.

One way to help with this, if you do have a tendency to be nervous, is to pretend that certain people in the audience are your friends, and you are talking mainly to them. Of course, you must establish eye-contact with more people than just one or two, but in the back of your mind is the idea that this is a conversation between friends.

You might have to talk a little louder than in a one-on-one situation, and to be clear, you might have to talk a little more slowly, but remember, everybody there is there because they want to hear you talk. And you are there because you want to talk. So everybody has the same wish for the success of your presentation. Nervousness (and its attendant tendency to rush) is not needed. Talk slowly and clearly. Vary the speed a little, to show your enthusiasm.

Your talk should be extemporaneous. You have prepared what it is you are going to say in outline; you have decided how to support that outline in detail. You have loaded your viewgraphs with reminders for your theme. Only the words will be decided at the last moment.

Early in the talk is the time to calibrate your talk against the audience, if you need to do so because you couldn't learn all about your audience ahead of time. It may be too late to do much in the way of changing your slides, but you can adjust the technical content somewhat as you talk.

Unless you are a trained actor, any way of giving your talk except naturally and extemporaneously will come across as artificial, and your credibility will suffer. But as you are talking more or less normally, and you are not reading a script, you have time to look at the audience, you have the freedom to move around a little, and in these ways you can draw the audience into your world. You set the right environment.

At the most, you should glance at the screen, taking your eye off the audience only for short periods. The computerized presentation may be a help in this. Back when the material on the screen came from a 35-mm slide, the only way you knew what was on the screen was to turn around and look. With a computer, you have a small version of what's up there right in front of you. A sort of "head down" display.

If it would help you, you can use some secret code in the slides to remind yourself of something. For example, you can change the color of the background, or of the line separating the header from the body, to remind you when you segue from the intro into the body, or to remind you that the wrap-up is approaching. You can use blank slides for the same purpose, or to have a nondistracting background when you need to spend some extra time expounding a point.

HK once used the slide shown in Figure 6.25 in a review, to remind himself to say that we were at an important point in the development of the system.

The right side of the screen is a representation of a game called Go, at a point in the game when strategic decisions have been made and tactics are coming into play. That was where the project was, the big decisions having been made and the little ones just starting. The slide was there because it was necessary to hear opinions and comments on the decisions that were already

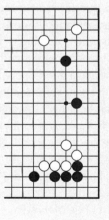

The purpose of this particular Review

• Ensure that the design work is on track
• If it isn't:
  • Identify areas of concern
  • Document them
  • Suggest corrective actions

Figure 6.25

made, on the grounds that if the review board agreed the project was on the right track, the rest of the work should be straightforward.

Mind you, it is important to know your audience if you intend to try anything quite as "off-topic" as a game of Go in a technical presentation! Normally, something like that might be reserved for a lunchtime seminar, but (as project manager) I didn't want too much formality, and this set the right tone. Besides, the talk was in Japan, where Go is well known.

FINISHING THE TALK

It seems like an obvious thing to say, but you must always have a finish to your presentation. It may be hard to believe, but sometimes you may encounter a speaker who *doesn't*. This speaker gets to the last slide and says, actually or implicitly, "Well, that's the last slide, so I'm going to stop now." Do not be that speaker!

You started the presentation with an objective, and you used that objective to decide on a structure. Use your closing remarks to finish that structure, and at the same time help you achieve that objective.

If you are confident that you have achieved your objective, you could end with something ultrasimple like "I would like to thank you for your attention. I'll be happy to answer any questions you may have." This is generic, and adequate if you're sure you "nailed it." But you may not be certain.

You should try to get a sense of how you are doing (in terms of moving toward your objective) as the presentation goes along, and then use the closing remarks to make or reinforce some final, persuasive, argument. You can get this feedback by looking at the audience, and by assessing the questions they ask as you talk. (This is why it is a good idea to allow questions during the talk, although you will need to manage them carefully if your schedule is not to be upset.)

For example, based on your judgment of how it went, you might say: "During the last thirty minutes or so, we have examined the *whatever the topic was*. I'd like to spend the last few minutes clarifying why *whatever you sensed didn't go well* is important. If you have further questions, I'd certainly be happy to address them."

Now, I would not recommend that you plant a "shill" in the audience. But depending on the circumstances, you may have the opportunity to stimulate a friendly question to get things going. For example, at a conference, before your paper presentation, you will almost certainly encounter people with overlapping interests who are going to attend your presentation. There is nothing wrong with talking to them beforehand. During this talk, you might encounter a question that would make a good opener. If so, why not say something like "You know, that's a good question. Why don't you ask it at the end of my talk, so the rest of the audience can hear my answer?"

If there is no obvious question coming up in conversation, you can propose one to a friend. "I've been thinking about *whatever you want asked*. Though it's not in my talk, I'd like to discuss it afterwards. Why don't you ask it as a follow-up question?" Sometimes even the session chairperson would appreciate the help that such a suggestion offers.

If none of these mechanisms for getting something going seems right, you could always start a question period with a question of your own for the audience.

This is particularly appropriate if you happen to know there is someone in the audience who is likely to have a question. But never ever ask a question that puts someone in your audience in an embarrassing situation. First, because it really doesn't help move you toward achieving your objective; and second: remember, one day, it will be *their* turn!

ETIQUETTE

It is surprisingly easy to create the wrong impression when you give a presentation. It is essential to treat your audience with respect, and to show good manners. Therefore, no matter how tense you are, do not grip the podium as if it is the only thing between you and falling on your face. No matter how relaxed you feel, do not lean casually on the podium while you give your presentation or afterwards, while you listen to the questions. Stand near the podium, walk around if you are permitted by the technology, and face the audience.

Laser pointers are a great invention, but they are better in support of putting up a shelf than making a presentation. With a laser pointer in your hand, you are tempted to use it to indicate things on your screen. That means you must look away from the audience. Not good. You may even find yourself waving it madly at the screen, circling a name or an important point. Generally, this will annoy most of the audience. So, as far as you can, do without a laser pointer.

Finally, when you are all done and the questions have stopped, or the chairperson has indicated it is time for you to depart, make some kind of statement thanking the audience for their attention, and continue to make eye contact until you have left the stage.

SUMMARY

Getting your presentation in good shape is a matter of remembering the fundamentals, and being organized. Start by writing an objective statement for the presentation. It should be based on what you expect of your audience.

- Remember the basic rules for viewgraphs:

    Use bulleted lists.

    Avoid sentences.

- Build up logically:

    Start with an outline.

    Use thumbnails.

- Adjust your presentation to your audience.

- Make your style decisions:

    Use a repeating background.

    Use sans serif font.

- Use color in a presentation even if you don't in the paper:

    Make the slide "work" first in B/W.

    Use color for appeal or for clarity.

Beware readability issues with colored lines and text.

Some colors have low contrast against other colors.

Fidelity of color is not maintained by all projectors.

■ Use the power of the software, such as auto-timing:

But don't overdo the fades and slide-ins.

Use branching, action buttons.

Use graphs and diagrams as part of presentation, but minimize equations.

Use the things you can do in a presentation that you can't do in print—such as animation or turning gridlines on and off.

■ Consider presentation integration and your final thoughts.

■ Check for rhythm and pace; count the slides on each subtopic.

EXERCISES

6.1. Say what would be needed to fix the viewgraph in Figure 6.26.

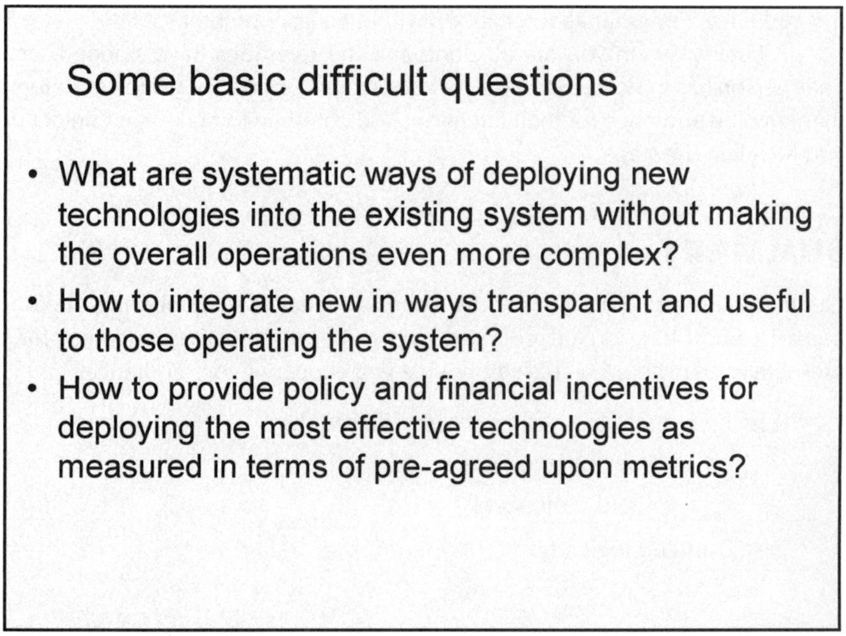

Some basic difficult questions

• What are systematic ways of deploying new technologies into the existing system without making the overall operations even more complex?

• How to integrate new in ways transparent and useful to those operating the system?

• How to provide policy and financial incentives for deploying the most effective technologies as measured in terms of pre-agreed upon metrics?

Figure 6.26

6.2. Comment on the activist viewgraph in Figure 6.27.

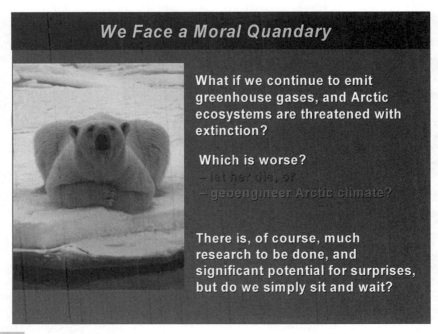

We Face a Moral Quandary

What if we continue to emit greenhouse gases, and Arctic ecosystems are threatened with extinction?

Which is worse?
– let her die, or
– geoengineer Arctic climate?

There is, of course, much research to be done, and significant potential for surprises, but do we simply sit and wait?

Figure 6-27   See insert for color representation of this figure.

6.3. The following is an outline for a "marketing" presentation aimed at getting funding for a new way of measuring current. Is it appropriate? If not, what should be changed?

Overview of current measurement technologies

   definitions

   voltage (differential)

   force

   magnetoresistance

   Hall effect

   magnetostriction

   Faraday effect

Examine the obvious approaches, and point out problems

   voltage (differential)

         problems with matching resistors

   Hall effect

         problems with leakage flux

   Faraday effect

         complexity

Propose two general-purpose solutions

   Improved differential voltage

   Closed-loop Hall

Request funding for new solutions

An introduction to spreadsheets

7

Spreadsheets are used for graphics mainly for one reason: they let you create a graph automatically and quickly from numerical data. While graphics was really a kind of afterthought in the early days of spreadsheets—in the early Lotus® 1-2-3® you had to close Lotus and open something called PrintGraph—these days it's all integrated. More or less. Well, actually, graphics still seems like an afterthought—or a nonthought—to the people who write spreadsheet software, and getting publication-quality graphs is sometimes such a chore that you wish for the good old days when you just gave the numbers to someone else to take care of.

However, the use of spreadsheets for graphics is very widespread. Not only do people enter manually gathered data into a spreadsheet, but data acquisition systems often boast that they produce data files that are compatible with common spreadsheets.

That, of course, makes a second reason for using spreadsheets: they define formats useful for the exchange of data. A third reason might be that they make it easy to do some sorts of statistical analysis of the data—linear regression, for example. This useful feature allows you to do a sanity check on the data.

Usually, the default graphs you get from spreadsheets are not publication quality. You will need to make improvements, unless you are publishing in a journal that will redo the artwork anyway. And it seems that the writers of the spreadsheet software are making it easier and easier to get the default, and harder and harder not to. But never mind. With a little effort you can get whatever you want.

The major aim of our chapters on spreadsheets is to show you how to reduce that effort: how to get what you want in terms of style and appearance without spending much time doing it. We have four whole chapter devoted to the topic, but do not be put off by that. It's just that there are two programs to consider and many kinds of graph to discuss, and we do need to examine them all.

The details of the process of using a spreadsheet to produce a graph vary from spreadsheet to spreadsheet, but the main steps are the same:

1. *Get the numbers*. They could be from a data acquisition system, or from other software, or from the spreadsheet itself. They may be in the spreadsheet in rows or columns.

The Right Graph. By Harold Kirkham and Robin C. Dumas
Copyright © 2009 John Wiley & Sons, Inc.

2. *Create the graph* by using the spreadsheet. This means telling the spreadsheet which of its rows and columns represent data, and telling it what kind of graph you want (bar chart, pie chart, scatter plot, etc.).

3. *Fix the graph* so it is presentable when inserted into your document.

Only the first of these steps will be covered (and that lightly) in this chapter.

GETTING THE NUMBERS

Spreadsheets started life as implementations of accounting balance sheets. The screen was arranged as a matrix of cells, and the numbers or letters that appeared in the cell could be simple numbers or they could be the result of calculations. One could visualize the spreadsheet display as the front layer of a three-dimensional stack of blocks, where the invisible cells behind the front were attributes that controlled the front layer.

Complicated mathematical expressions could be entered into a cell, so that the number that was visible was the result of some functional relationship with the contents of other cells.

Pretty soon, the attributes of a cell included the format of the display, and quite attractive printouts became routinely available. To go along with all this, some means of speeding up the process of creating a spreadsheet became necessary. Tools that allowed simple copying and editing of formulas were written, in addition to functions that allowed "what if" calculations to be done.

By the time Lotus® 1-2-3® came along, spreadsheets were capable of being operated via custom-written menus, using what were called macros.

Back then, a macro was essentially a recording of a series of keystrokes that would save the user the time and trouble of repetitive actions. These days, unfortunately, macros are the route by which many viruses enter your computer and perform their cheerful tasks of destroying your hard work.

Back then, a macro was an easily understood sequence of commands that echoed the instructions that you typed into the program. Everyone used them, because they were easy to generate—you just typed into the macro what you would have typed on the screen. These days, macros use obscure languages that are unrelated to your keystrokes (such as Visual Basic®), so practically nobody but "developers" use them. (Have you ever noticed that developers tend to write with other developers in mind? Worrying trend, that!)

Well, it may be that fewer people use macros than before, but the macro languages are quite powerful. It is possible to use a spreadsheet with macros to control automatic test equipment *and* to gather the results of the tests. Having such a process automated can save an enormous amount of work if the data you need represent the variation of a lot of parameters.

The first part of that process is beyond the scope of this book, and the second part, getting the numbers into a spreadsheet, is assumed to have happened. You might have a dozen or two numbers in a spreadsheet, or a couple of thousand. You might have calculated some of the numbers as a result of a Monte Carlo simulation, or you may have downloaded data from a test and measurement system. Since our interest really begins after this step, we will not address further the question of how you get the numbers you want to plot to appear in the spreadsheet.

SCOPE OF THE REMAINING SPREADSHEET CHAPTERS

There are four chapters addressing the use of spreadsheets. This brief chapter serves only as an introduction and a guide through the next three. Each of the next two chapters will be focused on one of two programs: Excel® from Microsoft (Chapter 8) and QuattroPro® from Corel (Chapter 9). Chapter 10 will deal with the use of a graphics program such as PowerPoint® or Presentations™ in conjunction with a spreadsheet.

The two application-specific chapters, 8 and 9, will show you not only how you can get what you (and your journal) want out of the spreadsheet, but also how to change the defaults in your software for the various kinds of graph, so that next time you make a graph, you can get the appearance you want automatically. While there is nothing you could not change in a graphics program, it is much more efficient to make the changes in the spreadsheet program because the program can remember them.

We are aware that there are specialist graphing programs that allow spreadsheet use: SigmaPlot, and a number of low-cost packages that you can find on the Web, and even Corel Presentations. These programs do not have exhaustive sets of spreadsheet features, but they generally do an outstanding job of allowing you to make graphs. We shall not deal with them here, as our intent is not to review software that was written for the purpose, but rather to help you with getting software that you already have to do a job for which it was not primarily written.

As far as spreadsheet use is concerned, please note that it is not our intent to give an exhaustive tutorial on the topic of using Excel and QuattroPro to generate graphs. A good tutorial would be a whole book in itself, a fact that would pretty well guarantee nobody would open it! What we intend to do is to point out some very useful features that can save time, and to make you aware that there are things you can learn by hacking around on your own time. We want to get you started, not graduated.

The two spreadsheet programs are broadly similar when it comes to graphs. Each has a set of not very applicable defaults that you need to change. Each has a useful feature or two that the other does not, and each has its own set of interesting bugs. Probably the biggest difference between the two programs in terms of how they handle graphics is this. Excel copes with presenting you most of the options by means of one pull-down menu and a lot of right-clicking, and a `Chart` toolbar that you have to add. Some of your options will be grayed-out, depending on context. A few things are available in a `Drawing` toolbar that you can also add to the screen, if you remember. In contrast, QuattroPro deals with all the options by adding context-sensitive toolbars to the top of the screen. (There is a `Drawing` toolbar, but it is not as useful as the one in Excel, as we shall see. However, you may find you don't need it anyway, because you can usually do more with the appearing and disappearing toolbars of QuattroPro than with the menus of Excel.)

Not only do new toolbars appear in QuattroPro, but a little window appears that will let you select what you want to edit by name (a pick list), so you don't have to go poking about with the mouse or the right arrow key. Excel has a similar looking window, but it doesn't work. The feature does work on the `Chart` toolbar, however.

Both the toolbar plus menu method of Excel and the toolbar method of QuattroPro do work, and by and large they have similar capabilities. For someone who has not memorized all the dialog-box sequences, the tool box approach of QuattroPro is more user-friendly, but there isn't really much to choose between them.

TIME AXES

Many years ago, when computer graphics was somewhat less advanced than it is now, HK was working on some software to present data on a computer doing data acquisition. The data represented the performance of a power transmission line being tested for power loss (due to corona in rain), and audible and radio noise generation (also principally in rain). A number of weather parameters were being measured, including the instantaneous rainfall rate. The rainfall rate instrument controlled the rate of data acquisition. The problem was to plot these various quantities in a graph on the computer monitor by scaling the vertical and horizontal axes suitably. After some deliberation, an algorithm was devised for vertical axes that used a minimum of 5 and a maximum of 11 tick marks, that put labels on only 5 or 6 of these (using every other one if the number was 7 or greater). The algorithm rounded the data *outward* in order to select the extremes of the axis. This way, it was possible to place the tick marks on the 1's, 5's, or the 10's of the axis, to give a convenient, readable scale.

Horizontal axes were handled similarly, but with somewhat greater latitude. Often the scale is physically longer and can accept more labeling without becoming cluttered. The same general rules applied, even though the scale was time.

First, time ranges of just a few minutes were dealt with in the obvious way: all the data points were plotted and the axis labeled for each one. For longer times, natural *breakpoints* were determined to exist at time ranges of 80, 120, 240, 480, 960, 1920, 3360, 5760, and 11,520 minutes, that is, 80 min, 2 h, 4 h, 8 h, 16 h, 32 h, 56 h, 96 h, and 192 h. (This is a large number of breakpoints, but the number is finite!) For any given range of data, beginning and end times for the axis were forced *outward* to values divisible by 10, 15, 30, 60, 120, 480, 720, or 1440 minutes (the last numbers correspond to 2, 4, and 12 h, and 1 day). The *new* time range was then divided by the force-value to yield the number of intervals for the x-axis.

For example, suppose the data were from 11:23 in the morning until 12:11 in the afternoon. The time range is 48 minutes, that is, between zero and 80. The axis would be forced out to the nearest interval of 10 minutes, with end points at 11:20 am and 12:20 pm. This is a range of 60 minutes, and would appear with 7 tick marks at 10-minute increments.

Suppose the data period had been longer, say, from 11:23 in the morning until 12:57 in the afternoon. The time range of the data is 94 minutes, that is, between 80 minutes and 120 minutes. The axis would therefore be forced outward to the nearest intervals of 15 minutes, with the endpoints at 11:15 am and 1:00 pm. This is now a range of 105 minutes, and the axis would still get divided into 15-minute divisions. There would thus be 8 tick marks. Note that the number of intervals will always be 5, 6, 7, 8, or 9. In the case of 8 or 9, only alternate tick marks were labeled. If the data from the example had continued for long enough that the forced axis endpoints had been 120 minutes apart, there would have been 9 ticks. A longer interval would have dropped the number by increasing the increments to 30 minutes.

This algorithm copes with time ranges from a few minutes to months. It was later extended to years.

We bring these examples up here to indicate that not only can simple scales always be devised, but the process can be automated! Spreadsheets already attempt automatic labeling, but they do a worse than mediocre job. Try the times listed here in your spreadsheet: Excel splits the axis up into unequal intervals, and QuattroPro just goes off to see the Wizard. If you have time graphs, you may have to fix this kind of problem in your graphics package.

But we are getting ahead of ourselves. We said we would look at Excel and QuattroPro in their own chapters. That is next.

SUMMARY

Spreadsheets allow ready analysis of data and allow graphs to be created quickly and easily.

For creating graphs, there are three steps:

- ■ Get the numbers.
- ■ Create the graph in the spreadsheet.
- ■ Fix the graph to be publication quality.

In the chapters that follow, the first step is assumed to have taken place, and the next two steps are examined. Most of the fixing of the graph will take place in the spreadsheet program (as described in Chapters 8 and 9). Some work may have to be done in a graphics program (Chapter 10).

Using spreadsheets: Excel® 8

BAR CHARTS

Bar charts are a very useful way to summarize statistical data. A table can be used to present the same kind of information, but the bar chart is sometimes more convincing. We introduce the graphing use of the spreadsheet by means of the bar chart, as its production is quite straightforward. Some of the techniques you will learn here for changing the appearance of the graph will be useful for other kinds of graphs.

Suppose you want to show the results of an evening-class vote for business classes on the different nights of the week. You might want to present something like the chart shown in Figure 8.1. Here's how to do it.

Open Excel. Type in the data, with the names of the days of the week in a column, and the number of votes each received next to their name, as shown in Figure 8.2. Select *all* the data. Click on `Insert` and `Chart`.

You will be invited to select from a number of `Standard Types` of chart. The default is called `Column`, and examples of it appear on the right of the dialog box. If you click on where it says `Press and hold to view`

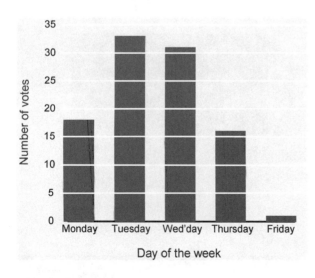

Figure 8.1

	A	B
1	Monday	18
2	Tuesday	33
3	Wednesda	31
4	Thursday	16
5	Friday	1

Figure 8.2

`sample`, the program will let you preview the chart with your data. Try it. Then, since this is the correct kind of chart, click `Next`.

You will be given the opportunity to add a title and to name the two axes. When you are all done, with all the defaults applicable, you get something like the graph of Figure 8.3.

Let's see what we need to do to get the result we're looking for. The graph is a little squashed in aspect ratio, and there's an unnecessary gray background. The bars are too thin and blue, not gray, and they are in front of the extended tick marks indicating number. There is a completely wasted label on the right telling us that the data are `Series 1`. (We put in a graph title, just to show where it would appear. Our target graph does not have one.)

There are many ways we can go about the business of adjusting the appearance of the graph. So many, it can get confusing. Here's how we fixed those aspects of the graph.

The first thing to do is to add the `Chart` toolbar to your Excel environment, by clicking `Tools / Customize` and checking the `Chart` box. This is probably the best approach of all to editing graphs, as it has a very useful pick list (called the `Chart Objects` list), and right next to it a `Format` option (with its icon to the right of the pick list), as in Figure 8.4. What it says when you hover the mouse over the icon depends on what is selected in the pick list, but it is generally something like "format the thing on the left." Using this list should get you in the habit of looking outside the graph as you make the changes.

If you were not looking outside the graph, you might miss the fact that when you click on the chart that you have inserted into the spreadsheet, a new pull-down menu item appears at the top of the spreadsheet just to the left of the window pull-down. It is the word `Chart`, and the pull-down options are `Chart Type`, `Source Data`, `Chart Options`, `Location`, `Add Trendline`, and `3D View`, as shown in Figure 8.5. (Some options may be grayed-out, depending on the kind

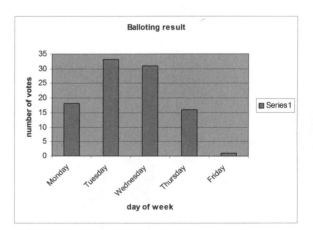

Figure 8.3

Figure 8.4

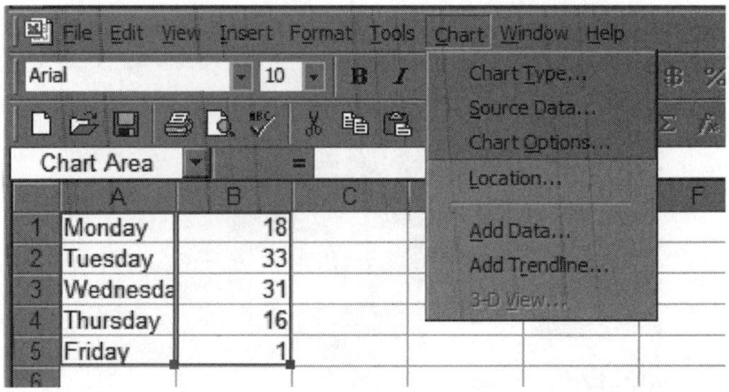

Figure 8.5

of chart.) With these functions, you can revise a good deal of the appearance of the graph. If you like this approach, there is a theory that you can have the Chart Menu permanently on the screen by clicking Tools / Customize and checking the Chart Menu Bar box. We can't make it work, however.

But notice this. On the left of the screen, where you normally see the address of the active cell in the spreadsheet, Excel is telling you which part of the graph you have selected. (In Figure 8.5 it is Chart Area.) It will also be telling you in the Chart toolbar that you added. The names may not be familiar to you—the horizontal lines across the background are called Value Axis Major Gridlines, for example—but you should get used to using these windows for help in selecting the right part of the graph. You will see the same information next to the cursor when you click to select, but of course, the cursor display varies as you move it over the screen, whereas the two windows remain locked to your selection. Note that the one in the left corner isn't a pull-down or a pick list. It might look that way, but it isn't.

The pick list that came with the Chart toolbar works. However, it is a sort of top-level look. If you want to see in detail what the items are that you can select on the graph, select something (anything), and then use the left–right and up–down arrows to cycle through all the elements of the graph. The left–right arrows will take you through everything in detail, for example, each column in the bar chart, whereas the up–down arrows will take bigger jumps, treating all the columns in your bar graph as Series 1, for example.

But for now, back to the Chart Menu pull-down. If you click on Chart Options, a dialog box opens that will let you edit the Title, Axes, Gridlines, Legend, Data Labels, and Data Table parameters. Most of these are simple or even binary options: do you have a legend or not, do you present the data values as labels or not, and so on. The Source Data option is similarly simple, and not very useful for our present purposes. (It can be valuable, however, when you need to select a different set of data to graph.)

However, if you *right-click* on the various parts of the graph, a whole new range of options appears. For example, if you right-click on one of the bars in your bar graph, a new dialog box opens, giving you the choices of (among other things) Format Data Series. Note that access to the Format dialog is also available through a pull-down menu item at the top of the screen, where it usually displays things like Cell, Row, or Column. When you have a data series selected, the Format menu displays Selected Data Series. You can get to the same place by typing Alt-o. These alternatives are worth remembering, as there are times when Excel just seems to change selection when you right-click, and you

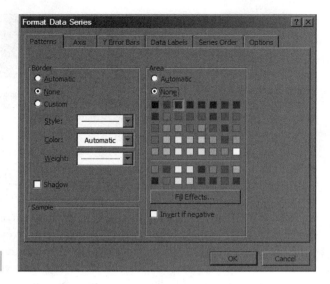

Figure 8.6

can end up formatting the wrong thing. (Perhaps this is due to a minute movement of the mouse at the instant of the click. No matter. The format pull-down (or Alt-o) provides a way around the problem.)

If you select that option, a new box appears (Figure 8.6) with the following tabs: Patterns, Axis, Y Error Bars, Data Labels, Series Order, and Options. These are what you need.

Note that you can get to exactly the same place by clicking on a bar in the chart (so as to select a series), and then clicking on the icon to the right of the Chart Objects pick list on the Chart toolbar. If you hover the mouse pointer over this icon, it tells you what it will allow you to do, as in Figure 8.4.

The Patterns tab (seen in Figure 8.6) opens up a window that allows you to select the bar color, and whether or not you want a line around it. If you do want a line, you can choose the style, weight, and color. For our target graph, we want a gray bar and no line around it.

The Options tab (seen in Figure 8.7) allows you to change the width of the gap between the bars, and to see the effect in real time on a scaled down version of your graph inside the dialog box.

With these two tools, you can make adjustments to get close to the target appearance. While you *can* change the color and thickness of the horizontal lines across the chart, the bad news is you *cannot* change the order in which things are stacked on the page, so you cannot put the horizontal lines in front of the bars, as in the target. Still, you can do a whole lot.

And while you might think the whole thing is rather tedious, *you can save your changes in a way that makes them available as the defaults next time you need a bar chart.* Once you have made all the changes you want, you click on the chart, so that it is highlighted with eight little black squares, and the Chart pull-down menu appears next to the Window pull-down.

Here you select Chart Type, as in Figure 8.8. ("WAIT A MINUTE," you say, "this is how we started!" You are right, but bear with us a moment longer please.)

Now, click on the Custom Types tab, as in Figure 8.9. Now, instead of selecting something from the default Built-in radio button, you click on the button for User-defined. Now here's something interesting: *you will be given the option to* Add *your chart.*

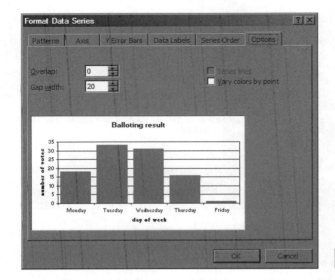

Figure 8.7

You must give the chart a name, and you can add a few words of description, as in Figure 8.10. The name will appear as in the main window where you select types, and the description will appear under the sample graph when you highlight it. Most of what the graph is about will be evident in the window, so for description it might be enough to add a few words about what is unusual or not obvious.[1]

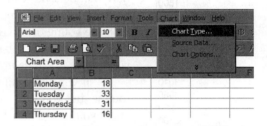

Figure 8.8

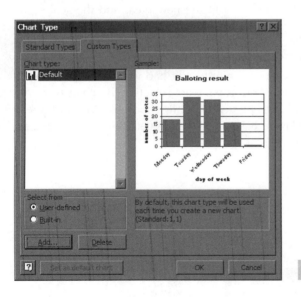

Figure 8.9

[1] Notice that you are also offered the opportunity to delete a chart type from the User-defined category. This is good, as the list may get cluttered as you converge on your style. Be warned, however. About half the time you try to delete a chart type, Excel crashes, and you get invited to send one of those error reports to the good folks at Microsoft. *Don't do it!* They have a competition to see whose bug can get the most error reports in a month. It *must* be! They surely never call a person back with help, or information that their software has been fixed!

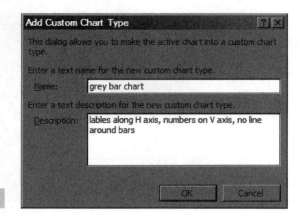

Figure 8.10

Try it! It only takes a moment to add the bar chart with gray bars, and give it a name. Make sure you have included a title and labels for the axes.

Now, just for fun, let's test the usefulness of the process. Add some new numbers back into the spreadsheet. There does not have to be the same number of numbers, and they don't have to have anything to do with the voting bar chart, but let us suppose they are numbers from which you want to create a bar chart that is largely similar in appearance to the voting one. The data from Figure 8.11 will serve as our example.

If you select all this in Excel, and go through the Insert Chart routine, but you select the Custom Types, User-defined, and pick the chart you just defined, what you get if you make no more changes is shown in Figure 8.12.

Now, we skipped a step to get here, but let's see what we have. The chart has been resized by Excel to fit the new data, and the scales are appropriate to the new data. The bars are gray, the background color has gone, the labels for the abscissa have been grabbed from the data, and the chart and axis labels have been picked up from the previous example. These we would have had the opportunity to edit had we not skipped a step and chosen Next instead of Finish as we were selecting the chart type. All in all, you can see this is going to be a very handy short-cut. With just a few edits, we have the graph of Figure 8.13, showing the data the way it appeared in the *L.A. Times*, January 17, 2007 (page A6).

J	1800
F	2200
M	2480
A	2320
M	2710
J	3180
J	3510
A	3050
S	3360
O	3750
N	3490
D	2900

Figure 8.11

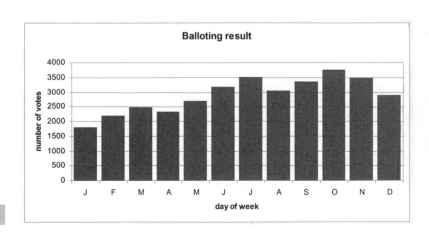

Figure 8.12

Note that in order to change the aspect ratio of the graph, all we have to do is select the entire graph (so that the eight little black squares appear) and use one of the middle ones on one side to squeeze the entire thing. Excel will change the font size of the x-axis labels if necessary to avoid them overlapping, or it will rotate them. The shape of the individual letters is maintained, so the text looks quite normal.

Once you have saved the graph as your own user-defined graph, the simplicity of implementing automatically another graph that looks similar fits with having your own *style*. More on style in Chapter 13.

One more little task. Our target graph had white lines in front of the bars, not black lines behind them. While Excel does not give us a way to move the lines in front, it does give us a way to add lines. However, such changes are not remembered by Excel in the user-defined styles. Since we will end up making the existing black lines go away, the user-defined version would have no lines at all, which would render it less useful. The only workable solution is to leave the lines black when you save the style, and fix them later. It is not hard, because Excel has several of the features of PowerPoint, in a very PowerPoint-like interface.

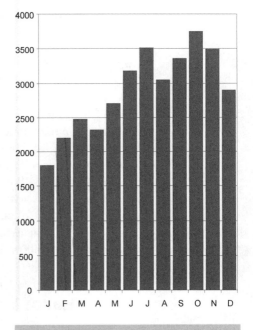

Figure 8.13

All you have to do is to add the Drawing toolbar. Under Tools, click Customize, and under the Toolbars tab, check the Drawing box. This will add a familiar series of icons at the bottom of the screen, just where they would be in PowerPoint. It *isn't* PowerPoint, but a sort of sawn-off version of it. Some of the options are grayed out. You cannot change the order of the elements of a graph, for example. Still, there are some handy capabilities, as we shall see.

For our present purposes, a capability to add white lines on top of the Value Axis Major Gridlines is useful. With the Drawing toolbar, you can add lines (any color) anywhere on the spreadsheet—half on and half off the graph, if you choose.

Strangely, while you can zoom the view in Excel, that option is grayed out when you have a chart selected. So choose the zoom you want *before* you get into the editing.

Then, using the Drawing toolbar, select the icon that lets you draw a line. Using your PowerPoint skills (i.e., using the Shift key to ensure that your line is truly horizontal), draw the line on top of one of those Value Axis Major Gridlines. The new line has been drawn *on top of* the earlier line, but that won't be obvious until you change color. If you can't quite line it up on top of the original line, press the Alt key, to release the snap feature while you are moving the new line. Then select a line color of white, and a line thickness of a point or more, and you should find the original line disappears.

Use Ctrl-c Ctrl-v to make a copy of the white line, and put the copy on top of the next Gridline.

Select *both* white lines, and `Ctrl-c Ctrl-v` them, and put them on top of the next two `Gridlines`. Repeat the process until you have all the lines white.

When you copy and paste the chart into your document, these lines will come with it.

Now because the process of scaling and transferring the graph to your word processor involves some behind-the-scenes approximations, it is worthwhile making the original black lines disappear, so none will leak through later. Click on something in the graph, and use the right or left arrow to cycle through all the things you can select until you get to `Value Axis Major Gridlines` in the little window in the top left corner (the one that looks as if it should be a pick list, but isn't). What you will see might be `Value Axis M...` because the window isn't long enough for more, but you'll know when you get there because all the lines in the set will get a little black square at their end, whether or not you covered the end of the line with a white line. Then right-click on one of the set of lines, and in the little dialog window that opens up either (a) click Clear (which will remove the lines for ever) or click `Format Gridlines`, and follow the boxes until you can turn the lines white.

Figure 8.14

You might wonder how you can right-click on a line that you have covered up. There are two options. Maybe you can leave a little of one of the lines showing, just so you can click on it. (Since you are going to turn it white, it won't matter if it shows.) Or maybe you notice that the cursor will tell you what it is hovering over. If you put the cursor just outside and to the bottom right of one of the black square boxes, it will give you `Value Axis Major Gridlines`, as in Figure 8.14. (Actually, you don't have to select the gridlines in order to right-click them, but it helps you to see where they are if you have them all covered.)

This way, you can get white lines on top of the gray bars, as in Figure 8.15, and reproduce exactly the target graph.

Note that the only real difference between Figure 8.15 and the target is that we left a title visible here.

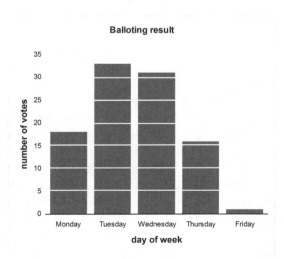

Figure 8.15

You can copy and paste this graph directly into Word. You can also insert a Text Box into your Word document, and paste the graph into that. (This sequence will have the beneficial result of reducing somewhat the frustrating tendency of the Word graphic to bounce uncontrollably from page to page as your document is edited. Reduce, mind you, not eliminate.)

HISTOGRAMS

While we are thinking about bar charts is a good time to mention histograms. The difference was discussed in Chapter 2, but it will bear repeating. A bar chart is the kind of thing we have just looked at: a presentation of several things each of which is different in some way, or of the same thing sampled at different times. A histogram shows the distribution of samples of a single variable: how many times in the sample the value was in this bin or that.

Excel does not have a chart option that makes it easy to produce a histogram from a set of data. There is an option to add it, but unless you are a frequent user of histograms, you probably have not done that. However, by adding a few cells with the appropriate function to the spreadsheet, you can get a bar chart that presents the data in exactly the same way a histogram would. (A histogram by any other name. . . .)

Here's an example. Suppose you have a collection of data that you want to summarize in a histogram. Call it a data array—it could be a column or a block. Suppose you know the bins that you wish to put the data in, metaphorically speaking. All you do is use the *array function* FREQUENCY on these two sets of information, and then make a bar chart of the values returned by the function.

To demonstrate the method, we set up an array of random numbers in a spreadsheet, to see how the random numbers are distributed. The data were generated using the Excel RAND function. In our example, we set up 1000 cells, A8:E207. We wanted to see how often the random numbers were between 0 and 0.1, between 0.1 and 0.2, and so on. The FREQUENCY function is written like this: =FREQUENCY(*data,bins*), where *data* defines the location of the cells with the data, and *bins* defines the location of the cells with the bin boundaries.

The way that the function uses the *bins* specification is that it expects the lowest limit to be zero, so you need only enter into your bin cells the upper limit of the first interval. Furthermore, the function returns the count of the data that are above the last limit in the bin definition, so you do not enter the value of the upper limit of the top bin. For random numbers between 0 and 1, you enter 0.1, 0.2, up to 0.9.

Now, the function will return 10 values (one more than the number of cells with the bin definitions), so you select a column of 10 unused cells somewhere, and *while they are still all selected* you type your array function. In the case of our example, we typed =FREQUENCY(A8:E207,G8:G16). When you do it, however, after the close-parens, *do not hit the enter key.* You must put the formula into the cells by pressing Ctrl-Shift-Enter. Excel will add curly brackets at either end of the formula in each cell in the selection, to remind itself this is an array function. When the spreadsheet calculates, the *value* in the cells of the FREQUENCY array will be the *count* from the data. But the formula will not be replaced—you can still see it if you click on one of the cells.

Figure 8.16 shows what the top of our spreadsheet looks like. Note that the bin array consists of the numbers from 0.1 to 0.9. This is all that is needed.

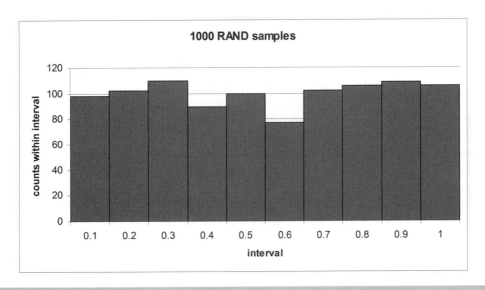

Figure 8.16

	B	C	D	E	F	G	H
1							
2							
3	Histogram of random numbers						
4							
5							
6		data array				bin array	
7							
8	0.005508	0.300519	0.611477	0.53182		0.1	93
9	0.789407	0.972676	0.866994	0.964128		0.2	82
10	0.140781	0.826107	0.406271	0.196286		0.3	86
11	0.179447	0.564814	0.245789	0.103622		0.4	93
12	0.598023	0.394786	0.406553	0.966334		0.5	111
13	0.43269	0.655623	0.080301	0.852104		0.6	97
14	0.540013	0.675627	0.5407	0.427772		0.7	119
15	0.310184	0.529615	0.56273	0.780259		0.8	111
16	0.476213	0.09944	0.815628	0.4892		0.9	112
17	0.618379	0.309516	0.971311	0.461484			96
18	0.323744	0.697545	0.797299	0.485035			

Cell H17 = {=FREQUENCY(A8:E207,G8:G16)}

However, unless you type "1" after the end of the array (in cell G17 to the left of the highlighted cell), you will not get a "1" under the last column in your histogram.

To make your histogram, all you need to do is make a bar chart of the array values by selecting the cells and doing the familiar `Insert Chart` routine. Figure 8.17 is one result of our example. The details change every time the spreadsheet recalculates.

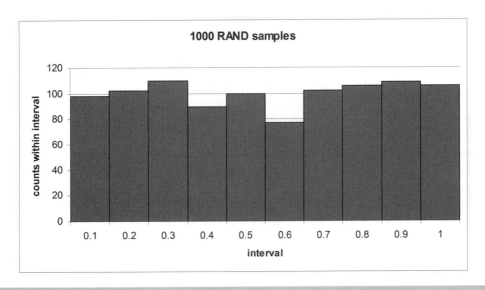

Figure 8.17

You will notice that, in keeping with common practice for histograms, we have adjusted the space between the bars to zero. With only modest effort in the spreadsheet, we have a passable histogram from Excel.

PIE CHARTS

As before, let's suppose we have a data set in a spreadsheet. Here we are pretending we have sampled some classes to find out who does the teaching, and come up with the data shown in Figure 8.18.

	A	B
1	Profs	12
2	Grads	130
3	Undergrad:	78
4	Temps	4

Figure 8.18

Our goal is to show the data in a pie chart so that we can see at a glance what fraction of the classes are being taught by qualified personnel (assuming we know what "qualified" means). Our objective is to get the result shown in Figure 8.19.

As before, type in the data, as shown above. Highlight the data, and click on `Insert / Chart`. Select `Pie`. With only a little effort, you get the result shown in Figure 8.20.

This graph shows the default colors and font sizes, as well as the default text placements. It's not bad, though it wasn't quite what we had in mind.

You can edit further in Excel; for example, you can edit the colors of the areas, and you can edit the font used for the labels. Fonts are straightforward: right-click on your selection.

Changing the color of the pie segments is not difficult. If you right-click anywhere in the pie, you are told you have the option to `Format Data Series`. In the pop-up window, select `Area`, and the shade of gray that most appeals to you. This change will have the effect of turning *all* the segments of the pie chart gray. To turn some of them to white, you need to click on them twice, fairly slowly, or use the left–right arrow once you have selected the series. Each segment is called a `Series Point` with a label (e.g., "Grads"), and you can see

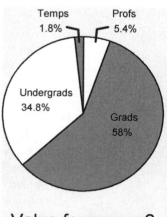

Value for money?
teaching duties

Figure 8.19

what you have selected displayed in the top left corner of the screen. Once you have selected the segments you want white, right-click and go through the process of changing the color via `Format Data Point`.

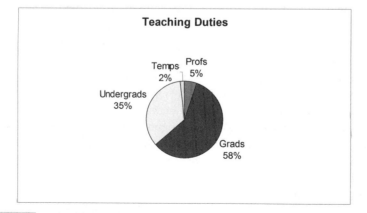

Figure 8-20 See insert for color representation of this figure.

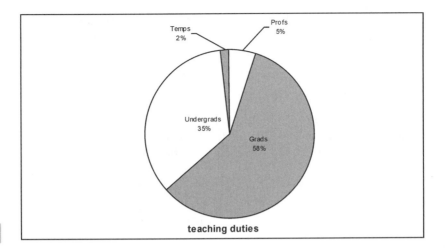

Figure 8.21

You can even edit the size of the pie itself, by selecting the `Plot Area` from the `Chart Objects` pick list, and stretching it to the desired size. As Figure 8.21 shows, you can then get quite close to our goal.

We are not actually stuck with the rectangle around the outside of the figure; we left it there to indicate the size of the pie compared to the size of the box that the pie comes in. (Our local pizza place does that, too.)

There seems to be no way to remove the box in Excel. You can make it white, or even transparent against the Excel cells, but in the Word document, this is how big the box is. While the proportions of the figure are about what we wanted, the actual size may be a bit useless if you have to accept all that white space.

We can, of course, bring the information into PowerPoint and make improvements to it there. We'll look at that in Chapter 10.

SCATTER GRAPHS

An important kind of graph that spreadsheets can do is variously called an *X-Y plot* or a *scatter graph*. Sometimes the word *trend* is used, though we would prefer to reserve this word for situations where the horizontal axis represents *time*. We will look at scatter graphs next. Before we do, however, a word of warning. For some reason, the writers of spreadsheet software perceive a difference between a graph with data spaced evenly along the abscissa and one where the data are not evenly spaced. Therefore, these are treated separately in the spreadsheet graphing software, and accessed separately. If you choose wrongly, you will find that you don't get what you expect.

You may not notice what is happening right away, so be on the lookout for it. Check that the graph you get is at least roughly what you expected. Does it have the right number of curves? Do the axes make sense? Is the slope what you expected? It might seem odd to ask these questions, but it is worthwhile to be sure. Excel will add data of its own if you let it!

Suppose you have two columns of data. Excel will give you two curves if you select `Line` as the type of plot you want, and only one curve if you select `XY` or `Scatter`. When you get the two-curve version (and expected only one), you

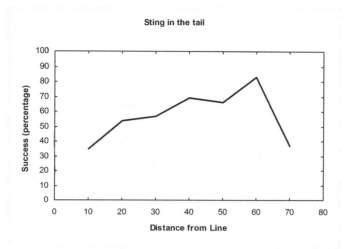

Sting in the tail

Figure 8.22

may note that the horizontal axis is simply derived from the *location* of your data in the column. The first row gets an *x*-value of 1, the second row a value of 2, and so on.

We honestly don't know if a Line Chart is useful. It certainly isn't useful to anyone we know, and it likely won't be useful to you even if your horizontal axis is time. So beware the trap, and check that you really are getting the kind of output you expected.

On with the example. Suppose we had looked at the survivability (in terms of percentage hatch rate) of mosquito larvae as a function of their distance from a high-voltage power line. The local power company, interested in allegations of biological effects of power lines, has done an investigation.[2] The results are shown in Figure 8.22, in a graph with a box around it.

As before, we can do this in two columns. Both programs start with these data in a spreadsheet, as in Figure 8.23.

	A	B
1	distance	success
2	(meters)	(percent)
3	10	35
4	20	54
5	30	57
6	40	69
7	50	66
8	60	83
9	70	37

Figure 8.23

In Excel, as before, you can highlight the numbers and click on `Insert Chart`. Our goal is similar to one of the `Standard Types`, called a `XY (Scatter)`. Under the `XY` option, you can choose a graph with lines or dots (or both) and with the lines simple straight lines joining the dots, or "smoothed." Our target was the lines and the dots,[3] so we select this option and click on `Next`.

With all the default values working, Excel generates the graph shown in Figure 8.24.

There are just a few things remaining to do. First, we didn't want a gray background, or colored lines and dots. Second, we don't need a legend. Oh, and we wanted the axes to form a box around the graph, with tick marks on

[2] Studies of this kind have actually been done. The data here are purely made up, however, and are not intended to illustrate the kind of result that a real experiment might obtain. We select the subject of the graphs in this book with two things in mind: first, we want to illustrate the kind of graph you might choose for a particular purpose; and second, we want to keep your interest!

[3] It is not obvious before you plot the data that maybe you should not be joining the dots. In this particular example, the straight-line joining looks to be inappropriate, but since you (as investigator) might reasonably have expected a smooth relationship, we'll leave the dots joined for the rest of the example.

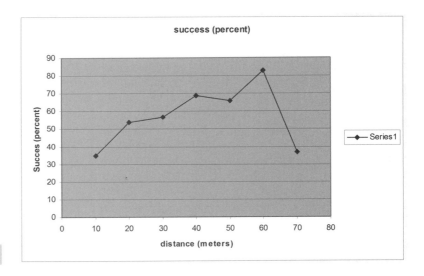

Figure 8.24

the inside, but we didn't want a line around the whole thing. We have a lot to change. *But not to worry, all the work can be remembered by Excel ready for next time.*

	A	B	C
1	distance	success	
2	(meters)	(percent)	
3	10	35	35
4	20	54	54
5	30	57	57
6	40	69	69
7	50	66	66
8	60	83	83
9	70	37	37

Figure 8.25

As it happens, we need a trick. In order to get Excel to allow an axis on the left *and* an axis on the right, Excel requires that there are two sets of data. If we can fool Excel into thinking that, we can then also add a copy of the horizontal axis at the top. In that order.

First, we make a duplicate copy of the last column of data, as in Figure 8.25. Select all the numbers, and create the usual X-Y (scatter) graph. You can let all the defaults apply—we're going to change them anyway. So now you've got a graph that looks about like it did before, because the two lines that are plotted are right on top of one another.

You start by clicking on the graph line. The words `Series 1` will appear in the little window at the top left.

With the cursor over a point on the line (the cursor will indicate the series number—either will do—and the data value), right-click. Up pops a window that will let you `Format Data Series`. Click on the `Axis` tab. Push the `Secondary axis` radio button, as in Figure 8.26.

A new scaled axis appears over on the right of the graph. *Once you have done that*, an easier way to control the secondary axes is made accessible. Right click on the `Plot Area`, which is indicated by the gray background. Now you get to select `Chart Options`, as in Figure 8.27, and then the `Axes` tab. Here you can turn on or off both the vertical and the horizontal axes, with the two `Secondary axis` boxes.

The result is the addition of two more axes, each with a set of numbers. Since the data are the same for each, the labels are the same, and you really do not need them. These labels can be removed by right-clicking on the axis whose numbers you want to remove, and following the `Format Axis` sequence.

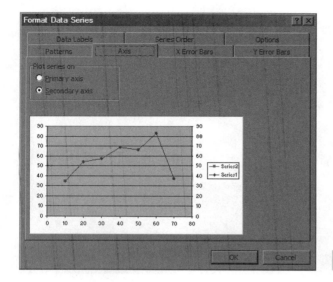

Figure 8.26

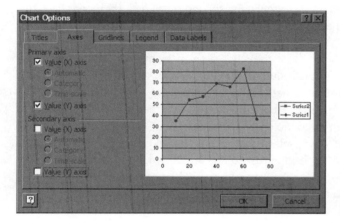

Figure 8.27

Before you close this window to do that, however, you can click on the `Gridlines` tab and remove the horizontal gridlines.

You need to format each axis, both to make all the tick marks appear in the appropriate place (`Inside`), and to remove the `Tick mark labels` on the top and the right. Start by right-clicking on the axis, and then make your selections in the dialog window shown in Figure 8.28.

After this, the gray background is easily turned white. Right-click on it (remember, Excel calls it the `Plot Area`) and follow the `Format Plot Area` sequence. You can remove the gridlines across the plot area in the same way—right-click on one and format it into the bit-bucket. The line around the outside of the chart is removed by right-clicking just outside the `Plot Area` and doing a format of the `Chart Area`.

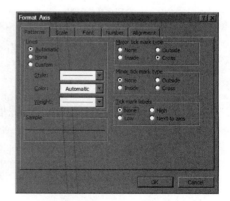

Figure 8.28

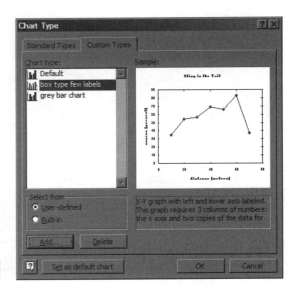

Figure 8.29

The last thing to do is to make the second set of data disappear from the screen. The defaults have placed the lines and dots from the middle column of data on top of the graph as Series 1, and those from the right column underneath as Series 2. However, since the shapes and colors of the markers are different, you can see bits of the underneath stuff showing.

With the mouse, it is hard to select the items you want to format. The easiest way is to use the `Chart Objects` pick list. Alternatively, select something else, and then cycle through the up and down arrows. Once you get the display telling you that you have selected `Series 2`, you can use the `Format` pull-down menu, or Alt-o, and make the line and the markers disappear.

Once you have the graph the way you want it, it is time to save it in Excel so you do not have to go through all these changes again.

Click on the graph, more or less anywhere, so that the `Chart` pull-down appears at the top of the screen. With this menu, select `Chart Type`, `Custom Types`, and `User-defined`, and fill out the brief name and description, as in Figure 8.29.

As you can see from the example in Figure 8.29, it is a good idea to include a reminder that the graph requires a spare copy of the data in order to fool Excel into allowing the box to be created.

You may have noticed that we did not change the scale on the vertical axis. There was a reason for that. If we force the upper scale mark to be 100, and then save the graph type, *all other graphs created with this starting point* will default to 100 for the scale. If we defer that change, future graphs will autoscale.

So far, we've looked at what you might call the "standard product" of spreadsheet software. Next, we are going to go a little further, into what may be unexplored territory.

MULTIPLE SERIES

We have seen the expression "`Series 1`" a few times on these charts. We have generally ignored or deleted the label, as it seems redundant when we have only one set of data. Now it is time to see what happens if we need to show more. The

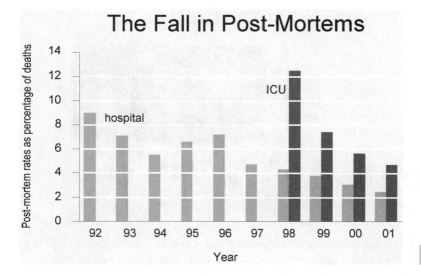

Figure 8.30

example in Figure 8.30 is adapted from a graph in *New Scientist*, February 21, 2004. The graph showed that fewer postmortems were being performed in the United Kingdom over the decade of the 1990s. (The article it accompanied argued that evidence of wrong diagnosis was being missed as a result.)

This graph is interesting in that data are not available for identical times. Perhaps statistics just were not collected for the ICU category prior to 1996. Apart from that little quirk, the graph is pretty straightforward.

There is a little problem with the horizontal axis, however. Strictly speaking, the numbers shown do not increase monotonically in the order given. Let's have a look to see how our software handles this. We start by entering the numbers into a spreadsheet (the numbers in Figure 8.31 are somewhat of a rough estimate of the values that might have been in the original graph, but they are adequate for demonstration purposes).

If you select the three columns beginning with row 2, and going down to row 12, your first attempt at a chart will be as seen in Figure 8.32.

	A	B	C
1	year	rate	
2		hospital	ICU
3	92	9	
4	93	7.1	
5	94	5.5	
6	95	6.6	
7	96	7.2	
8	97	4.7	
9	98	4.3	12.5
10	99	3.8	7.5
11	0	2.1	4.5
12	1	1.5	3.6

Figure 8.31

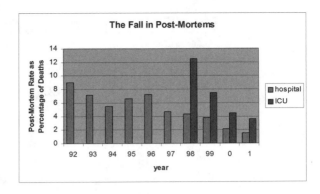

Figure 8-32 See insert for color representation of this figure.

The usual problems are there—the color, the gray background, the lines, the explanatory key over on the right—but really this is pretty close. All these aspects of the graph can be edited in Excel. The long label of the vertical axis was automatically split and looks good. Even the numbers on the horizontal axis are close to our target.

If you had selected the top row of data (the labels "year" and "rate"), Excel would have given you the graph of Figure 8.33 as the first try.

The numbers we actually want are down in the noise because the program decided to show the year as another set of values, and plot everything against its position in the spreadsheet relative to the first number. Go figure!

We can see from each of these graphs that there are problems with the numbers in the first column. In Figure 8.32, the software is treating them as the numbers they are, rather than as the year. In Figure 8.33, they are being treated as data to be plotted against cell position in the column of the spreadsheet.

We might try to add "19" in front of the first few numbers, and "20" in front of the rest, but that wasn't what the target graph was. In fact, there seem to be two solutions.

First, you can make the numbers into *labels* by beginning each of them with a single-quote mark. (Recall that the first column of the voting graph consisted entirely of labels: the days of the week.) If the first column is all labels, Excel® will use the information as labels for the *x*-axis.

Second, you can head the column of numbers with an empty cell. If you do this, the column can be a mixture of numbers and labels. With that capability,

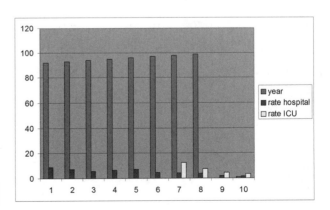

Figure 8.33

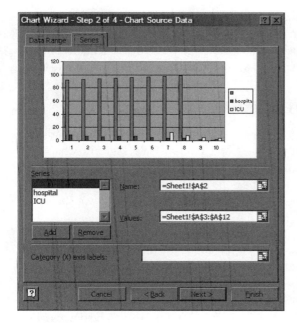

Figure 8.34

you need to make only the 00 and 01 labels nonnumeric, in order to still have Excel use the first column to label the *x*-axis.

However, if you forget to make either of these changes in the spreadsheet, you are given an opportunity after you have selected the kind of graph you want. If you select the `Series` tab, as in Figure 8.34, you will see that there is a blank space for the *x*-axis labels.

This last method is not recommended, however, as it almost certainly will mean that you will have to edit the other series as well. Get used to checking for the appropriate cell references in that `Category (X) labels` window. If you don't see the right information, go back to the spreadsheet and do something drastic, like adding an empty cell at the top of the first column.

More editing is needed to get the result shown in Figure 8.35.

Once you have the *x*-axis labels fixed, the graph is starting to look a lot like our target. Starting with our already-saved gray bar chart style, a few minutes

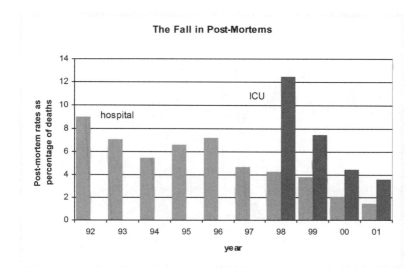

Figure 8.35

editing (change the color and line of the added bar, add a couple of text boxes) will give us the version shown in Figure 8.35.

This is close to our target. Many people would find this acceptable and not bother with the final step that covered the horizontal lines with added white lines.

A COMPLICATED EXAMPLE

Here's an example that demonstrates the power of finishing the spreadsheet process in a graphics program. The graph that we saw back in Chapter 2 showing the Earth's temperature and the carbon dioxide level looked straightforward enough, and while it could not be done in a spreadsheet alone, much of the exacting assembly of the curves *can* and *should* be done there, as we shall see.

We reproduce our version of the graph in Figure 8.36 as a reminder.

In this particular case, the data are from two separate files, both of which were available for download from the Web at the time of this writing. Originally they were found at the website `http://www.yourplanetearth.org/terms/details.php3?term=CO2+and+Global+Warming`. The data are from J. M. Barnola, D. Raynaud, C. Lorius, and N. I. Barkov, "Historical carbon dioxide record from the Vostok ice core" (1999), and can be downloaded from `http://cdiac.esd.ornl.gov/trends/co2/vostok.htm` and J. R. Petit, D. Raynaud, C. Lorius, J. Jouzel, G. Delaygue, N. I. Barkov, and V. M. Kotlyakov, "Historical isotopic temperature record from the Vostok ice core" (2000), with data available at the time of this writing at `http://cdiac.esd.ornl.gov/trends/temp/vostok/jouz_tem.htm`. (Feel free to go to the websites to see how ice core samples are obtained and dated, and why their information is so useful.)

We have no intention here of getting into the discussion of global warming (except to observe that the carbon dioxide data here do not include about the last 5000 years). Our aim is to show how this kind of graph can be generated from the data.

The first step in the process is to get the data and transform it into a usable form. On the Web, there is simply a button to click to see the data. The numbers

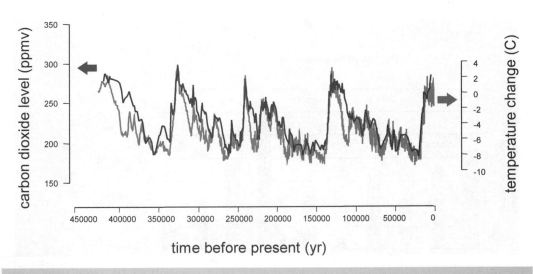

Figure 8.36

are in columns on your screen, but when you copy and paste them into (say) Excel, the information is all jammed into the first column. This won't be useful.

The solution is to copy the data and everything to a file by clicking `File / Save Page As` (pick a name and location) on your browser. (If the file does not save as a text file, open the file into a word processor or WordPad, and save as a text file (`File/Save Page As/Text Document`).) Now, for one of the files, if you copy and paste from there into Excel, the tab delimiters in the original are interpreted as cell delimiters, and you have the data in handy columns. Alternatively, you can open the text file in Excel, and you will be asked about delimiters. But either way, it works, and you now have the data in usable form.

The other data set was not saved as a tab-delimited file; it was space-delimited. You should be able import it as a space-delimited file, treating consecutive delimiters as one. (Also, be aware that with this file, about the first 999 entries have a leading space, which means you will have to slide them all over one column once you have the data in the spreadsheet.)

The problem is that you have actually now got *two* sets of data, in separate files, and *with different lengths*. There are somewhat less than 400 data points in the CO_2 record, and more than 3300 in the temperature data. Each can be plotted readily enough on its own, but there seems to be no way to get a spreadsheet program to deal with the two data sets from two files in a single graph. So it is necessary to combine data sets, if you want to try to make the graph in a spreadsheet.

Now, you *could* combine the data in PowerPoint, after getting Excel to draw you two separate graphs. However, there is a small chance that you could accidentally change the relative alignment of the two data sets. It is much safer to do the combining in the spreadsheet program.

Ideally, you would want three columns: one for the time, one for the CO_2, and one for the temperature. Since, in general, the time numbers from one data set are not exactly the same as in the other, the time numbers would have to be interleaved somehow. As it happens, Excel can deal with that problem. However, for producing the graph, that is not the end of the story.

When you assemble the data, with the time data made negative for each data set, you have *four* columns, not three. The first column should be the time data for the longer (temperature) curve, the second column the data for that curve, the third column is the time data for the shorter (carbon dioxide) curve, and the fourth column the data for that curve. You *select all four* columns of numbers and tell Excel to `Insert` a `Chart`. Of course, since Excel selects a rectangular section of the spreadsheet, you have a section of the selection that is blank. No matter.

Excel *has to be told* that there are two sets of data for the x-axis. Thus, you begin by choosing an `X-Y (Scatter)` graph. The next dialog box allows you to select the `Series`, and you need to do this in order to choose the appropriate data for the x-axis. The first series that Excel produces is correct: it uses the first column for the x-axis and the second column for the data. However, the program uses the first column as the time axis for *all* the other curves, so they are incorrect. You can easily delete either series and set the appropriate columns for the (new) second curve. Figure 8.37 shows what you get.

The data are there, but they share a vertical axis. Because the numbers are so different, the temperature data are rather compressed! We need to edit the graph to add a second vertical axis. We have already seen how that could be

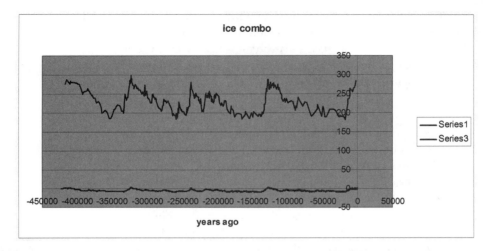

Figure 8.37

done when there are two—or more—sets of data. While Excel is capable of coping with the secondary axis issue for small data sets, it seems the ice data overwhelms the program, and there are problems. (Yes, Virginia, there *are* bugs in Excel!) It takes some work to sort things out, but it can be done.

In Figure 8.37, the top curve is the carbon dioxide curve, with numbers between 150 and 300. The temperature data are shown below, with numbers between about zero and minus 10 or so. Whichever series you select for the secondary axis, the visible axis on the screen is scaled for the *other* data set, and only one axis is given on the screen. Figure 8.38 shows what happens if temperature is supposed to be on the secondary axis.

This is close to acceptable in terms of relative sizes of the curves, but the scale for temperature is in the wrong place. However, if you select the vertical axis and *force* it to cross the *x*-axis at the left end, you can get the two axes to appear at the same time, as in Figure 8.39.

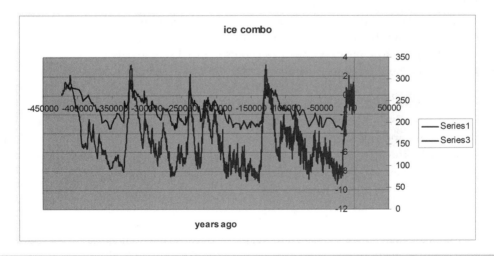

Figure 8.38

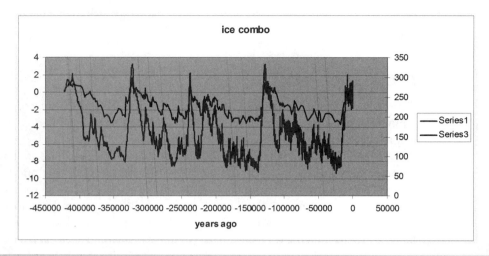

Figure 8.39

From here, it is a simple matter of editing the graph to get the appearance you want. Of course, you are stuck with having the axes touch one another, so you cannot quite reproduce the graph as it was in Figure 8.36. Those final touches were added in a graphics package and are discussed in Chapter 10.

INSERTING DATA

As well as presenting graphical data, you might want to present data numerically, in the form of a table, *in the graph*. There are two ways to proceed. You can get the spreadsheet to add some information, or you can do it yourself. In an Excel bar chart, you can add what is called a Data Table that will appear under the horizontal axis, in the fashion shown in Figure 8.40.

For the monthly report, this sort of thing is excellent. Your management might expect the dots to be joined, but since the hours and the costs are monthly summaries, it hardly makes any sense to join the dots. (Still, managers

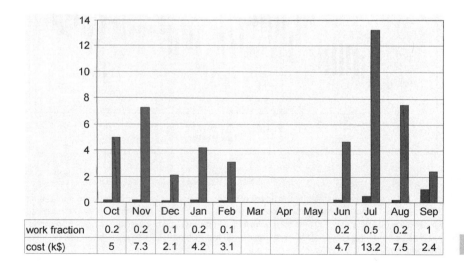

	Oct	Nov	Dec	Jan	Feb	Mar	Apr	May	Jun	Jul	Aug	Sep
work fraction	0.2	0.2	0.1	0.2	0.1				0.2	0.5	0.2	1
cost (k$)	5	7.3	2.1	4.2	3.1				4.7	13.2	7.5	2.4

Figure 8.40

being simple souls, it may be as well to humor them by selecting a different basis chart for the Data Table.)

FLOATING BARS

Figure 8.41 is another uncommon spreadsheet product. You might think of this as a modified bar chart—the data of interest are floating bars in the middle of the graph area.

In this example, the links are the numbers assigned (somewhat arbitrarily) to the nodes in a power delivery system. The voltages are the upper and lower limits of voltage at each node under certain conditions. The data were calculated by a program written in C++, but all that was available when the graph was needed was a table in a Word document. However, the table seemed rather large, and in the end not very informative, as the reader would have been required to do a good deal of studying to understand the results. A graph (a narrower version of this for putting into a single column) was the preferred alternative. So we transferred the numbers to a spreadsheet.

There were originally three columns: the link number, the lower limit, and the upper limit. It was easy to add a fourth column for the difference.

The graph was produced as what Excel calls a Column, subtype Stacked Column.[4] After selecting all the data, Excel was told to produce the chart with all defaults applicable. Then, with the chart still in the spreadsheet, we edited the colors so the visible bar in the middle was gray, and the bar underneath was white, and so not seen.

However, the graph had axes at the left and at the bottom, and we wanted a box around the whole thing. This turned out to be an interesting challenge.

Naturally, the use of a `Secondary axis` is called for, but as soon as you add it, the data become "unstacked," and the graph is not what is wanted. Once again, trickery is needed to get Excel to do what is needed. A whole column of

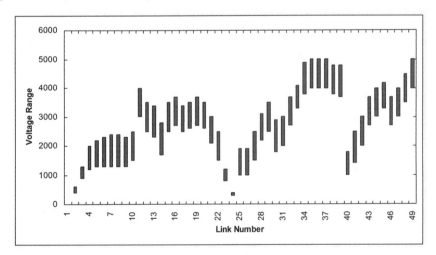

Figure 8.41

---

[4]Yes, dear reader, we are aware that there is an Excel option under `Custom Types` for a similar floating bar. The use of this option would have been no simpler, it seems, because more defaults would have had to be overridden (the default is 3-D and has a colored background), and because the bars are horizontal, the end result would have required a 90° rotation as well as a horizontal flip to get it the way we wanted it, something possible only in a graphics program.

dummy data is added, purely to support the second axis. In this case, a column of numbers that would force the axis to be the same as the primary axis makes sense. A number like 5700 will do, copied all the way down the column.

It is a simple matter to select all the data (there are now five columns), and choose a stacked column. Rather than add the axis labels during the chart creation process, we headed the first column with an empty cell. The other columns were headed with identifying labels. Before you put the graph into the spreadsheet, however, you might as well remove the `Series` corresponding to the third column of numbers. The third column was headed `Max V` and is a calculated value in the stacked column chart. Select the `Series` tab and click on `Remove` for this series. (If you forget, you can click later in the `Plot Area`, select `Source Data`, and find the `Series` tab there.)

Once the graph is created, you assign the column of dummy numbers to its own secondary axis and force the corresponding columns to go away by giving them no color and no border.

As always, it is worthwhile saving the graph in the user-defined types, with a note about the need for a column of dummy data. When you reuse the `Type`, Excel will show you the preview in color, but will make the graph in gray. While it will remember to give the dummy data a secondary axis, it will not remember to *not use* the first series as data to be plotted unless you have headed the column with an empty cell.

For some reason, Excel will also not remember that you formatted the colors of the `Max V` column into invisibility. Still, it's not a big edit that you have to repeat.

MULTIPLE TYPE: COMBINING TREND AND HISTOGRAM

Once in a while, you will see a graph that combines a line graph and a bar chart. There seem to be two occasions when you might need this. One is to show two quantities that vary in time, but one quantity must be plotted as a bar chart (because the guidelines of Chapter 3 make it clear that a line graph would be an error) whereas the other can be plotted as a line graph without risking misinterpretation. We examined an example of this in Chapter 3, when we saw the support for capital punishment graphed as a line above a bar chart of the number of executions carried out each year.

The other kind of graph that requires two kinds of data is the Pareto chart. This is a chart named after an Italian economist who observed (in 1906) that 20% of the population (of Italy) owned 80% of the wealth. Of course, if wealth was evenly distributed among a population, 20% of the population would have 20% of the wealth. Given that wealth is not evenly distributed, it may be supposed that the more unevenly distributed it is, the smaller the first number and the larger the second.

But when you think about it, it becomes clear that the 20/80 split is arbitrary: the richest 20% of the population of a country is bound to own some fraction greater than 20% of the wealth. It is said that in the United States, 20% of the population owns 50% of the wealth. There is no reason to choose two numbers that add up to 100 apart from their being more easily memorable.

The major application of the 20/80 split is as a management tool. You may consider it quick and dirty, but it often works adequately as a rule of thumb: 80%

TABLE 8-1

Reason	%	Cumulative
Leaves on the line	38	38
Switch jammed by heat	23	61
Signal error	10	71
Driver overslept	6	77
Cafeteria running late	5	82
Funny noises in train	4	86
Fire at trackside	3	89
Power failure	3	92
Animal on line	2	94
Presidential motorcade	2	96
Funny smell in carriage	2	98
Accident on line	1	99
Unidentified	1	100

of customer complaints can be dealt with by addressing 20% of the known problems, and so on.

Given that you are going to pick two integer numbers that sum to 100, then at one extreme distribution (a uniform distribution) 50% of problems would cause 50% of complaints, and at the other extreme (basically there is only one important problem) 1% of problems would cause 99% of complaints. It is reasonable to assume that a 20/80 split is somewhere in the middle. Fine, 20/80 is convenient, and unless the distribution is unusually close to one of the extremes, 20/80 is likely to be a reasonably accurate estimate of the situation *even if you know nothing at all about the statistics*. Accurate enough for some management decisions to be based on, anyway.

Table 8.1 is an example of some data you might assemble to explain why your train to work was late. Note that the right column is the cumulative total of the numbers in the middle column. This is only necessary for a Pareto chart. For any other chart that combines bar chart with line graph, this constraint is not applicable.

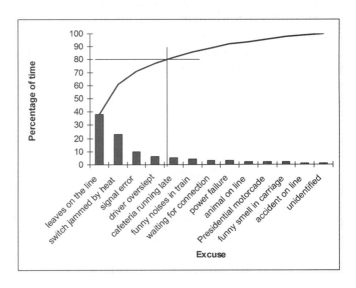

Figure 8.42

You could plot this information in the way shown in Figure 8.42. This is straight out of Excel. Note that there is a line drawn across the graph at the level of 80%. From where it intersects the cumulative line, another line is drawn down the graph, separating four excuses for lateness from the rest. These four excuses are responsible for a little less than 80% of the problems. Since there are altogether 12 identified causes, and one unidentified, we can see that the four are about 31% of the number. Evidently the distribution is not skewed enough to get the 20/80 split: never mind, the diagram is still a Pareto diagram.

The question is: How do we draw it? You type in the data, and with it all selected, you go through the `Insert Chart` routine. Excel generates labels for the horizontal axis, but with as much text as this example, they overlap and would be illegible. Excel therefore rotates the labels to 45° and solves the problem. (You could adjust the angle, but we left it at this value for the example.)

You select what Excel calls a `Line-Column` from the `Custom Types`. As always, to get our target you will need to fix it up a little.

If you have remembered this chapter, that will be straightforward.

SUMMARY

Excel has many convenient features as far as creating graphs is concerned. Editing graphics in Excel is much more efficient than doing it in PowerPoint because Excel can remember the changes you made and reproduce them later for another graph.

- Add the `Drawing` toolbar to your custom version of Excel.
- Select the zoom level you want before you edit a chart. Changing zoom is not an option once you have a chart element selected.
- Excel does not do a good job with pie charts. Anticipate editing them in PowerPoint.
- To have Excel automatically choose your first column of data as x-axis labels in a bar chart, put an empty cell on the top of the column, or make all the cells into labels by starting them with a single quote mark.
- Get accustomed to using the `Chart` pull-down menu, and to looking at the window in the top left of the screen that indicates what you have selected.
- You can select elements in turn on your graph by using the up–down arrows (for broad selections, such as an entire series of data) or the left–right arrows (for more detail, such as a single data point in a series).
- Right-clicking on selected items brings new capabilities to change the appearance of your graphs.
- If you override the automatically selected scale on an axis, the change will be remembered. Usually, it is worth deferring this step until you have saved your chart `Type`.
- To create a box around your graph, with tick marks on all sides, you need to add a `Secondary axis`. Often, it is best to add a dummy set of data for the axis. Start with the vertical axis. You can select a data series and use the `Format` pull-down, or you can right-click on

a data series, and go through the `Format Data Series` sequence to add another vertical axis.

- Once you have added a second vertical axis, you have two ways to add tick marks to a top horizontal axis: click in the `Chart area` (outside the `Plot Area`), and select `Chart Options` from the pull-down menu, or right-click in the same area, and select `Chart Options` from the dialog box.

- Format the color and border of the dummy data so they do not show on the graph.

- You can add a data table under a bar chart.

- When you have finished making a graph the way you want it, consider adding it to the selection of user-defined `Types` under the `Chart` pull-down. The time saving next time around will be considerable.

The following changes to a graph will not be remembered:

- To add callouts and comments, use the text and arrows available in the `Drawing` toolbar.

- To make white (or any other color) lines on top of your graph, use the `Drawing` toolbar.

EXERCISES

8.1. From the following data produce the graph shown in Figure 8.43.

Age	Males	Females
0–9	90	97
10–19	85	94
20–29	95	90
30–39	82	80
40–49	80	78
50–59	70	75
60–69	60	80
70–79	50	70
80–89	40	60
90–99	14	24
100–110	3	4

(*Hint*: Note that the numbers under the left side of the graph are colored red.)

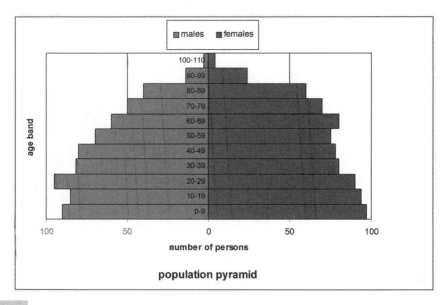

Figure 8-43   See insert for color representation of this figure.

8.2. Some readings were taken of the output of power supply, with a digital instrument with a three-digit display. Generate a histogram.

9.75	9.78
9.84	9.80
9.81	9.76
9.82	9.80
9.77	9.75
9.79	9.76
9.85	9.81
9.85	9.78
9.75	9.77
9.77	9.78
9.87	9.76
9.78	9.80
10.1	9.76
10.1	9.79
10.2	9.79
9.82	9.76
9.78	9.77
9.80	9.79
9.91	9.81
9.87	9.75
9.84	9.81
9.82	9.78

From the histogram, does this look like a power supply you would think was functioning properly?

8.3. How do you rotate a pie chart in Excel without changing the order of the pie segments?

Using spreadsheets: QuattroPro®

9

BAR CHARTS

Bar charts are a very useful way to summarize statistical data. A table can be used to present the same kind of information, but the bar chart is sometimes more convincing.

Suppose you want to show the results of an evening-class vote for business classes on the different nights of the week. You might want to present something like the graph shown in Figure 9.1. Here's how to do it.

Open QuattroPro®. Type in the data, with the names of the days of the week in a column, and the number of votes each received next to their name, as in Figure 9.2. Select *all* the data. Click on `Insert` and `Chart`.

The chart "wizard" will start by assuming you want a pie chart, but after one key click, you can select from a number of standard types of chart. The bar chart is up near the top of the list. If you select that, fill in the titles and axis labels, but default to everything else; you get something like Figure 9.3, only with color.

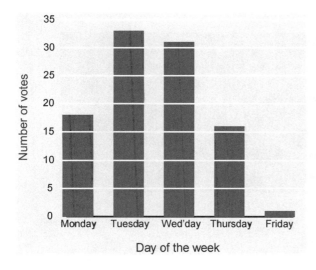

Figure 9.1

	A	B
1	Monday	18
2	Tuesday	33
3	Wednesda	31
4	Thursday	16
5	Friday	1

Figure 9.2

In what happens next, QuattroPro differs from Excel. When you click inside a chart, for example, on one of the bars of the chart, the top of the computer screen changes. Not only does a new pull-down menu for `Chart` appear next to the `Tools` pull-down, *but a bunch of new tools automatically appear on a new toolbar.*[1] If you click on a bar, you get a color and line toolbar, that are reminiscent of the way these things are controlled in Presentations. The screen shown in Figure 9.4 is an example of the start of the transition.

The screen changes to the layout in Figure 9.5, as soon as you click on a bar. Note the shaded box on the left—here you can select not only color, but also fill type, if you want a shaded look or some such frivolity.

Not only do new toolbars appear, but a little window appears that will let you select what you want to edit by name (from a pick list), so you don't have to go poking about with the mouse or the right arrow key.

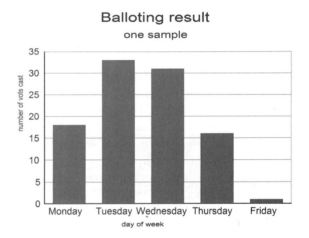

Figure 9.3

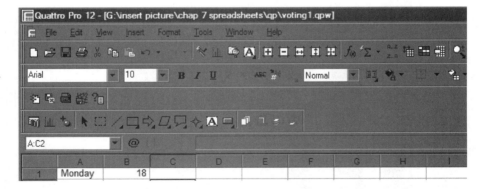

Figure 9.4

---

[1] If this does not happen when you try it, you need to pull down the `Tools` menu and go through a sequence of `Customize...`, `Customization`, `Toolbar`, and make sure `Property Bar` is checked.

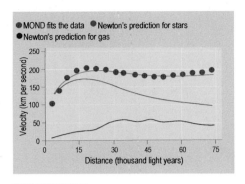

Figure 1.57   See page 28 for text discussion of this figure.

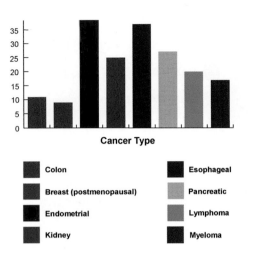

* Extra percentage of cancer risk, by type, attributable to obesity in the U.S. population

Figure 1.59   See pages 29–30 for text discussion of this figure.

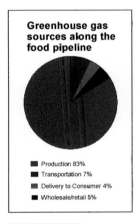

Figure 1.63   See page 33 for text discussion of this figure.

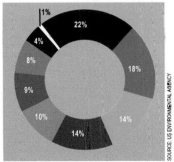

Figure 2.15   See page 43 for text discussion of this figure.

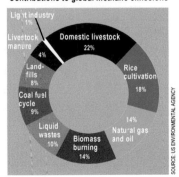

Figure 2.16   See page 44 for text discussion of this figure.

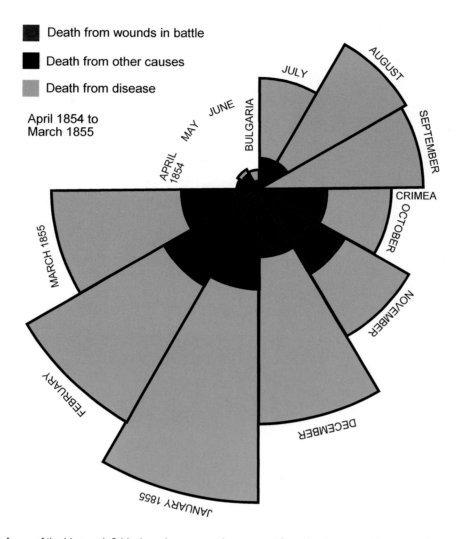

Death from wounds in battle

Death from other causes

Death from disease

April 1854 to
March 1855

The Areas of the blue, red, & black wedges are each measured from the centre as the common vertex.

The blue wedges measured from the centre of the circle represent area for area the deaths from Preventable or Mitigable Zymotic diseases, the red wedges measured from the centre the deaths from wounds, & the black wedges measured from the centre the deaths from all other causes.

The black line across the red triangle in Nov. 1854 marks the boundary of the deaths from all other causes during the month.

In October 1854, & April 1855, the black area coincides with the red, in January & February 1855, the blue coincides with the black.

The entire areas may be compared by following the blue, the red, & the black lines enclosing them.

Figure 2.34   See pages 54–55 for text discussion of this figure.

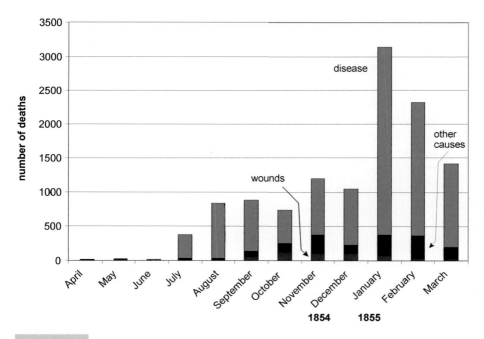

Figure 2.35   See page 55 for text discussion of this figure.

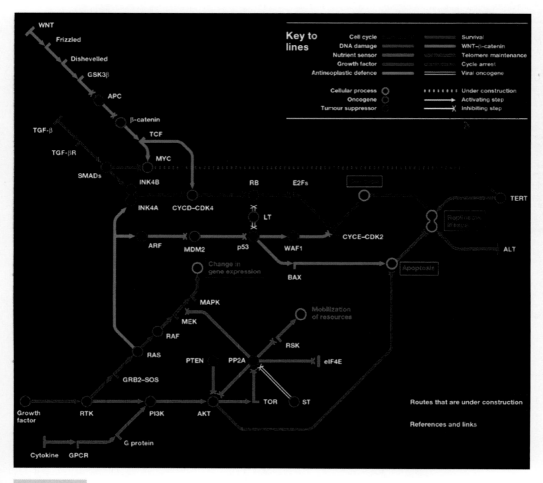

Figure 2.38   See pages 57–58 for text discussion of this figure.

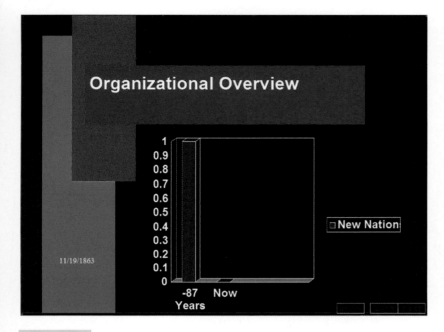

Figure 6.10   See pages 156–157 for text discussion of this figure.

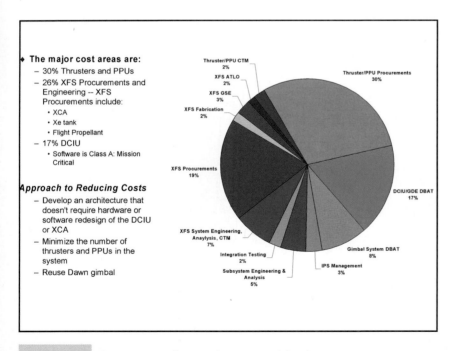

Figure 6.15   See page 160 for text discussion of this figure.

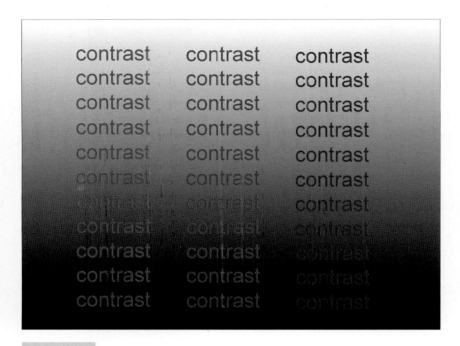

Figure 6.16   See page 161 for text discussion of this figure.

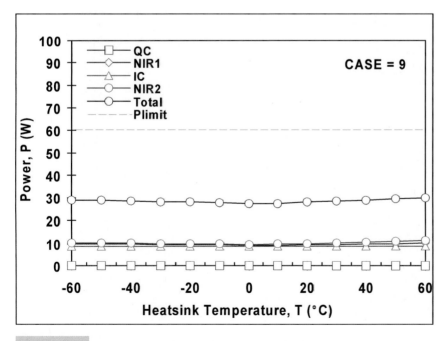

Figure 6.17   See page 161 for text discussion of this figure.

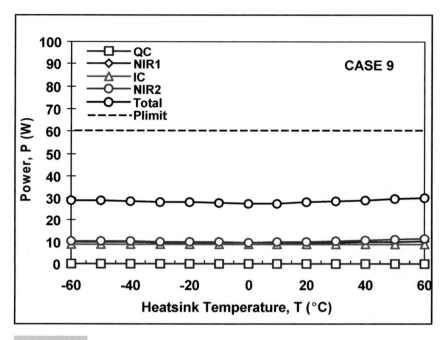

Figure 6.18   See page 162 for text discussion of this figure.

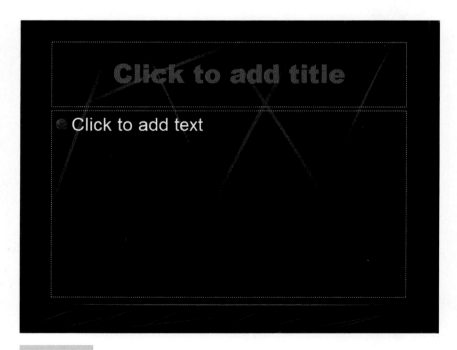

Figure 6.20   See pages 163–164 for text discussion of this figure.

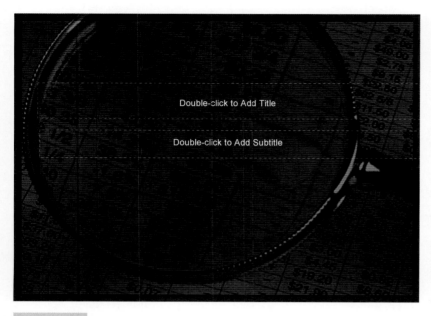

Figure 6.21   See pages 163–164 for text discussion of this figure.

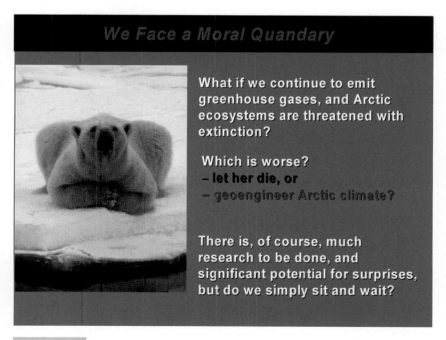

Figure 6.27   See page 173 for text discussion of this figure.

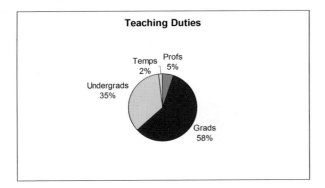

Figure 8.20   See page 191 for text discussion of this figure.

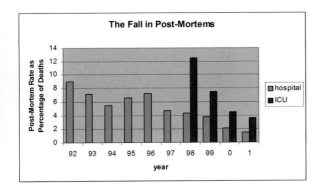

Figure 8.32   See pages 197–198 for text discussion of this figure.

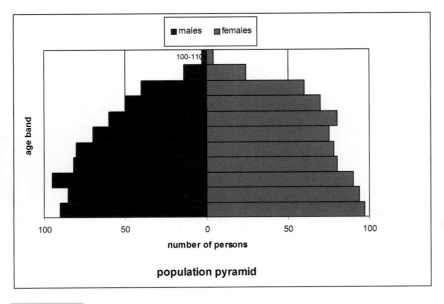

Figure 8.43   See pages 208–209 for text discussion of this figure.

Teaching Duties
Value for Money?

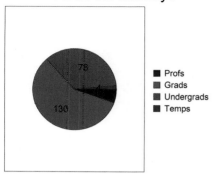

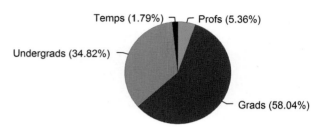

Figure 9.24 See page 220 for text discussion of this figure.

Figure 9.26 See page 221 for text discussion of this figure.

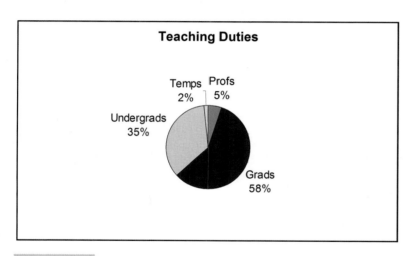

Figure 10.13 See page 249 for text discussion of this figure.

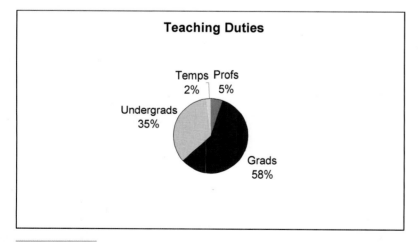

Figure 10.14 See page 250 for text discussion of this figure.

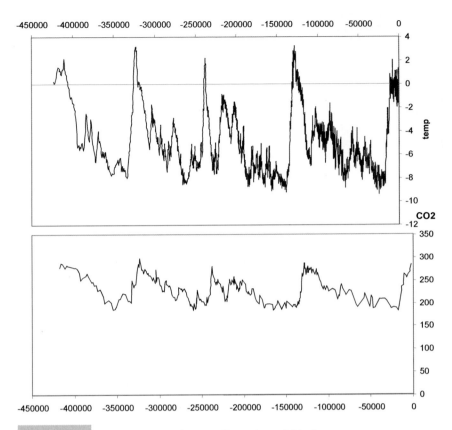

Figure 10.25   See page 256 for text discussion of this figure.

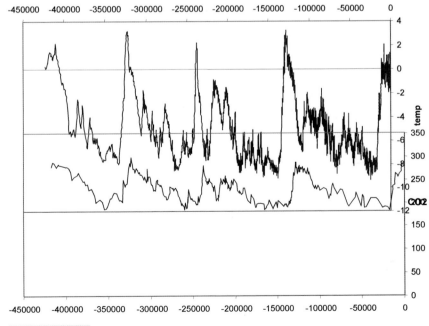

Figure 10.26   See pages 256–257 for text discussion of this figure.

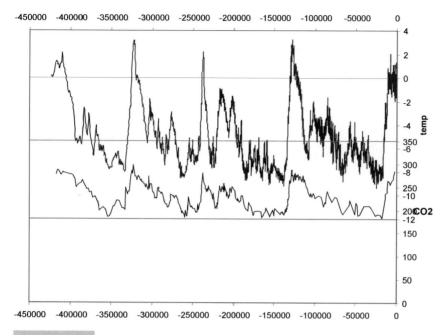

Figure 10.27   See page 257 for text discussion of this figure.

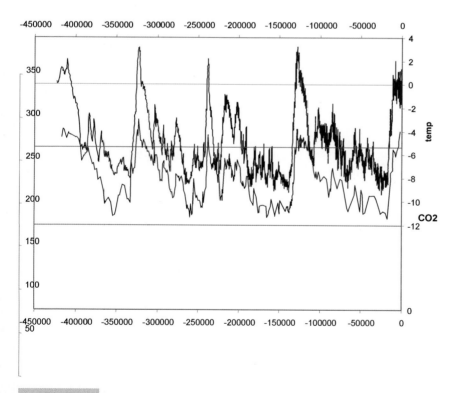

Figure 10.28   See pages 257–258 for text discussion of this figure.

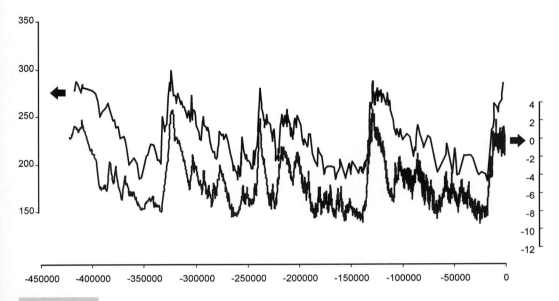

Figure 10.29   See page 258 for text discussion of this figure.

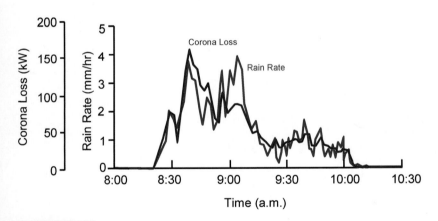

Figure 10.52   See pages 272–273 for text discussion of this figure.

Figure 11.1　See pages 279–280 for text discussion of this figure.

Figure 11.2　See pages 280–281 for text discussion of this figure.

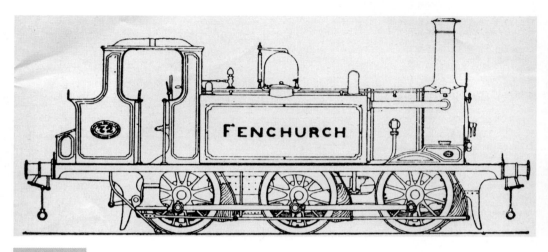

Figure 12.2   See page 302 for text discussion of this figure.

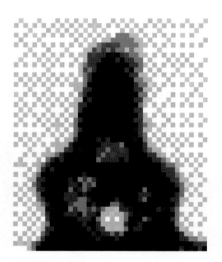

Figure 12.3   See page 302 for text discussion of this figure.

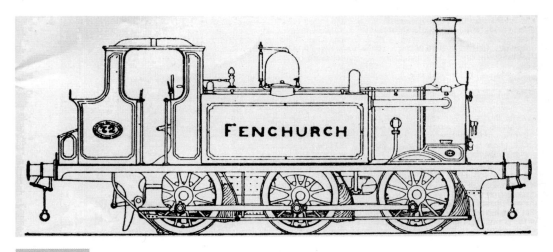

Figure 12.8   See pages 303–304 for text discussion of this figure.

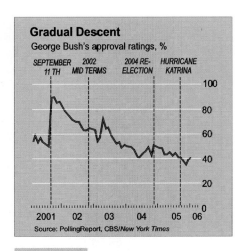

Figure 13.1   See page 321 for text discussion of this figure.

Not the way to present the data

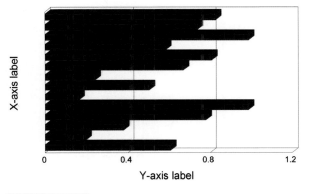

Figure 13.22   See page 330 for text discussion of this figure.

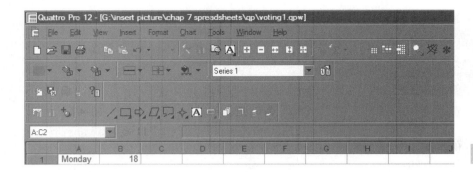

Figure 9.5

If you choose the chart title or subtitle, you get a text toolbar that will let you change the formats of these entities. The same is true for the axes, and everything else. A consequence of the flexibility of the approach is that the pick list (oh so useful!) moves about up there at the top of the screen, so you may have to hunt for it. In Figure 9.6 it has moved over to the right. A minor inconvenience.

Suppose you want to change something about the *y*-axis. Without the pick list, you can right-click on the axis, and open a dialog box with `Axis Properties` as an option. Choose this, and you will see that one of the tabs you can select is `Grids & Ticks`. Under this tab, you have the option not only to change their color, width, and style, but also a check-box to put `Grids On Top`. If you use this option and make the grid lines white, it greatly speeds the process of getting to the target graph. This is fortunate, because although the `Drawing` toolbar (visible in the previous figures) will let you add lines and so on, there is evidently no way to copy such lines out with the graph. Good work Corel!

Another, and very convenient, way of making this change is to use something called the `Property Selector`, which is indicated by the little icon to the right of the pick list for choosing what you wish to edit. Figure 9.7 shows the `Property Selector`.

Notice in Figure 9.6 that near the middle of the pick list is `Y1-Axis`. Use the pick list to select `Y1-Axis`, use the `Property Selector` to access `Grids & Ticks`, as before.

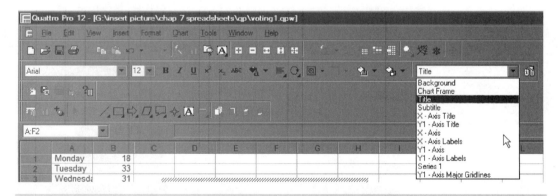

Figure 9.6

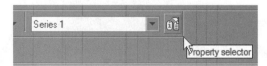

Figure 9.7

Figure 9.8

To change the width of the bars, you choose `Series 1`; then you have a couple of methods. If you right-click on the bar back in the graph, a little window opens as in Figure 9.8 that allows you to select as shown. You select `Series Properties`, of course, and another window opens that has tabs showing `Series Options`, `Trendline`, `Error Bars`, `Data Labels`, and `Type Options`. (It turns out that tabs for `Line` and `Fill` are hidden but can be slid into view by arrows under the question mark at the top right of the window. You can see the arrows in Figure 9.9.)

However, you can get to the same place by clicking on the `Property Selector`.

Figure 9.10 shows the way you can change the bar width. By using the pick list to select elements of the graph that you want to change, and by using

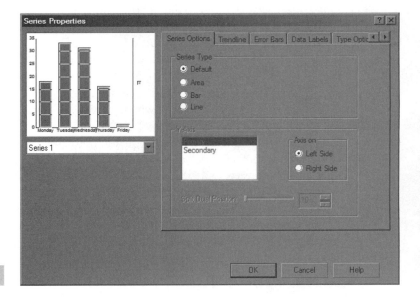

Figure 9.9

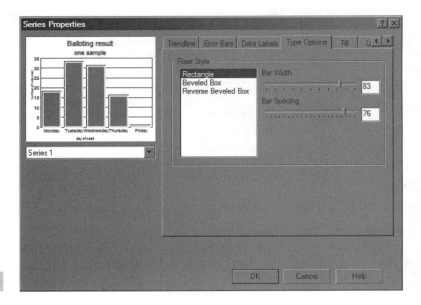

Figure 9.10

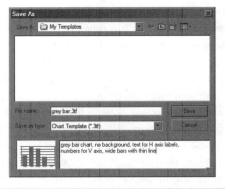

Figure 9.11

the `Property Selector` to change them, you can very easily make all the changes you want to the graph.

An important feature of QuattroPro is the ability to change the gridlines to be on top of the bars. Select the `Y1-Axis;` then in the `Property Selector`, choose `Grids & Ticks`, and check the box for `Grids on Top`. After you have the lines on top, you can change the color to white and adjust the thickness to suit your taste. See Figure 9.11.

Our target is obtained within a few minutes.

Once you have the graph the way you want it, you will of course save it, by copying it out to a graphics program or inserting it into a document, and of course by saving the spreadsheet. But you should also add it to the `Templates`.

This you do by clicking on a bar or some such, so that the `Chart` pull-down appears, and then clicking on the pull-down so that (down at the bottom) you have the option to `Save Template`. You will get the option to add a name and (though it isn't pointed out) a few words of description. Figure 9.12 shows the dialog box.

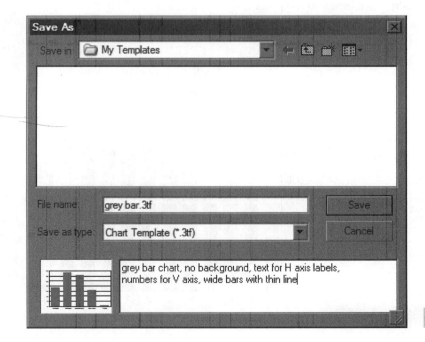

Figure 9.12

J	1800
F	2200
M	2480
A	2320
M	2710
J	3180
J	3510
A	3050
S	3360
O	3750
N	3490
D	2900

Figure 9.13

The picture of your graph is small, but it is recognizable. The big blank window in the middle is where the other templates you have created will appear. It is presently empty because this will be the first. Note that the template is being saved (by default) into a folder called `My Templates`. This is important to remember when you want to retrieve it. (You could, of course, put it in another folder if that suited you.)

Now if we type in some data to test the idea of applying our own defaults, we can get a feeling for the speed with which a graph *in our style* can be created. Suitable data are given in Figure 9.13.

First, you create the bar chart, using all the defaults (so it's very fast). Next, you click on a bar, or whatever you wish in the chart, just so that the `Chart` pull-down appears, and you click on `Retrieve Template`, near the bottom of the menu. Click on `Custom`, then `My Templates`, then `OK` (since at this point you should have a choice of only one custom template). The result is quite impressive, as in Figure 9.14. All the desired properties have been incorporated, and a few undesired ones (such as the various labels). Although there are a few things that need to be fixed, from here on it's simple.

A very few minutes of further editing will get the graph looking like a hybrid between our style (with the white lines in front of the bars) and the way it appeared in the *L.A. Times*, January 17, 2007 (page A6), as in Figure 9.15.

We note in passing that QuattroPro allows the export of charts in file form. Normally this would be ideal, as you could then grab the file and put it into your document, or your graphics package. This production of a graphic file must be considered one of the advantages of using QuattroPro. You can even select from a menu of file formats, so compatibility with your software shouldn't be an issue. However, Corel has not adequately debugged the process, and when you export the file things change. In some versions, an export to cgm format (the one and only true open standard graphic format) or wpg format (Corel's own format) results in all the text disappearing! You are left with something like Figure 9.16. Very nice, and quite editable, but why bother with all the labeling in QuattroPro if we lose it in the process of saving to a file?

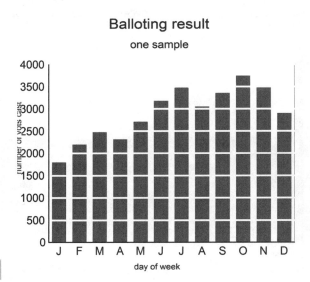

Figure 9.14

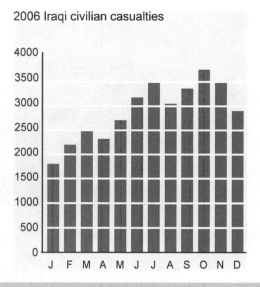

2006 Iraqi civilian casualties

Figure 9.15

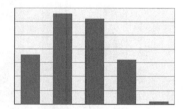

Figure 9.16

We'll look at some ways to deal with this kind of issue when we deal with file format problems in Chapter 11.

HISTOGRAMS

While we are thinking about bar charts is a good time to mention histograms. The difference was discussed in Chapter 2, but it will bear repeating. A bar chart is the kind of thing we have just looked at: a presentation of several things each of which is different in some way, or of the same thing sampled at different times. A histogram shows the distribution of samples of a single variable: how many times in the sample the value was in this bin or that.

QuattroPro makes the process simple.

Suppose you have a collection of data that you want to summarize in a histogram. Call it a data array—it could be a column or a block. All you have to do is select the data array, click on the `Chart` pull-down menu, and choose `Histogram`. To demonstrate the method, we set up an array of random numbers in a spreadsheet, to see how the random numbers are distributed. The target graph is shown in Figure 9.17.

The data were generated using the QuattroPro @RAND function. In our example, we set up 1000 cells, A8:E207. We wanted to see how often the random numbers were between 0 and 0.1, between 0.1 and 0.2, and so on. Figure 9.18 shows what QuattroPro did, using all the defaults.

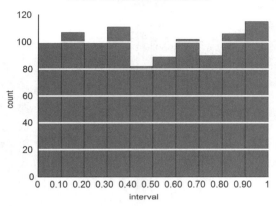

Figure 9.17

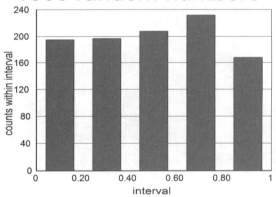

Figure 9.18

The usual changes need to be made: the bar color, the grid lines in front, the spacing of the bars, the font size, and so on.

The editing is similar to the work we did on the bar graph, and so is the result, in Figure 9.19.

As before, we save this as a template, called gray histogram. However, the result is not quite the same as the target, as the intervals are not as we wanted.

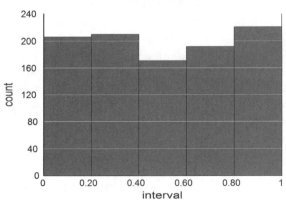

Figure 9.19

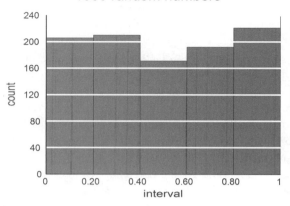

Figure 9.20

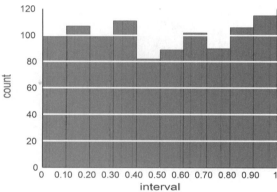

Figure 9.21

Using the pick list, we select the X-axis, and the Property Selector to adjust the scale, as seen in Figure 9.20.

We changed the Increment value to 0.1, and Figure 9.21 shows the result. This is quite close to our target, but there is a serious problem. A rough check of the average value of the count shows right away that the average is not 100, but somewhat less.

In fact, although the axis labels are at values of 0.1, 0.2, and so forth, the bins are narrower—there are 11 of them, instead of 10. The problem goes away if you force the x-axis to start at zero (YES, it said zero on the Property Selector, and it seems to *be* zero on the graph, but *forcing* it to zero apparently enables QuattroPro to get the right number of intervals.) This sort of nonsense is why it is always worth doing a sanity check on what your software is doing.

Remember, software is written by people who are convinced that if you eat a candy bar and follow it with a diet cola, the calories cancel out.

PIE CHARTS

Let's suppose we already have a data set in a spreadsheet. Here, we are pretending we have sampled some classes to find out who does the teaching, and we have come up with the data given in Figure 9.22.

	A	B
1	Profs	12
2	Grads	130
3	Undergrad:	78
4	Temps	4

Figure 9.22

Our goal is to show the data in a pie chart so that we can see at a glance what fraction of the classes are being taught by qualified personnel (assuming we know what "qualified" means). Our objective is to get the result seen in Figure 9.23.

As before, type in the data, as shown earlier. Highlight the data, and click on `Insert / Chart`. With no editing, and all the defaults, you get the pie chart of Figure 9.24.

Well, that's what you get if you use QuattroPro 12, anyway. You can edit it to make it all grayscale, and you can remove the box around the outside, but you have to do quite a lot of work to make it like the target. You can produce Figure 9.25.

You can move the labels about, and get the percentages calculated, but while the version in the spreadsheet includes the leaders from the two labels at

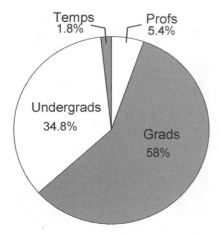

Temps
1.8%
Profs
5.4%
Undergrads
34.8%
Grads
58%
value for money?
teaching duties

Figure 9.23

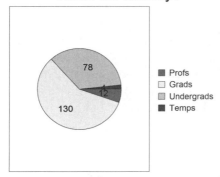

Teaching Duties
Value for Money?
78
12
130
Profs
Grads
Undergrads
Temps

Figure 9-24 See insert for color representation of this figure.

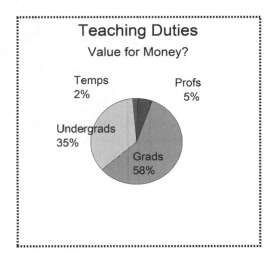

Figure 9.25

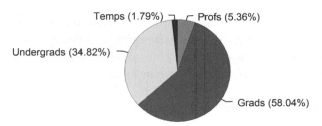

Figure 9-26   See insert for color representation of this figure.

the top, there seems to be no way to include them in the process of getting the graph into the word processor. The unacceptable amount of white space—here shown by the dotted line that we added in Figure 9.25—can be reduced a little by choosing `Chart Frame` from the `Object Selector` and stretching appropriately.

Ironically, if you were lucky enough to have left QuattroPro 9 on your computer when you upgraded, you would get something editable, as in Figure 9.26. At least the leaders got included this time. Actually, on the screen of QuattroPro 9, the left end of the label `Undergrads` does not show. It reappears when you paste the chart into Presentations, however.

We will cover in Chapter 10 a process whereby you can use Presentations or PowerPoint to improve the appearance.

SCATTER GRAPHS

An important kind of graph that spreadsheets can do is variously called an *X-Y plot* or a *scatter graph*. Sometimes the word *trend* is used, though we would prefer to reserve this word for situations where the horizontal axis represents *time.* We will look at scatter graphs next. Before we do, however, a word of warning. For some reason, the writers of spreadsheet software perceive a difference between a graph with data spaced evenly along the abscissa and one where the data are not evenly spaced. Therefore, these are treated separately in the spreadsheet graphing software, and accessed separately. If you choose wrongly, you will find that you don't get what you expect.

You may not notice what is happening right away, so you need a sanity check to look out for it. Check that the graph you get is at least roughly what you expected. Does it have the right number of curves? Do the axes make sense? Is the slope what you expected? It might seem odd to ask these questions, but it is worthwhile to be sure. QuattroPro will add data of its own if you let it!

Suppose you have two columns of data. Depending on whether or not you include the words heading the columns, QuattroPro may give you two curves if you select `Line` as the type of plot you want, and only one curve if you select `XY` or `Scatter`. When you get the two-curve version (and expected only one), you may note that the horizontal axis is unlabeled but seems to be derived from the *location* of your data in the column.

We honestly don't know if a Line Chart with an unlabeled axis is useful. It certainly isn't useful to anyone we know, and it likely won't be useful to you even if your horizontal axis is time. So beware the trap, and check that you really are getting the kind of output you expected.

On with the example. Suppose we had looked at the survivability (in terms of percentage hatch rate) of mosquito larvae as a function of their distance from a high-voltage power line. The local power company, interested in allegations of biological effects of power lines, has done an investigation.[2] The results are shown in Figure 9.27.

We can do this with the data in two columns, as in Figure 9.28.

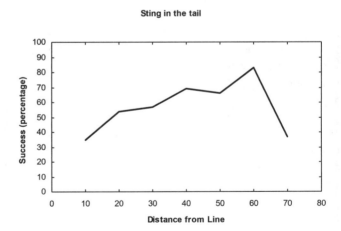

Figure 9.27

	A	B
1	distance	success
2	(meters)	(percent)
3	10	35
4	20	54
5	30	57
6	40	69
7	50	66
8	60	83
9	70	37

Figure 9.28

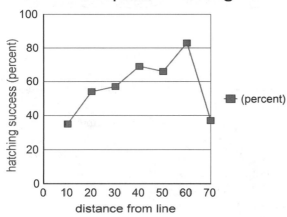

Sting in the tail
mosquito hatching

Figure 9.29

In QuattroPro, you can highlight the numbers, and click on `Insert Chart`. Several choices are automatically made for you in the program's first guess at what you want, but you can overrule them if you need to.

For example, the initial guess is that you want two three-dimensional pie charts. What an odd idea! However, by clicking on `Next`, you have some options. If you tell the program you want a `Scatter` diagram (which you do) you get the message "`Insufficient data for this chart type!`"

The exclamation point at the end is particularly grating, don't you think? However, if you *first* unclick where it is giving you 3-D, and *then* click on `Scatter`, you get the reasonable result seen in Figure 9.29.

As expected, you can change colors and choose to have the dots connected by a line, or not.[3] Time to see how to make the changes.

First, we can have QuattroPro add axes at the right and the top of the graph. Click on anything in the graph so that the pick list appears, and use it to select an axis. Figure 9.30 shows the options for the horizontal axis.

---

[3] It is not obvious before you plot the data that maybe you should not be joining the dots. In this particular example, the straight-line joining looks to be inappropriate, but since you (as investigator) might reasonably have expected a smooth relationship, we'll leave the dots joined for the rest of the example.

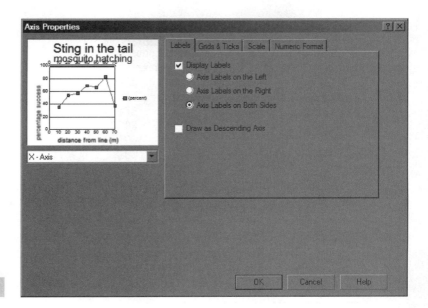

Figure 9.30

As you can see, by selecting `Axis Labels on Both Sides`, the top of the box appears, along with labels and tick marks. When you click on the `Grids & Ticks` tab, you can choose to have the tick marks on the inside (of both axes at the same time), as in Figure 9.31.

It is simple to repeat the process for the vertical axis, and to edit the line and box color and to remove the legend. The pick list and the `Property Selector` make it very straightforward and fast. The result is shown in Figure 9.32. The target does not have the tick marks labeled on the top or the right, however. Removing these labels is possible, but a bit tedious.

As soon as you select (with the pick list) the axis labels, you select *all* the axis labels. Edits to any one of the properties affects all the labels in the same way. The property selector/formatter shown in Figure 9.33 offers no help.

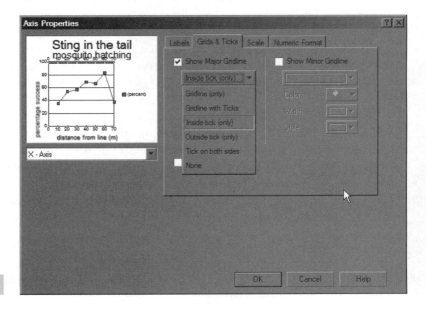

Figure 9.31

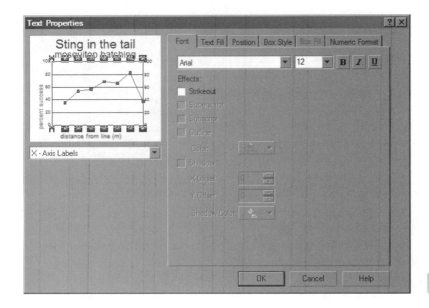

Figure 9.32

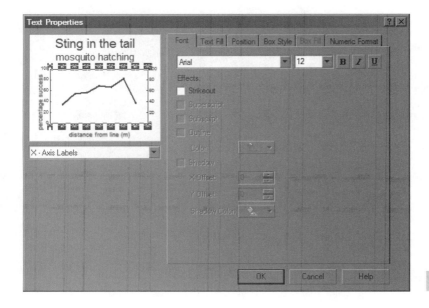

Figure 9.33

The only solution, it seems, is to click (left-click) on each individual label in turn, then left-click again (not quickly enough to count as a double-click, as that opens the Property Selector) and change the properties of each label one at a time by right-clicking.

The sequence is to click on a number, slowly click on it again, then right-click on it. A dialog box (Figure 9.34) opens that will let you change the text properties for just that label.

Once you have the dialog box for the Text Properties, you can make the color white, as in Figure 9.35, for example, to have the labels disappear.

Some simplification results if, instead of having the vertical axis labels on both sides, you add a duplicate copy of the data and create a secondary axis instead. Then you can make all the labels for the secondary axis disappear at once. However, the same simplification does not apply to the x-axis.

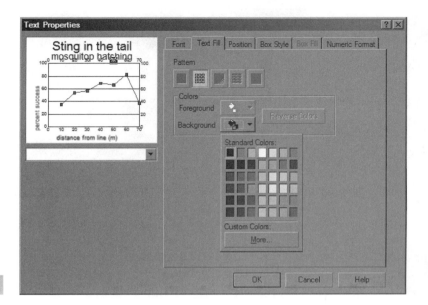

Figure 9.34

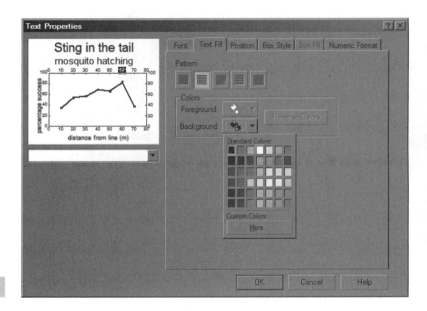

Figure 9.35

Tedious it may be, but with these various edits, you can get as close to the target as you wish, editing every parameter of the graph more or less at will.

Now let's save the template.

MULTIPLE SERIES

We have seen the expression "Series 1" a few times on these charts. We have generally ignored or deleted the label, as it seems redundant when we have only one set of data. Now it is time to see what happens if we need to show more. The example in Figure 9.36 is adapted from a graph in *New Scientist*, February 21, 2004. The graph showed that fewer postmortems were being performed in the United Kingdom over the decade of the 1990s. (The article it

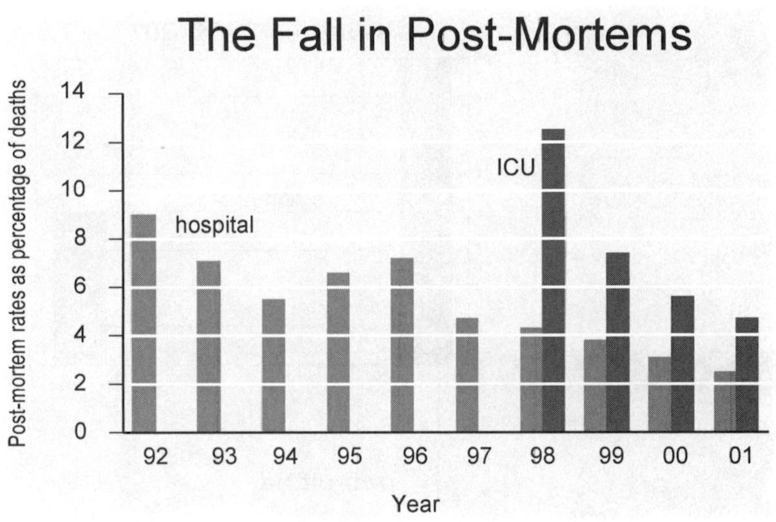

The Fall in Post-Mortems

Figure 9.36

accompanied argued that evidence of wrong diagnosis was being missed as a result.)

This graph is interesting in that data are not available for identical times. Perhaps statistics just were not collected for the ICU category prior to 1996. Apart from that little quirk, the graph is pretty straightforward.

There is a little problem with the horizontal axis, however. Strictly speaking, the numbers shown do not increase monotonically in the order given. Let's have a look to see how our software handles this. We start by entering the numbers into a spreadsheet, as in Figure 9.37. (These numbers are somewhat of a rough estimate of the values that might have been in the original graph, but they are adequate for demonstration purposes.)

If you select the three columns beginning with row 2, and going down to row 12, if you select Histogram as the type of graph you want, your first attempt at a chart will be as shown in Figure 9.38.

Pretty silly. So silly, in fact, that it's almost impossible to figure out what went wrong. (But not *quite* impossible. The program counted how many of the numbers were between 0 and 20, and how many between 20 and 40, and so on. In other words, you are generating a *real histogram*, and not a bar chart.

	A	B	C
1	year	rate	
2		hospital	ICU
3	92	9	
4	93	7.1	
5	94	5.5	
6	95	6.6	
7	96	7.2	
8	97	4.7	
9	98	4.3	12.5
10	99	3.8	7.5
11	00	2.1	4.5
12	01	1.5	3.6

Figure 9.37

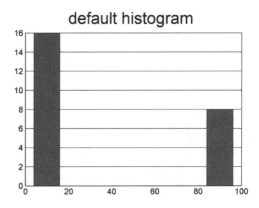

default histogram

Figure 9.38

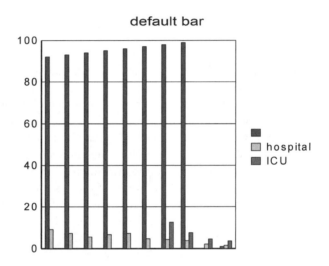

default bar

hospital
ICU

Figure 9.39

Oops! *We made a wrong choice.* Be aware that QuattroPro has a histogram option, and it is not the same thing as a bar chart!)

On the other hand, if you select Bar as the type of graph you want, you get closer, with Figure 9.39.

Not much better, but at least we can see from the graph of Figure 9.39 that there is a problem with the numbers in the first column. The software is treating them as the numbers they are, rather than as the year. This is the same as Excel did if you selected the top line of text. We might try to add "19" in front of the first few numbers, and "20" in front of the rest, but in fact the solution is to make the numbers into *labels* by adding a leading space or by beginning each entry with a single quote mark. Once this is done, the default bar is a lot like our target, and a few minutes of editing will give us the graph of Figure 9.40.

This is close to our target. Many people would find the version in Figure 9.40 acceptable and not bother with the final step that moved the horizontal lines in front of the bars and made them into white lines. But it's so easy. Use the pick list to select Y-Axis, and with the Property Selector use the Grids & Ticks tab to put Grids on Top. Then with the pick list, select the Y1-Axis Major Gridlines, and use the Property Selector to change the line color to white and adjust the thickness.

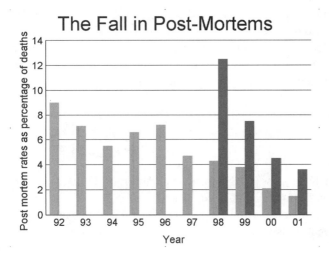

The Fall in Post-Mortems

Figure 9.40

A COMPLICATED EXAMPLE

Here's an example that demonstrates the power of finishing the spreadsheet process in a graphics program. The graph that we saw back in Chapter 2, showing the Earth's temperature and the carbon dioxide level, looked straightforward enough, and while it could not be done in a spreadsheet alone, much of the exacting assembly of the curves *can* be done there, as we shall see.

We reproduce our version of the graph in Figure 9.41 as a reminder.

In this particular case, the data are from two separate files, both of which were available for download from the Web at the time of this writing. Originally, they were found at the website `http://www.yourplanetearth.org/terms/details.php3?term=CO2+and+Global+Warming`. The data are from J. M. Barnola, D. Raynaud, C. Lorius, and N. I. Barkov, "Historical carbon dioxide record from the Vostok ice core" (1999), and can be downloaded from `http://cdiac.esd.ornl.gov/trends/co2/vostok.htm`

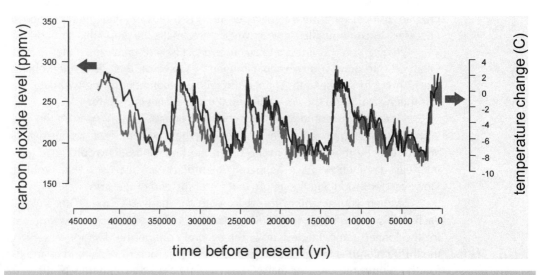

Figure 9.41

and J. R. Petit, D. Raynaud, C. Lorius, J. Jouzel, G. Delaygue, N. I. Barkov, and V. M. Kotlyakov, "Historical isotopic temperature record from the Vostok ice core" (2000), with data available at the time of this writing at `http://cdiac.esd.ornl.gov/trends/temp/vostok/jouz_tem.htm`. (Feel free to go to the websites to see how ice core samples are obtained and dated, and why their information is so useful.)

We have no intention here of getting into the discussion of global warming (except to observe that the carbon dioxide data here do not include about the last 5000 years). Our aim is to show how this kind of graph can be generated from the data.

The first step in the process is to get the data and transform it into a usable form. On the Web, there is simply a button to click to see the data. The numbers are in columns on your screen, but when you copy and paste them into your spreadsheet, the information is all jammed into the first column. This won't be useful.

The solution is to copy the data and everything to a file by clicking `File/Save Page As` (pick a name and location) on your browser. (If the file does not save as a text file, open the file into a word processor or WordPad, and save as a text file (`File/Save Page As/Text Document`).) Now, for one of the files, if you copy and past from there into QuattroPro, the tab delimiters in the original are interpreted as cell delimiters, and you have the data in handy columns. Alternatively, you can open the text file in QuattroPro, and you will be asked about delimiters. But either way, it works, and you now have the data in usable form.

The other data set was not saved as a tab-delimited file; it was space delimited. You should be able import it as a space-delimited file, treating consecutive delimiters as one. (Also, be aware that with this file, about the first 999 entries have a leading space, which means you will have to slide them all over one column once you have the data in the spreadsheet.)

The problem is that you have actually now got *two* sets of data, in separate files, and *with different lengths.* There are somewhat less than 400 data points in the CO_2 record, and more than 3300 in the temperature data. Each can be plotted readily enough on its own, but there seems to be no way to get a spreadsheet program to deal with the two data sets from two files in a single graph. So it is necessary to combine files, if you want to try to make the graph in a spreadsheet.

Now, you could combine the data in Presentations, after getting QuattroPro to draw you two separate graphs. However, there is a small chance that you could accidentally change the relative alignment of the two data sets. It is much safer to do the combining in the spreadsheet program.

Ideally, you would want three columns: one for the time, one for the CO_2, and one for the temperature. Since, in general, the time numbers from one data set are not exactly the same as in the other, the time numbers would have to be interleaved somehow. As it happens, QuattroPro can deal with that problem. However, for producing the graph, that is not the end of the story.

When you assemble the data, with the time data made negative for each data set, you have *four* columns. The first column should be the time data for the longer (temperature) curve, the second column the data for that curve, the third column is the time data for the shorter (carbon dioxide) curve, and the fourth column the data for that curve. You select all four columns of numbers and tell QuattroPro to `Insert a Chart`. Of course, since QuattroPro selects a

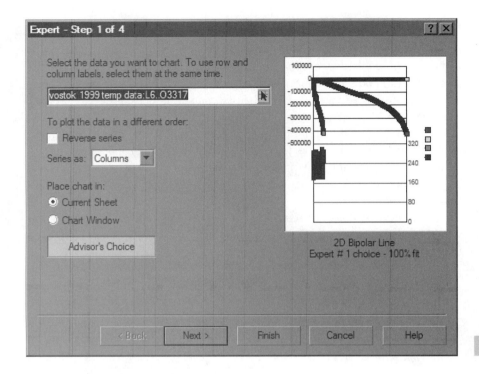

Figure 9.42

rectangular section of the spreadsheet, you have a section of the selection that is blank. No matter.

QuattroPro does not have to be told that there are two sets of data for the x-axis. Perhaps it senses that the two sets of data, although in columns of different length, cover the same range. Whatever it does, it works. The first guess is not very good, mind you, as you can see on the right side of Figure 9.42.

However, as soon as you choose a Scatter graph, QuattroPro catches on and gives you the graph in Figure 9.43 right away.

There are a few things to edit, but nothing challenging. You must change the graph to have no markers for the data points, get rid of the legend, edit the axis labels, and so on, but we have seen how to do that kind of thing already.

Of course, you are stuck with having the axes touch one another, so you cannot quite reproduce the graph as it was in Chapter 2. Those final touches were added in a graphics package and are discussed in Chapter 10.

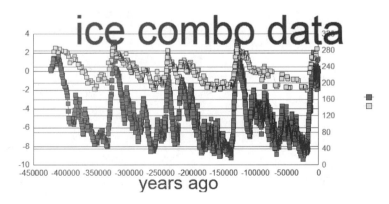

Figure 9.43

INSERTING DATA

As well as presenting graphical data, you might want to present data numerically, in the form of a table, *in the graph*. There are two ways to proceed. You can get the spreadsheet to add some information, or you can do it yourself. It would sometimes be useful to add a Data Table that will appear under the horizontal axis, as in Figure 9.44.

For the monthly report, this sort of thing is excellent. Your management might expect the dots to be joined, but since the hours and the costs are monthly summaries, it hardly makes any sense to join the dots. (Still, managers being simple souls, it may be as well to humor them by selecting a different basis chart for the Data Table.)

QuattroPro does not seem to have exactly this kind of feature, so you would have to construct it in a graphics program. This kind of thing is discussed in the next chapter.

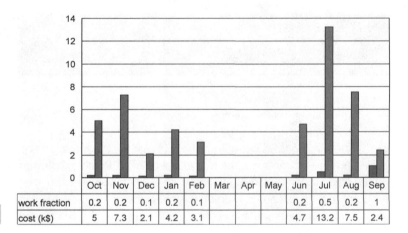

	Oct	Nov	Dec	Jan	Feb	Mar	Apr	May	Jun	Jul	Aug	Sep
work fraction	0.2	0.2	0.1	0.2	0.1				0.2	0.5	0.2	1
cost (k$)	5	7.3	2.1	4.2	3.1				4.7	13.2	7.5	2.4

Figure 9.44

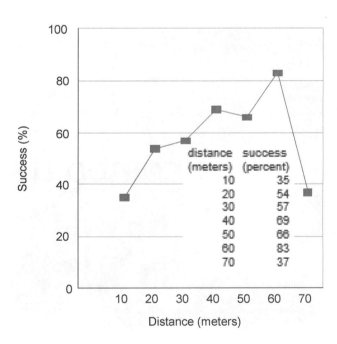

Figure 9.45

A feature that QuattroPro *does* have will allow you to do an *insert* anywhere you want in the graph area, and the insert can be anything you want out of the spreadsheet. Figure 9.45 is an example of the mosquito data.

Figure 9.45 shows what it looks like on the computer screen. Pretty spiffy! And if the folks at the factory ever debug this, it will be really useful. Unfortunately, there seems to be no way to get this graph into a document with the table intact! All attempts to copy-and-paste, or to export the chart to a file, result in the *removal* of the Data Table. To generate this image, we had to reinsert the table by a separate copy-and-paste operation in a graphics package. Once again, we are adding that fourth wheel after we paid for the car.

It's not the end of the world, because, in general, the process of adding such a table into a graphics program is simple with copy-and-paste: just select the numerical area you want in QuattroPro, copy it, and paste it into your graphics program in an empty area of the graph. There was an example in Chapter 1 (Figure 1.5), where we "inserted" a bar chart in the middle of a graph. We deal with spreadsheets and graphics packages in the next chapter.

FLOATING BARS

Here is another uncommon spreadsheet product. You might think of Figure 9.46 as a modified bar chart—the data of interest are floating bars in the middle of the graph area. In this example, the links are the numbers assigned (somewhat arbitrarily) to the nodes in a power delivery system. The voltages are the upper and lower limits of voltage at each node under certain conditions. The data were calculated by a program written in C++, but all we had available when the graph was needed was a table in a Word document. However, the table seemed rather large, and in the end not very informative, as the reader would have been required to do a good deal of studying to understand the results.

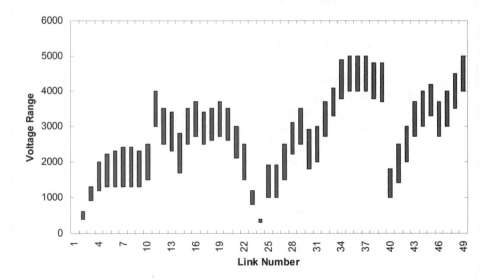

Figure 9.46

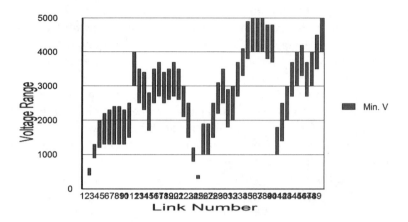

Figure 9.47

A graph (a narrower version of this for putting into a single column) was the preferred alternative. So we transferred the numbers to a spreadsheet.

There were originally three columns: the link number, the lower limit, and the upper limit. QuattroPro differs from Excel in that there is no need to add a fourth column for the difference, as there is a graph type called High-Low.

Remembering that in order to get the program to treat the first column as data for the x-axis, we tried replacing the numbers by numbers with a leading blank. That worked, but there were a lot of numbers, so we tried a shortcut. We just replaced the first number by a number with a leading blank. That *didn't* work. But when we replaced the first number by a number with a leading single quote, that *did* work, and the result of selecting a High-Low graph and letting the program default to everything is shown in Figure 9.47.

The overall chart occupied the entire width of the page, even though there is no information anywhere near the edges. As a matter of fact, the chart did not look quite like this on the computer screen—this is the result of copying and pasting. On the screen, all the axis labels ran from bottom to top, and the words "Voltage Range" were in a much larger font. Those interested in pointing fingers would likely point at the screen driver interface. Very little editing is actually required to change the color of the bars, remove the legend, and put the gridlines on top.

However, our target graph had tick marks on the top and at the right, and it did not have all the numbers jammed into one another on the x-axis. Getting there is not so simple.

Putting the tick marks on all the sides is simple enough—we saw how to do that with the mosquito graph. All you do is use the pick list to select the two axes in turn, and use the Property Selector to get the axis labels on both sides. However, there seems to be no way to make the labels on the top and at the right go away and leave the tick marks except to go through each label in turn and make it white.

There is good news, however. There are two ways to get the x-axis labels "thinned out." One way is to select the X-Axis, and use the Property Selector to instruct the program to skip some labels. The first label will be used, and you can then skip one or two or more.

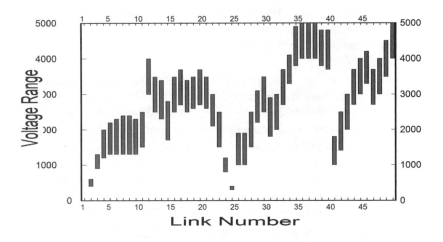

Figure 9.48

The other way is to remove from the column of data the numbers that you don't want on the axis. This method gives you complete control. Removing all the even numbers halves your white-out problem. Removing all except the fives and tens, as in Figure 9.48, is even more parsimonious.

Editing this into an acceptable version of our target is not difficult. And it is worth saving as a Template.

MULTIPLE TYPE: COMBINING TREND AND HISTOGRAM

Once in a while, you will see a graph that combines a line graph and a bar chart. There seem to be two occasions when you might need this. One is to show two quantities that vary in time, but one quantity must be plotted as a bar chart (because the rules of Chapter 3 make it clear that a line graph would be an error) whereas the other can be plotted as a line graph without risking misinterpretation. We examined an example of this in Chapter 3, when we saw the support for capital punishment graphed as a line above a bar chart of the number of executions carried out each year.

The other kind of graph that requires two kinds of data is the Pareto chart. This is a chart named after an Italian economist who observed (in 1906) that 20% of the population (of Italy) owned 80% of the wealth. Of course, if wealth was evenly distributed among a population, 20% of the population would have 20% of the wealth. Given that wealth is not evenly distributed, it may be supposed that the more unevenly distributed it is, the smaller the first number and the larger the second.

But when you think about it, it becomes clear that the 20/80 split is arbitrary: the richest 20% of the population of a country is bound to own some fraction greater than 20% of the wealth. It is said that in the United States, 20% of the population owns 50% of the wealth. There is no reason to choose two numbers that add up to 100 apart from their being more easily memorable.

The major application of the 20/80 split is as a management tool. You may consider it quick and dirty, but it often works adequately as a rule of thumb: 80% of customer complaints can be dealt with by addressing 20% of the known problems, and so on.

Given that you are going to pick two integer numbers that sum to 100, then at one extreme distribution (a uniform distribution) 50% of problems would cause 50% of complaints, and at the other extreme (basically there is only one important problem) 1% of problems would cause 99% of complaints. It is reasonable to assume that a 20/80 split is somewhere in the middle. Fine, 20/80 is convenient, and unless the distribution is unusually close to one of the extremes, 20/80 is likely to be a reasonably accurate estimate of the situation *even if you know nothing at all about the statistics.* Accurate enough for some management decisions to be based on, anyway.

Table 9.1 is an example of some data you might assemble to explain why your train to work was late. Note that the right column is the cumulative total of the numbers in the middle column. This is only necessary for a Pareto chart. For any other chart that combines bar chart with line graph, this constraint is not applicable.

You could plot this information as in Figure 9.49.

Figure 9.50 is a monochrome version of what you get if you allow QuattroPro to use all the defaults for a chart it calls Mixed.

There are a few things here that need fixing, but in spite of the horrible appearance of the default graph, things are not too out of control.

First, we select the X-Axis Labels and rotate them by 45° or 90°. Then we fix the overlapping labels all over the bottom axis by using the Property Selector on the X-Axis (counterintuitive—not the X-Axis Labels) and find the Labels tab there, and have the labels occupy a large number of lines. We chose 7 in the dialog box shown in Figure 9.51. It turned out that 11 lines were needed to make sense of the labels at 90°. They take up less space at 45°.

Note that in the target there is a line drawn across the graph at the level of 80%. From where it intersects the cumulative line, another line is drawn down the

TABLE 9.1

Reason	%	Cumulative
Leaves on the line	38	38
Switch jammed by heat	23	61
Signal error	10	71
Driver overslept	6	77
Cafeteria running late	5	82
Funny noises in train	4	86
Fire at trackside	3	89
Power failure	3	92
Animal on line	2	94
Presidential motorcade	2	96
Funny smell in carriage	2	98
Accident on line	1	99
Unidentified	1	100

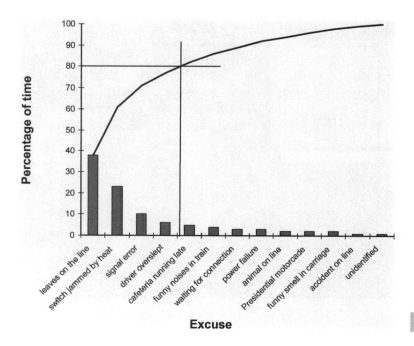

Figure 9.49

graph, separating four excuses for lateness from the rest. These four excuses are responsible for a little less than 80% of the problems. Since there are altogether 12 identified causes, and one unidentified, we can see that the four are about 31% of the number. Evidently the distribution is not skewed enough to get the 20/80 split: never mind, the diagram is still a Pareto diagram.

Editing the color away so that the cumulative curve has no fill is simple enough: so is removing the legend and so is making the lower curve into a bar chart (just use the properties selector, and under `Series Options` push the `Bar` radio button). The result is shown in Figure 9.52. In fact, the only part of the target that presents a problem is the matter of the additional lines artistically drawn to mark off the four excuses on the left.

There is seemingly no way to draw the lines in QuattroPro and bring them out into your document. That must be done in a graphics program. We will cover that sort of editing in the next chapter.

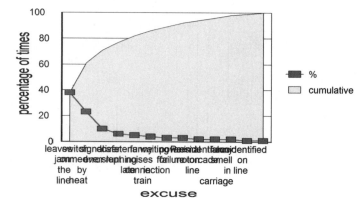

Figure 9.50

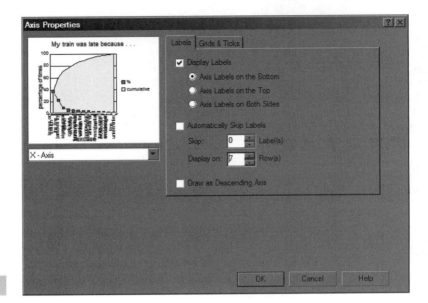

Figure 9.51

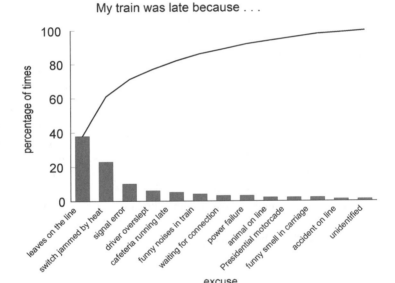

Figure 9.52

SUMMARY

QuattroPro has many convenient features as far as creating graphs is concerned. Editing graphics in QuattroPro is much more efficient than doing it in Presentations because QuattroPro can remember the changes you made and reproduce them later for another graph.

- It may not be worth adding the Drawing toolbar to your custom version of QuattroPro, because the changes you make cannot be copied out of the spreadsheet environment into your word processor.

- Select the zoom level you want before you edit a chart. Changing zoom is not an option once you have a chart element selected.

- QuattroPro does not do a good job with pie charts. Anticipate editing them in Presentations or PowerPoint.

- To have QuattroPro automatically choose your first column of data as x-axis labels in a bar chart, make all the cells into labels by starting them with a space or make the first one into a label by starting with a single quote mark.

- Get accustomed to using the Chart pull-down menu, and to using the Properties Selector on the Property toolbar. This is by far the easiest and most consistent way to move through the graph.

- Right-clicking on selected items is often an alternative way to get to the Properties Selector.

- If you override the automatically selected scale on an axis, the change will be remembered. Usually, it is worth deferring this step until you have saved your chart Template.

- To create a box around your graph, with tick marks on all sides, you can choose to have axis labels on both sides. This works for the x-axis and y-axis. However, you will get what you asked for: *labels* on both sides. Removing them to leave only the tick marks is tedious.

- When you have finished making a graph the way you want it, consider adding it to the selection of user-defined Templates under the Chart pull-down. The time saving next time around will be considerable.

The following changes to a graph will not be remembered:

- To add callouts and comments, use the text and arrows available in the Drawing toolbar.

- To make white (or any other color) lines on top of your graph, use the Drawing toolbar.

EXERCISES

9.1. Attempt the population pyramid question in Exercise 8.1 using QuattroPro.

9.2. Do the histogram question in Exercise 8.2 using QuattroPro.

9.3. How do you rotate a pie chart in QuattroPro without changing the order of the pie segments?

Fixes using graphics programs

10

In the last two chapters we saw that neither of our popular software packages are inclined to produce exactly the pie chart we want by default. However, with some effort on our part in either program, we can usually get the desired result. Either of the two spreadsheet programs we looked at will save you enormous amounts of time editing graphs, once you have saved examples of the styles you prefer.

However, neither of these programs is capable of producing *all* the changes to *all* the graphs you might want. There just are some graphs and some details that require the additional capabilities of a graphics package. For example, you might just need to add some labels to a graph, possibly with leader lines identifying some particular feature, or you might want to convert the graph into a colorful viewgraph.

In this chapter we will deal with the products of either Excel® or QuattroPro®, and with the two graphics programs PowerPoint® and Presentations™. However, the intention is to deal with principles as far as possible, so unless there is some compelling reason to get involved in the details of which software we are considering, we will try to be generic.

The first question to answer is how to get a graph out of the spreadsheet and into the graphics program.

GETTING THE GRAPH INTO A GRAPHICS PROGRAM

The first part of this turns out to be quite simple. Although there are sometimes errors of transcription, for the most part the ubiquitous clipboard is all you need. You click on the graph so that the eight little black squares appear, and copy with `Ctrl-c`.[1]

---

[1] In Excel, you can also hold down the Shift button while you click on the `Edit` pull-down menu. This will allow you to do a `Copy Picture`. The rest of the pasting process is the same. Just why you would want to do this is not clear. We think that a *picture* in Microsoft-speak was once a version of something that did not have all the features of the complete item whose image you were seeing on your screen. It was, we think, enough to reproduce the image, and nothing more. As we shall shortly see, an Excel graph in PowerPoint has quite a complicated structure. Perhaps originally that complexity was

The Right Graph. By Harold Kirkham and Robin C. Dumas
Copyright © 2009 John Wiley & Sons, Inc.

Then, in your graphics package you drop the image onto a blank page. In PowerPoint, you can use `Crtl-v` or you can select `Paste` or `Paste Special`, from the `Edit` pull-down menu, as a `Microsoft Excel Chart Object`, `Picture (Windows Metafile)`, or `Picture (Enhanced Metafile)`. In Presentations, you must paste as a `Metafile Picture`, *not* as a `Device Independent Bitmap` or a QuattroPro `Chart`. `Ctrl-v` does not allow the image to be ungrouped.

Now the graph, looking more or less as it did in the spreadsheet program, is yours to control completely. Some care is needed to maintain certain of the essential relationships. You could accidentally commit mayhem with the data, anything from causing some data to disappear entirely, to changing its relationship with the space it occupies. Very likely the first thing you need to do in your graphics package is to ungroup the elements of the image, so you can perform your edits. As soon as possible, consider regrouping some of the image just so as to be sure the appropriate relationships are maintained. More on this in a moment.

Both Excel and QuattroPro have elements of the image grouped. When you paste into Presentations, an ungrouping occurs automatically. In PowerPoint, you may have to ungroup more than once.

The elements of the chart are arranged in a similar way in the two programs. Thus, we will describe only one, the Excel structure.

An Excel graph is constructed, when viewed from PowerPoint, as a great many pieces, all ordered from front to back in such a way that the scales and the graph line and so on are all properly visible. Imagine each of the elements of the graph to have been drawn on a transparent playing card, and all the cards stacked up one on top of the other. The stack may be thousands of cards high.

At the bottom of this imaginary stack (referred to as the *back* from the viewpoint of the graphic software, which imagines your computer screen as the front), is a white rectangle. On top of that is a smaller gray rectangle that makes the background for the graph, and then, going toward the front, the various straight lines that cross the graph area, the lines that surround the graph area, and the little marks that indicate the scale divisions, whether or not there are lines across the graph.

Next, we consider the various line segments that make up the graph itself (and there could be thousands of these, depending on how big the original data columns were, and the various scale numbers). On top of everything is a frame, consisting of a rectangle with no fill, sized to match the white rectangle at the bottom.

Many of these elements are grouped in nonobvious ways. For example, in a large graph with many data points, some of the data lines are grouped with the gray rectangle and some with the long lines across the graph. In general, the data lines are grouped, but there are many groups, and not all of the data are so grouped. It may be that the groups are formed by counting up (or down) through the stack of cards, and when the number of pieces counted reaches the maximum number for a group, a group is created.

reduced. Such a thing might then have allowed the use of software like Word® to load files with embedded images more rapidly on the slow hardware, such as was available for the early PCs. Processor speeds and disk access speeds have increased about two orders of magnitudes since the days when such tricks were necessary, and it may be that the original idea has fallen by the wayside. Now, whether you copy by `Ctrl-c` or by `Copy Picture` the end result seems to be the same.

The details of this grouping approach may be true only for Excel, but the general principles are the same for QuattroPro. One difference is that QuattroPro seems to have the graph line itself as a single element. Another difference is that for some reason once they are outside the spreadsheet program the titles and axis labels in QuattroPro are incorporated as bitmaps (useless for editing), whereas in Excel they are ordinary text.

MAINTAINING RELATIONSHIPS

Now, why is it important to know about the grouping and regrouping of bits of the graphic images? Consider the graph in Figure 10.1, where the solid lines that indicate the location of the axes, and the little tick marks, have been removed.

This is a picture of the essential information in the graph. Without the axis lines, it looks as if the graph is floating in a two-dimensional space. That's fair enough—it is! *It is the relationship between the line and that space that is important.*

If you are going to have (say) a vertical axis, tradition would have you place it on the left of the curve, aligned above the zero of the horizontal axis. *But it doesn't have to be there.* Provided you do not change the vertical relationship between the graph and the axis, you can put the axis itself anywhere you like (or leave it out, as in the example of Figure 10.1).

You could also have put the axis in any (one) of the four places shown in Figure 10.2. (Note that you will have to put the axis labels back in—we took them out to save the clutter of having four sets of them.) In fact, you can surely put it anywhere in the same vertical position. Right through the middle of the graph if you wish, to emphasize that that is the location of the zero, for example.

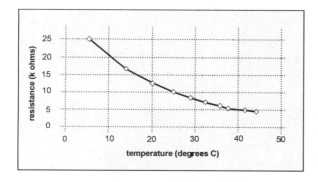

Figure 10.1

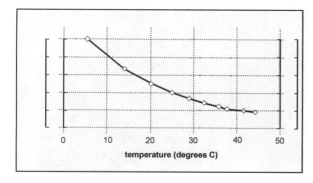

Figure 10.2

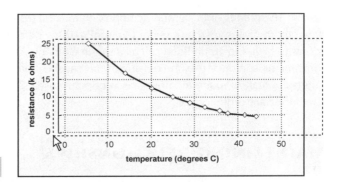

Figure 10.3

What is more, if you lasso the graph line, the horizontal lines, and the vertical axis, as shown in Figure 10.3, where the lassoing, shown in the figure by the ⟨symbol⟩ symbol and the dotted rectangle, started at the top right (outside the box) and does not include the vertical lines or the box, you can stretch things vertically, as in Figure 10.4.

Almost all graphics programs will let you stretch the set of things you lassoed and will let you do it to all of them together. The various programs differ a little in how they deal with the stretching question.

PowerPoint will let you do this even if you include the axis numbers, and it will not distort the number font. But you must group all the objects first. Adobe Illustrator® has trouble grouping things like this, where there are already groups. CorelDRAW® will let you stretch or compress without grouping, but it distorts the font. Corel Presentations will not let you include text.

Now, the example of Figure 10.4 looks pretty silly, but we did it to make a point. Yes, you need to go put the axis labels back in the right place, and yes, the box around the outside of the graph got too small all of a sudden, but the graph is still fair and true. That will be clearer after you put the numbers in the right place, but this kind of distortion is *fair*!

The point is, there is nothing to stop you distorting the vertical scale, or doing something similar to the horizontal scale, either, provided you keep the relationship between the graph line and the space in which it is presented the same.

Consider the example of the Figure 10.5, which shows the corona current generated when a pair of wires are added to some high-voltage equipment as a function of voltage.

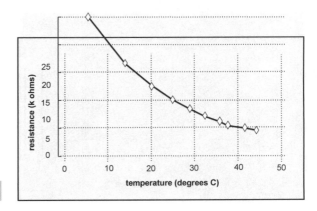

Figure 10.4

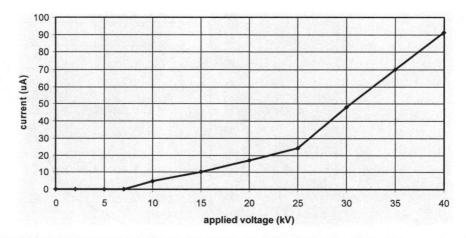

Figure 10.5

Now, you are certainly entitled to squish the graph[2] as in Figure 10.6.

You can stre-e-etch things vertically, too, if you wish, as in Figure 10.7.

All these edits are "legal" because the relationship between the graph line and the space it occupies is not changed. We have left lines all across the graph area to make this clear.

The edits of Figure 10.8 might look funny, but they are legal, too.

Now, you wouldn't actually want to do this, with the lines all over the top of the graph, and the vertical grid lines sticking out the bottom like so many fence posts. We did this to show that the horizontal axis can be moved vertically away from the graph, just as the vertical axis can be moved in the horizontal direction.

Without the lines, as in Figure 10.9, it does not look *quite* so silly. In fact, if you moved the vertical axis to the left, as in Figure 10.10, this wouldn't look too bad at all.

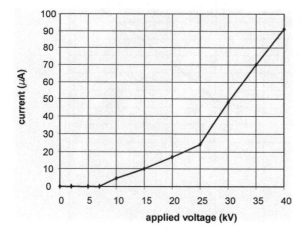

Figure 10.6

[2] You might note in the versions in the graphics package that the vertical axis label has been changed. The label "uA" has been replaced by "μA." We have been able to find no way of doing this in a spreadsheet program, but it can be done in your graphics package without great difficulty.

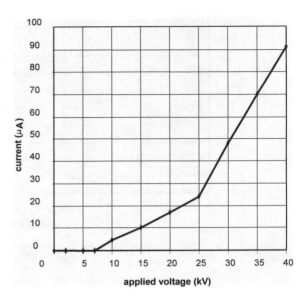

Figure 10.7

Now, if you had just dragged the axes down (without moving the graph itself), you would get the result shown in Figure 10.11.

The edit of Figure 10.11 would *not* be correct. If you look, you can see that the current values are now all different. Putting all the lines back in, as in Figure 10.12, makes it a bit more obvious.

In Figure 10.12 the space has been stretched while the graph line has not. If you must increase the separation of the graph line from the horizontal axis, you can certainly move the axis line, but *just the axis line!* Do not move anything

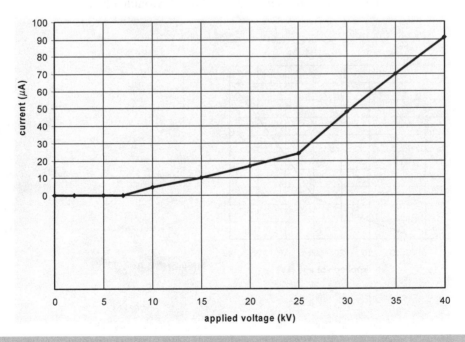

Figure 10.8

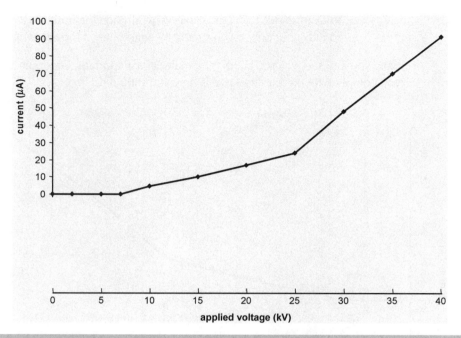

Figure 10.9

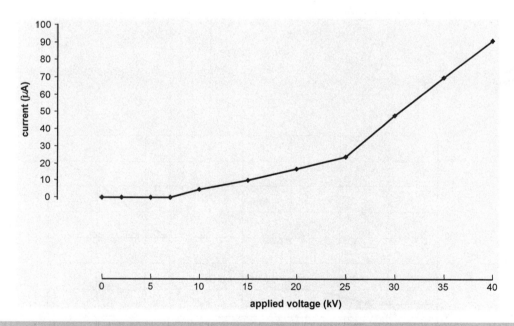

Figure 10.10

corresponding to those "value axis major gridlines" unless you move the graph line as well.

Two simple rules can be derived from these considerations.

1. You can *move* the horizontal axis vertically, and the vertical axis horizontally, but make these moves without involving the graph line.

2. You can *stretch* the vertical axis vertically and the horizontal axis horizontally, but only if you stretch the graph line at the same time.

Now you can see why regrouping the material is important. Grouping will help you comply with these rules. Here is how you should do it.

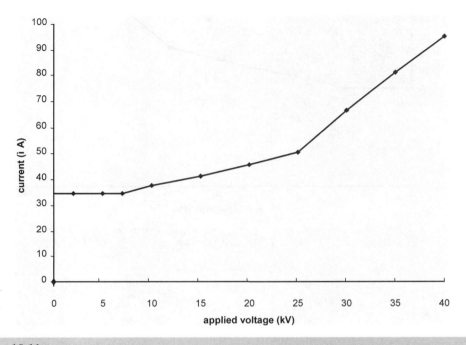

Figure 10.11

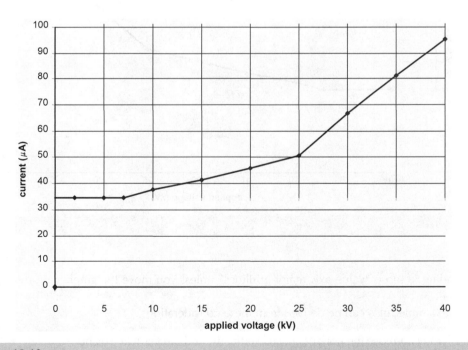

Figure 10.12

You should group the horizontal axis (and its labels and tick marks and any lines across the graph), the vertical axis (and all the stuff that goes with it), and the graph line. Now you have three groups.

Group all three together, and you can distort the graph however you wish. Separate an axis, and you can move it—with the restrictions we stated above: vertical movement for the horizontal axis, and horizontal movement for the vertical axis. This will let you move the vertical axis from the left side of the graph to the right side, for example. Remember in many applications, you can constrain an object's movements by holding down the shift key while you drag.

Stretching (or compressing) the graph requires grouping the graph and the axis associated with the direction of the stretch. You can therefore make a graph narrower (perhaps for two-column format), but you must group the graph and the horizontal axis before you do the squeeze.

We will see applications of these rules in what follows. But first, a final word of warning, and a suggestion. It is important to have all the parts that you need in each of the three groups. If you miss a little, for example, the end of the graph line (perhaps because it was already grouped with something else, and it wasn't included when you lassoed it), you might not notice that fact until it is too late to correct any errors it has caused. Therefore, it is a good idea to check, as soon as you create your three major groups. Take each group, one at a time, and do something with it that will allow you to see what it contains. You could delete it, stretch it, compress it, translate it, or change its color. Do something obvious. Then, check that you have the group correctly defined. Then—do a `Ctrl-z` to put things back the way they were. Don't forget!

FIXING A PIE CHART

The reason we need to understand all this is, of course, so we can "improve" a graph without cheating. Let us look first at how to improve a pie chart. Pie charts did not come off well from either of the spreadsheets we have looked at—let's see what we can do in the graphics packages.

Figure 10.13 shows the result of our endeavors in Excel.

Our original target was monochrome (which we could have achieved without leaving the spreadsheet program), and the pie was a little bigger

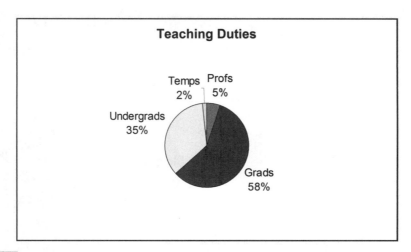

Figure 10-13 See insert for color representation of this figure.

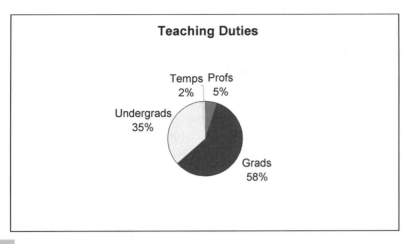

 See insert for color representation of this figure.

compared to the text, and a lot bigger compared to the box it came in. When the image is ungrouped, odd things may happen. Depending on how you pasted and what version of the software you have, the straight radial lines between the segments of the pie may all disappear, or the color may vanish from the segments of the pie, as in Figure 10.14!

It seems that it is best to `Paste` as an `Enhanced Metafile`.

Of course, it is straightforward now to enlarge the pie to engulf some of the words, to convert to monochrome, and to add the color or the radial lines back if necessary. Just remember to group the pie segments and hold down the `Shift` key when you are stretching the pie to make it larger, or you will get some odd-looking distortions, such as a noncircular pie. The main thing, of course, is to remove the big white rectangle and the black line that marks its edge. Once these are gone, the text of your document can be wrapped appropriately close to the figure.

In QuattroPro, we obtained the result shown in Figure 10.15. Apart from adding the leaders, and reducing the size of the white space around the pie, there is not much to do here.

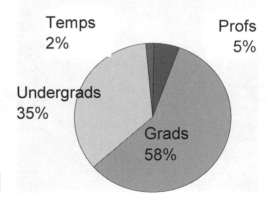

Figure 10.15

SOME EXCEL/POWERPOINT TIPS

We mentioned earlier that there are "data legality" reasons to group things appropriately before you change the aspect ratio of a graph. There are also artistic reasons. PowerPoint will allow you to squeeze or stretch a mixed selection of things, but the relationship between them usually needs to be fixed by grouping. If you have the Iraqi casualty graph we saw earlier (copied here as Figure 10.16), and you try to stretch it in PowerPoint to make it wider, unless you have things grouped, if you simply select everything (by lassoing) and grab one of the little squares (thinking it is a sizing handle), you will end up with the mess shown in Figure 10.17.

You have selected the horizontal lines, and the vertical bars, and made them wider, but you have not achieved your objective. To do that, you must group things appropriately. Then you can make it as wide as you like, as shown in Figure 10.18.

In general, if you squeeze or stretch a mixture that includes text, the text will remain the same size and will not get distorted. Its position relative to everything else may change, but the text font and style remain constant.

2006 Iraqi civilian casualties

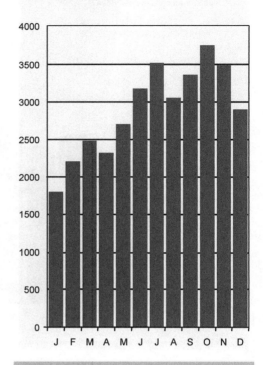

Figure 10.16

2006 Iraqi civilian casualties

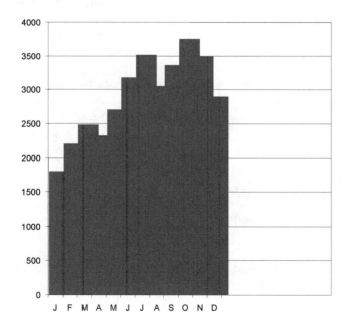

Figure 10.17

2006 Iraqi civilian casualties

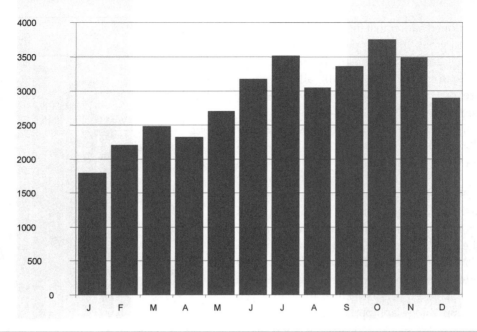

Figure 10.18

The text box that holds the text, however, will change in size and that may make the text wrap differently.

By grouping and ungrouping things, and by deleting things and changing the properties of what remains, you can get precisely what you want. You can save the image as a PowerPoint file, and you can copy-and-paste the image into a text box in Word.

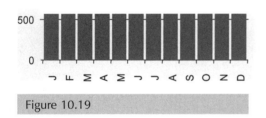

Figure 10.19

One more little trick. In Excel and PowerPoint, you may have a graph in which the abscissa labels appear facing up, rather than to the right. Perhaps the previous graph had the axis labels as shown in Figure 10.19.

In PowerPoint, it is simplicity itself to fix this. You select all the numbers by lassoing them and, making sure this time that they are *not* grouped, use `Draw/Rotate` and `Flip/Rotate Right`. Each number will execute a 90° turn on the spot, leaving it in the right place.

SOME QUATTROPRO/PRESENTATIONS TIPS

Let's look briefly at what you might need to fix in Presentations if you are bringing a graph in from QuattroPro.

The main deficiency in QuattroPro was that if you added lines or leaders, you could not get them out of the spreadsheet. Of course, it is easy to make these changes in Presentations. However, during the transfer from spreadsheet to graphics package, some changes occur. For example, the bar chart of the previous example arrives looking as seen in Figure 10.20.

2006 Iraqi civilian casualties

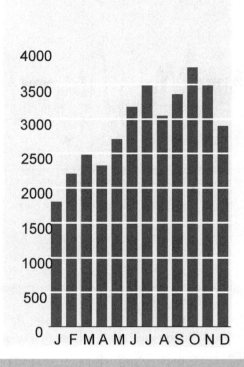

Figure 10.20

It's not bad, but the labels for the vertical axis have all shifted to the right and are overlapping the first bar. The problem is readily fixed, and at the same time you can align the right sides of all the numbers instead of the left if you wish. The point is that it pays to do a sanity check on what is on the screen. As with most software, it seems, you must do it, because they are not doing it very well back at the factory.

DATA MANIPULATION: THE CARBON DIOXIDE DATA

Here's an example that demonstrates the power of finishing the spreadsheet process in a graphics program. The graph that we saw back in Chapter 2 showing the Earth's temperature and the carbon dioxide level looked straightforward enough, and a good deal of the assembly of the graph could be done in Excel or in QuattroPro. The start of the process was described in Chapter 8 for Excel and Chapter 9 for QuattroPro. But the graph, reproduced here as Figure 10.21, could not be done in a spreadsheet alone.

It was created with a spreadsheet and a graphics program, and the graphics part was essential. To demonstrate the power of the graphics approach, we will assume that our starting point is that all the data are *not* in one spreadsheet. Combining the data in your spreadsheet is the best way to start as it handles the axes and lets you group the data for safe manipulation—but you can actually do the combining even if you never combine the data on a single

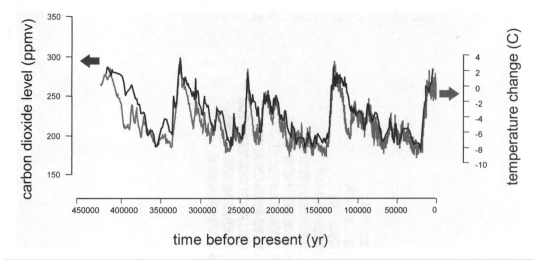

Figure 10.21

spreadsheet. You plot the data, in two separate plots, and copy and paste them into one.

Figure 10.22 is the temperature plot from Excel. By a little work in the spreadsheet, it was possible to move the locations of the axes to the positions shown, and (unexpectedly) the numbers appeared away from the graph.

Note that the time axis is a little unusual. A separate column was added to the numbers, containing the age of the ice data, multiplied by minus 1. When this is used as the abscissa, the result is that time is still shown in the customary direction.

Figure 10.23 is the CO_2 graph.

Now all that remains is to combine the two graphs in a way that is both fair and clear. First, you start your graphics package going. We recommend that you reserve three blank pages, one for each of the original graphs, and one for the combination.

Next, you copy and `Paste Special` each of the graphs shown in the spreadsheets into the appropriate blank page in your graphics program. Save

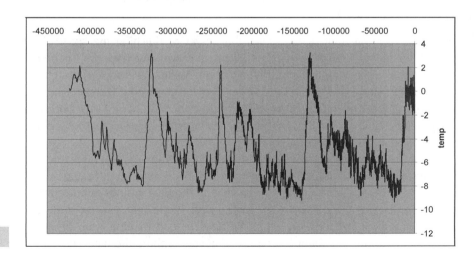

Figure 10.22

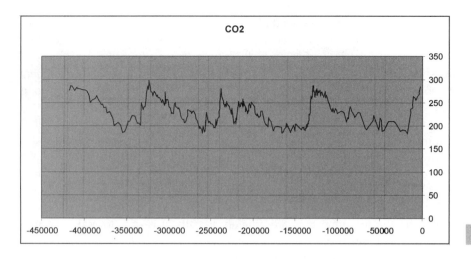

Figure 10.23

the file at this point just in case things get out of hand. (Perhaps in some distant and happy future, there will be an operating system and software that does not crash when called upon to do something a little out of the ordinary, like handle a graphics file. It isn't here yet, but we can all dream, can't we?) These copies are your backup versions, now kept in a handy location.

Next, you break each image apart, carefully. In PowerPoint, this will mean performing the ungroup operation twice, as there are groups within the groups of the image. However, ungrouping the graph twice seems to split it up into elements small enough that you can make your changes well enough. (Perhaps with some graphs a third ungrouping may be needed. It is worth remembering.)

Having ungrouped the graph, the first thing to do is to create groups that make sense from *your* point of view. *In particular, you need the data and each of the axes to be grouped.* From time to time, you may need to group some of your groups together and ungroup them, as we shall see, but each will stay as a group.

If you cannot create a group "cleanly," that is, if you cannot simply lasso everything you want in a group because your lasso would include something you do not want, you can create smaller groups out of things that you *can* lasso, and make the final group by grouping these groups. And remember, in PowerPoint and in Presentations, unless you lasso something completely, it will not be selected. With this in mind, you can do things like this to be selective about selecting elements. Look at the lasso line shown in Figure 10.24.

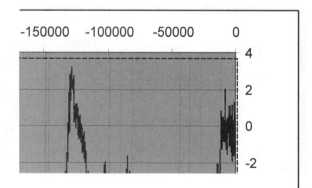

Figure 10.24

Here the lasso (the dotted line) will select the graph line, but it will not select the numbers, the gray rectangle, the vertical axis line, or the little tick marks just outside the vertical axis. If you use this method to group the data lines first, you can then group the vertical axis with its tick marks and numbers without the need for particular care to avoid any of the data lines. Having done that, you can group all the horizontal axis elements easily.

Having grouped all the required elements of each of the two original graphs in this way, you now combine them. Suppose you just insert one graph below the other on your third blank page, as in Figure 10.25. Here we have removed some extraneous material from the graphs. However, we deliberately left a box around the graphs. This will come in useful soon.

Two problems are immediately obvious: the two curves are the same color, and the two time axes are different lengths. It is easy to make the line color different, since that is just a group. We will make the temperature line red. Then, with the entire bottom graph grouped together, we will slide it up and line up the left ends of the boxes, as in Figure 10.26.

Now the bottom graph can be squeezed to match the size of the top one. The lines of the boxes must be matched on both sides. This must be done with great care, on both left and right sides, iteratively, and at maximum magnification.

Since the graph and the axes are all grouped, they should stay in the same relationship. That means this is a fair and honest thing to do. The result is quite confusing, because both vertical axes are labeled on the right. Figure 10.27 shows the result.

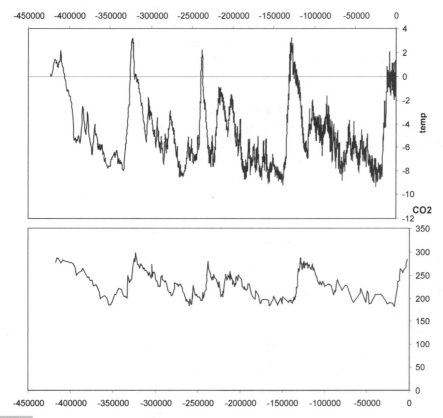

Figure 10-25   See insert for color representation of this figure.

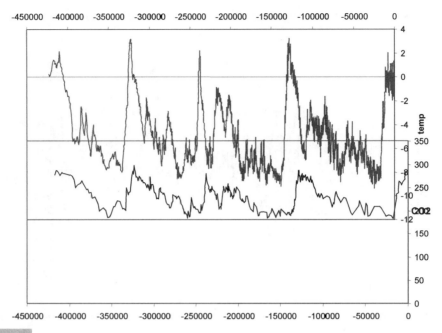

Figure 10-26   See insert for color representation of this figure.

Obviously, one of the vertical axes has to get moved. We choose the lower one, the one for CO_2. In PowerPoint, we can simply ungroup the entire axis and its labels from the rest, and move it over to the other side of the screen. That is, you can use the arrow keys to move the material or you can drag with the shift key depressed. Since the axis is now ungrouped from the graph line, it is essential to move it *only horizontally*, so as not to change that crucial relationship. The use of the arrow keys (or the shift key) ensures that. As we have seen, the axis

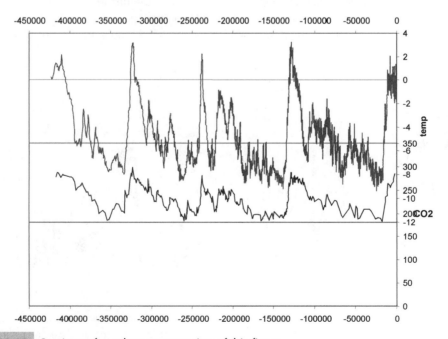

Figure 10-27   See insert for color representation of this figure.

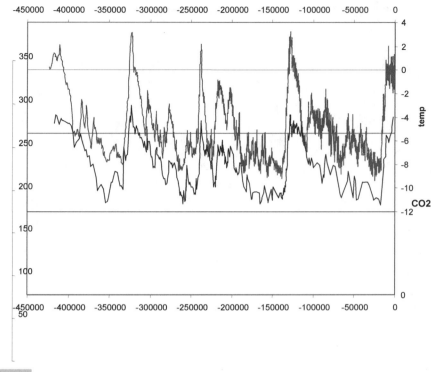

See insert for color representation of this figure.

can be relocated anywhere on the same horizontal line. Figure 10.28 shows a suitable location.

At this point, we can regroup the axis with the graph line, and add some vertical gain, in an artistic attempt to match the spread of the temperature graph. In PowerPoint, stretching like this will not cause distortion of the text.

From here on, it is just a matter of tidying up. One of the horizontal axes must go, the vertical axis labels must be fixed, and some means of identifying

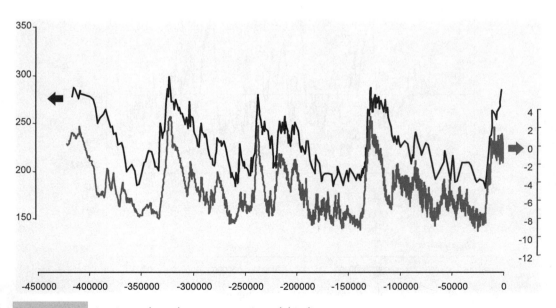

See insert for color representation of this figure.

each graph with its axis must be added. The result, as we have already seen, is quite neat (Figure 10.29).

Provided we keep the relationship between the red curve and the right scale fixed, and that between the blue curve and the left scale fixed, we can move either of them up or down, and squeeze or stretch them vertically. You can separate the curves, as we just did, or overlap them, as we did earlier. If the curves are separate, the presentation could be monochrome. We could stretch the whole thing out to fit across two columns. We could even compress into one column—and offer a magnifying glass to our readers.

SCANNING AND TRACING

Scanning is a way of using the computer as a copier, only the product is a file instead of a piece of paper. However, there are a couple of problems. First, the file may not be ideal for what you need, so you need to edit it. Second, because scanning produces bitmap files, the file may be quite large and may slow down the process of working on your document. A solution to both of these problems is to trace over the scanned information. You get a vector file that you can edit. (See Chapter 12 on file conversion for an exposition on file types.) Thus, for graphics, scanning and tracing are useful together.

If the scan is just words, the software you need probably came with your scanner software, or your graphics package. It is termed Optical Character Recognition software, or OCR. OCR software will reproduce text but will likely not get all the formatting right. And it may add some interesting features of its own. An OCR of this paragraph scanned at 200 bpi produces this:

> If the scan is just words, the software you need probably came with your scanner software, or your graphics package. It is termed Optical Character Recognition software, or OCR. OCR software will reproduce text, but will likely not get all the formatting right. And it may add some interesting features of its own. An *OCR* of this paragraph scanned at 200 bpi produces this:

Not bad: apart from gaining a ragged-right justification, the only mistake was that the third "OCR" came out in italic. We will say no more about OCR here, as it is a reasonably mature and very useful feature. And with text, we will assume you know enough about what you're aiming for that we would not be helping.

Our interest here is scanning and tracing *graphics.* While there are some programs that will do that for you and produce a very good representation of the original, the capability of editing the copy is not always what you would expect.

For example, the diagram in Figure 10.30 is the (much reduced) result of a trace operation on a very old blueprint—a radio design from 1923.

In the image in Figure 10.30, the "lines" are mostly made up of a large black area with two white areas placed on the top. If you click on an area and move it, the result is to modify lines all over a large area of the diagram. In Figure 10.31, the outline of one large white area has been made dotted, so it can be seen. It covers the entire width of the diagram, and the black lines are just the spaces between its parts.

There are other tracing options, but none seem satisfactory. For example, you could end up with the spidery thing seen in Figure 10.32. At least the lines are *lines* and not spaces between acres of white. That means you could edit the line width, style, direction, and so on. But what a lot of work before you get something usable in a publication!

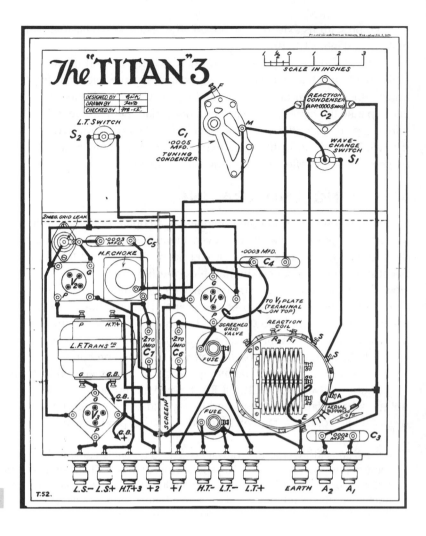

Figure 10.30

It seems that if you want *only* an exact copy (no editing), tracing is satisfactory. The traced image is a good copy of the original, and the file is a lot smaller. However, if you want to *edit* the diagram, you will have to do the tracing by hand. This is hard work, but you still might want to do it.

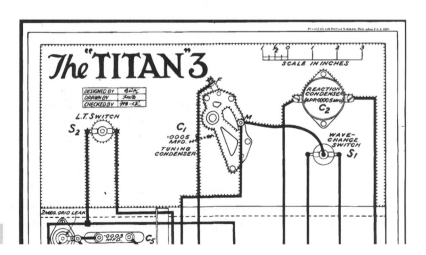

Figure 10.31

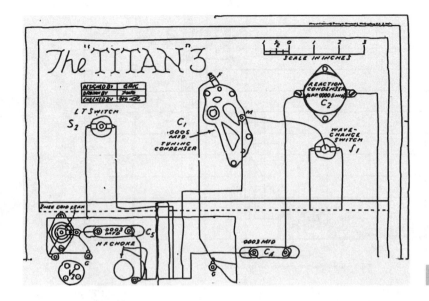

There are two reasons you might want to scan and trace something by hand. First, you might need an up-to-date and electronic version of an old figure that you have only on paper. Once you have the graph in vector form, you can edit it however you wish and embed the result in a paper or report.

Second, you might wish to find out whether some particular equation was a fit to some measured data. Note that this is not quite the same as performing a curve fit: the assumption is that you have the equation from some theoretical considerations and want to check its accuracy.

We will look first at how you can solve the problem of getting the graph from paper to electronic form. Then we will show how to adapt the technique to solve the second problem, without using a curve-fitting program.

TRACING EXAMPLE 1: ELECTRON VELOCITY

As our example, suppose you needed a new, electronic, version of the graph in Figure 10.33.

Figure 10.33 is taken from the *Radio Engineers' Handbook*, by Frederick Terman, McGraw-Hill, 1943 edition. It shows the velocity of an electron (as a fraction of the speed of light) as it falls through a given potential. The figure is a bitmap image, from a scanner.

The curve here is a tad wobbly. There seem to be a couple of kinks that likely don't belong there: one where the abscissa reaches 10^4 and one just before it reaches 10^6. The original seems to be a little awry.

So rather than just tracing the curve, or artistically fitting a Bezier[3] curve, let's improve on it. Let us see if we can produce a curve that does not have the defects of the original by using a spreadsheet. However, we will retain the rather special graph paper that Terman used, and that can only be done by using a graphics package.

---

[3] This is the U.S. spelling of the last name of the creator of the thing. Pierre Étienne Bézier (1910–1999) was a French engineer and mathematician who was at Renault when the work was done in the 1960s. As used in graphics programs, a Bezier curve is a third-order curve with a very user-friendly way of controlling the parameters.

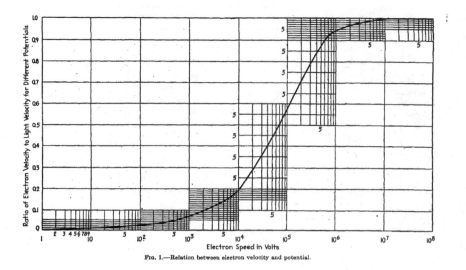

Fig. 1.—Relation between electron velocity and potential.

Figure 10.33

According to Terman, the equation that the graph shows is

$$v/c = \sqrt{1 - \frac{1}{(1 + 1.94 \times 10^{-6}\, E)^2}}$$

The curve of that equation will not be difficult to generate in a spreadsheet.

What about the "graph paper" that Terman used?

For that, the first thing you have to do is decide what software you will use. Most graphics packages will handle both vector and bitmap formats (see Chapter 12 for more on what this is all about), so you should be able to use your "usual" program. Your version of PowerPoint might let you down here, though, if you don't have the right conversion module (called a "filter") installed. In this case, try opening the bitmap in a bitmap-capable program (and there are some good ones available free on the Web), and copy the information to the clipboard. PowerPoint will then let you paste-in the information.

If you have a program that will let you handle your images in *layers*, so much the better. Adobe Illustrator and CorelDRAW will both allow the use of layers. The advantage is that you can put the bitmap on one layer, and create the vector on another without the inconvenience of occasionally selecting the bitmap (which you don't want) instead of a feature on the vector layer (which you do). Having layers is a convenience, then, but not essential. (We usually just call the added layer "vector," a name associated only with the file you are working on, so we can reuse the same name with any file.)

Let's start tracing the graph paper.

If the image came from a scanner, the lines might not be exactly parallel to the directions that the software calls vertical and horizontal. You need to check this very carefully, as getting everything parallel can save a lot of time. If you are using PowerPoint, you are out of luck for a simple way of fine-tuning the angle. Nevertheless, you must persevere and make this correction as soon as you have opened the scanned image.

As it happens, the scanned image above is *distorted*. While the top and bottom lines are horizontal, and the right side is vertical, the left side is not. It points about 0.3° to the west of north. The distortion may have been caused by the page not being flat against the scanner. It is not a large error, and we will deal with the problem "artistically."

Since we know that the top, bottom, and right sides are lined up correctly, we can start by drawing them. Draw a line across the top. Your screen should look somewhat as shown in Figure 10.34.

Here we have zoomed in on the top left of the screen. You can see some of the individual pixels of the bitmap. The new line you have just added is at the very top, selected, and not quite lined up with the top line of the bitmap. It is also not quite horizontal yet—it has a little "jog" in it. Very carefully, line it up. And then go line up the other end of the line, with as much zoom as you can for maximum accuracy.

You now have the top line in place, though very likely the line weight is wrong. It can wait.

Make ten copies of this line. As we have seen, depending on which program you are using, the copies will be in the same place as the original, or they will be distributed somewhat across the screen.

Either way, use the `Align and Distribute` function of your program to line up the right or left ends of the copies, and to spread them vertically to cover the top and bottom lines of the bitmap.

It is probably convenient to *group* them. In PowerPoint, grouping at this point is a necessary defense against illogical object selection for editing, given the program's limited magnification capability and its tendency to select something you didn't want. In all programs, grouping is a convenient way to allow selection of whole or partial images, depending on whether you "click on" the element or "lasso" it.

So, group the lines, and use the stretch capability of the software to make sure that both the top and bottom lines are nicely on top of their counterparts in the bitmap. Your screen should look as shown in Figure 10.35, depending on what your graphics program is.

Here, you can see the whole area around the graph selected (black squares), and the individual lines across the screen also selected (open

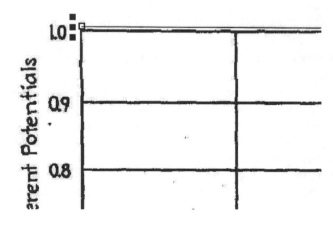

Figure 10.34

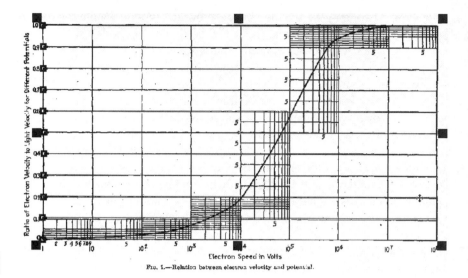

Fig. 1.—Relation between electron velocity and potential.

squares). You will also notice that the line corresponding to 0.1 on the ordinate does not line up with the line we drew.

Since our lines were computer generated, we can take it they are accurate, and the original is a little off. It is likely that the original draftsman traced over parts of some semilog graph paper to produce the image in Terman's book. Are we to retain the error of the original, or correct it as we generate the vector file? Correct it, of course. This is not an error that has any merit. It does not deserve to be retained.

At this point, we have eleven horizontal lines across the screen, evenly spaced. Now we have a choice. We can make a copy of the group, rotate the copy 90°, and stretch the group to fit the vertical lines. Or you can repeat the procedure that created the original set of lines, only matching the vertical lines. Copying and rotating is simpler. When we are done, we have a nice rectangular grid of lines. (If your software does layers, you can make the bitmap layer invisible and check the appearance of the grid. If you don't have this capability, you can select the bitmap and delete it, to see your progress. *Only don't forget to restore it right away afterwards!*) In the next several pictures (Figures 10.36 to 10.41) we will be disappearing the bitmap so you can see the progress on the vector layer.

Now we can add some of the detail. Let's start with the additional vertical lines for the scale. We selected some in the middle to trace over, as in Figure 10.36.

With only a little effort, copies of those new lines can reappear at various lengths, to match the original. Selecting them as a set is easy since we have grouped the other lines, because we are lassoing these lines *inside* the groups of the other lines. Remember, if the things you lasso are not *completely* surrounded by your lasso, they will not be selected, so by enclosing only the lines you want, you can pick them out and make them available on the clipboard for copying. They can easily be squished, as a set, to fit the various places you need them. Figure 10.37 shows the result.

Now you can add some of the horizontal detail. We note that there are sets of five long horizontal lines occurring a few times, sets of four once or twice, and

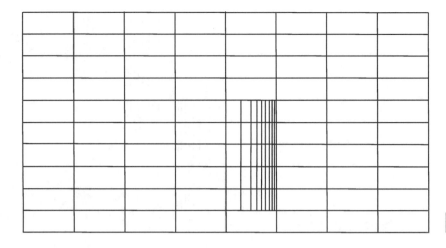

Figure 10.36

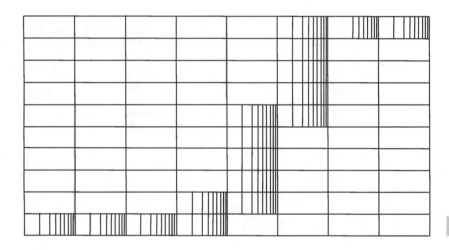

Figure 10.37

single lines here and there. We start by creating a set of five horizontal lines, evenly spaced, in the bottom left. Figure 10.38 shows the result.

It is easy now to replicate these lines and also to convert them to four-line groups, replicate them, and make them evenly spaced.

You can arrange things to allow you to be efficient. For example, you will notice that in the vertical spaces at 0.25 through 0.85, there are single horizontal lines, indicating the $x.5$ location. The longest of these, occurring only once, is 2 spaces wide.

We start with this one, by copying one of the 2-space horizontal lines from the groups of four we just added. We put it in the space at (vertical) 0.55. Now, we copy this single 2-space line into the locations at 0.45, 0.35, and 0.25. Then we select all three together, and squeeze them down to single-space width.

Next, we copy those three shorter lines and move the copies up to fill in the spaces at 0.65, 0.75, and 0.85.

In nothing flat, we have the grid exactly as it is in Terman. See Figure 10.39.

With a modest effort, we can add the text. It is all but impossible to match the font, as the original seems to have been hand-done. While fonts that are close

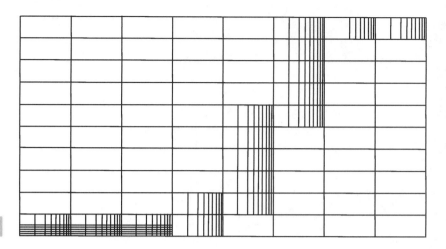

Figure 10.38

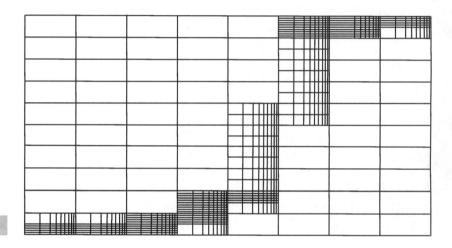

Figure 10.39

to handwriting do exist, that really is too "cutesy" for serious work, and so we go with a conventional choice like Arial. The sizes seem to be 9 point and 7 point. (We saw in Chapter 4 that it is never wise to go much smaller than 7 point.) The result is shown in Figure 10.40.

Now it would be a simple matter to add a nice curved *Bezier* line and push it around to make it match the original. We have done this in Figure 10.41. However, you would still have the kinks that were in the original line, and the point was to remove them.

So rather than just tracing the curve, or artistically fitting a Bezier curve, we will use a spreadsheet to improve on it. As we mentioned earlier, the equation that the graph shows is

$$v/c = \sqrt{1 - \frac{1}{(1 + 1.94 \times 10^{-6} E)^2}}$$

Figure 10.42 is a table of values from the spreadsheet. Note that we generated a logarithmic version of the values of E, so we could use them directly as the

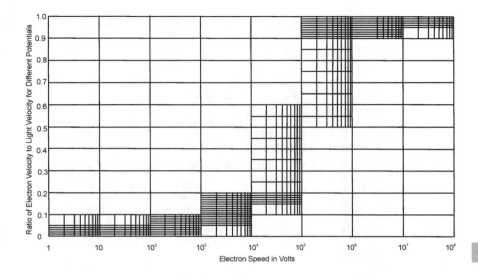

Figure 10.40

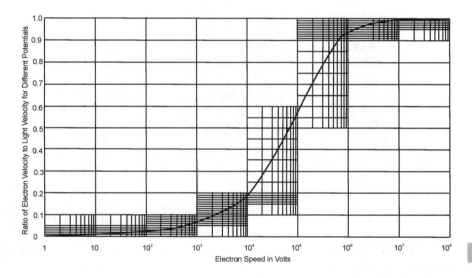

Figure 10.41

x-axis. Strictly speaking, we need not have done that, as the spreadsheet would let us use a log scale.

Now, we can get the software to generate a graph for us. All we will need from the spreadsheet is the curve, which we will fit to our new graph paper. In our case, we get the result seen in Figure 10.43 from Excel with all the original defaults.

It does look like the curve in Terman's book. If we edit the curve in the spreadsheet so as to remove the squares indicating the data points, the curve will be more useful, as there are no dots on the original curve. For reasons that will become apparent, we should also rescale the vertical axis so that the limits are zero and one. Once this is done, we select the chart, and copy it to the clipboard, and `Paste Special` it into our favorite graphics package.

At this point, we have on one page of the graphics program the scanned image from Terman's book, with a vector version of the graph paper, and on

equation 8 on page 277 of Terman, Radio Engineers Handbook, 1943				
values of E	root of denom	Ratio v/c	x-axis	
1	1.00000194	0.001969769	0	
3	1.00000582	0.00341173	0.477121255	
6	1.00001164	0.004824893	0.77815125	
10	1.0000194	0.006228874	1	
30	1.0000582	0.010788412	1.477121255	
60	1.0001164	0.015256453	1.77815125	
100	1.000194	0.01969485	2	
300	1.000582	0.03410256	2.477121255	
600	1.001164	0.048207278	2.77815125	
1000	1.00194	0.062199183	3	
3000	1.00582	0.107420509	3.477121255	
6000	1.01164	0.151260534	3.77815125	
10000	1.0194	0.194163419	4	
30000	1.0582	0.327067598	4.477121255	
60000	1.1164	0.44458575	4.77815125	
100000	1.194	0.546405233	5	
300000	1.582	0.774877646	5.477121255	
600000	2.164	0.886824068	5.77815125	
1000000	2.94	0.940376236	6	
3.00E+06	6.82	0.98919177	6.477121255	
6.00E+06	12.64	0.996865581	6.77815125	
1.00E+07	20.4	0.998797816	7	
3.00E+07	59.2	0.999857322	7.477121255	
6.00E+07	117.4	0.999963722	7.77815125	
1.00E+08	195	0.999986851	8	

Figure 10.42

another page the graph from the spreadsheet. But at least they are in the same graphics program. We need to start matching them up.

First, in the spreadsheet graph, ungroup to separate the elements of the chart. You may be able to select just the curve *and a box surrounding it*, but it may be easier instead to delete the parts you do not want. The idea of the box being kept with the curve is to help you locate exactly the corners of the graph

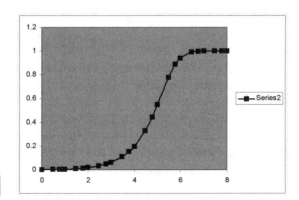

Figure 10.43

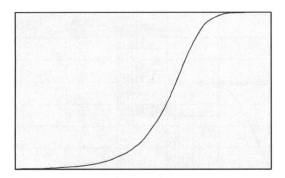

Figure 10.44

area. Not all curves neatly define the locations of two diagonal corners of the graph area the way this one does, but the method we are demonstrating here will still work. Figure 10.44 shows what you are left with.

If the spreadsheet was Excel, group the parts of the line together. Then, group this group together with the box. In what follows, we have made the box a lot thicker, so you can see what is going on. In practice, make the lines of the box as thin as you like. It is easier to do the next steps accurately with thin lines.

Now, copy the box-and-curve group, and paste it into the scanned and traced version of the graph. You should get something like Figure 10.45. The new material appears on the page in a position that bears no particular relationship to the curve we are trying to match. We can move the whole thing, the four lines of the box and the new curve, so that the intersection of the lines at the bottom left overlays the intersection of the axes, as in Figure 10.46.

Next, we can enlarge the new material until the intersection matches up at the top and the right, checking mainly in the top right corner. We choose these two diagonally opposite corners for our match because they are where the graph itself is. Given that there could be some distortion in the image on the original page, these would be the corners where alignment is most important. Here's the result, in two steps, beginning with the top line, in Figure 10.47.

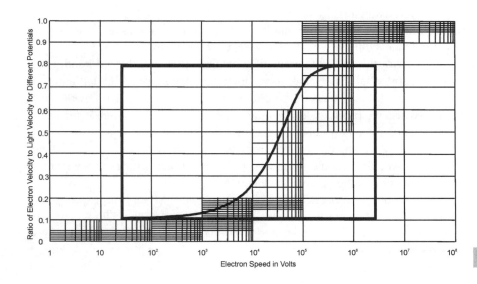

Figure 10.45

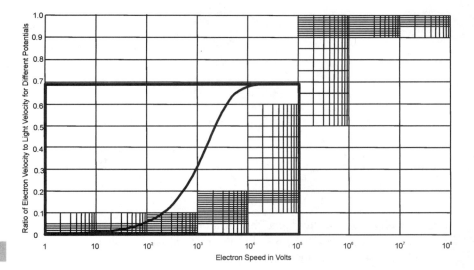

Figure 10.46

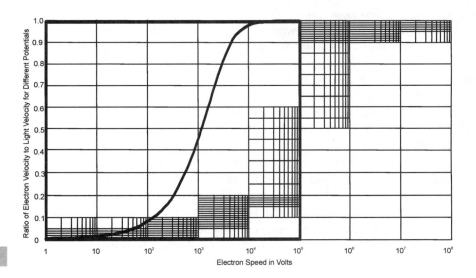

Figure 10.47

Now a horizontal stretch is needed, as in Figure 10.48. Some iteration may be required to get both corners lined up. It is important to be as accurate as possible (and that is difficult with fat lines like the ones here).

Obviously, the new computer-generated curve and the old hand-drawn one are substantially the same. If we bring back the scanned bitmap and look closely, we see the defects in the old curve. (Provided there is no evidence of error because we provided too few data points in the spreadsheet, it seems fair that we assume the new curve is accurate.)

Now, if we ungroup the box from the spreadsheet curve, and delete it, we have an all-vector version of what was in Terman's book, with a somewhat more accurate version of the curve, as in Figure 10.49.

It would have been impossible to generate this graph in a spreadsheet program alone. We could have made a simpler version, with no lines across the graph area, or a more cluttered one with lines all over it, but not a half-way version like Terman's. What Terman wanted to show was that there is a kind of square-law increase at the low-velocity end, and a limiting due to the relativistic

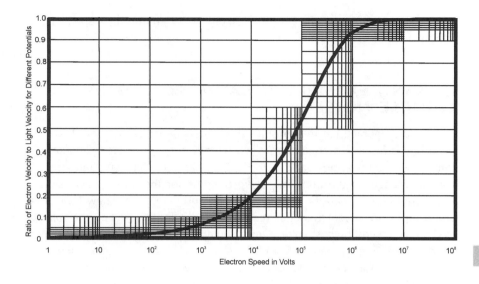

Figure 10.48

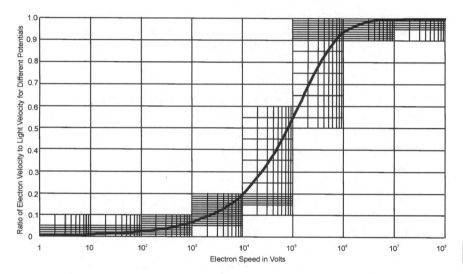

Figure 10.49

mass effects at the high-velocity end. He managed that with an attractive and clear graph. Not bad for an era that did not yet know about the atom bomb!

Of course, you might have saved a few steps if what you wanted was something simple, as in Figure 10.50.

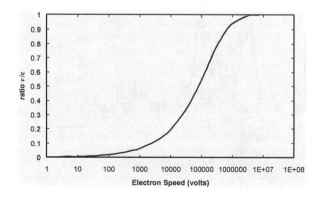

Figure 10.50

You can generate the image in Figure 10.50 entirely in a spreadsheet, without any use of the graphics program.

TRACING EXAMPLE 2: RAINFALL AND AUDIBLE NOISE

Keeping the geometrical relationships the same when there is more than one graph line is no less essential than when there is only one. Here is an example from power engineering. Back in the 1970s, some power companies sensed that the voltage on the lines they were building for the bulk transmission of power could be increased maybe once more, and then certain new limitations would appear to prevent further increases. Typically, as time went by, the voltages used on the newest highest-voltage lines had approximately doubled as the new designs were made: a company with lines at 345 kV would overlay them with lines at 765 kV, for example. Double that and the voltage is in the neighborhood of 1500 kV. Such a line may be technically and economically feasible. There was some evidence that hinted that the next jump after that (to around 3 or 4 MV) was likely not. So engineers were trying to understand how best to design lines for these "ultrahigh" voltages. Of particular interest was the interaction between the line and the environment.

One of the factors was the influence of the intensity of rainfall. There were a number of effects that were influenced by rain: the audible noise produced by the line, the radio interference, and the corona loss. A line could be designed that resulted in none of these factors being important in good weather, but when it rained, these parameters increased in complex ways that depended also on the age of the conductor in the line. Data like those in Figure 10.51 were obtained in research facilities.

Actually, the computer-generated graphics were not quite that good back then; this graph has been tweaked a little in a graphics program. It is clear from this presentation that the corona loss and the rain rate are connected in some way, but the curves are a bit cluttered.

One approach to making a clearer graph would be to get the graphics program to use color, as in Figure 10.52. This looks much more attractive, but some journals discourage (or even forbid) the use of color, so you might not be

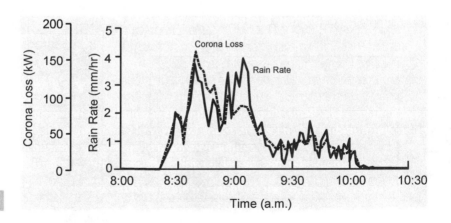

Figure 10.51

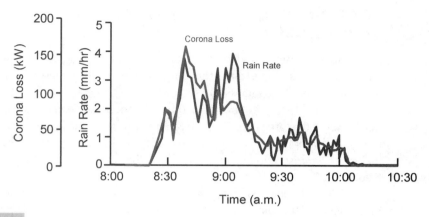

Figure 10-52 See insert for color representation of this figure.

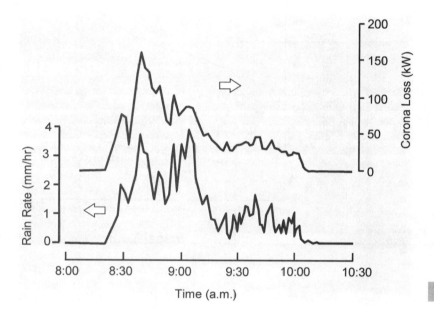

Figure 10.53

able to use this in a publication. Another way of clarifying the picture is to separate the curves by grouping them with their axes, and sliding them vertically (maintaining the horizontal relationship), as in Figure 10.53. This is not hard to do in a graphics package.

The graph of Figure 10.53 is actually quite close to how the data were presented back in the 1970s. However, back then, to get this separation one instructed the data acquisition computer to force the vertical axis of the corona loss parameter to go to −150 kW in order to obtain the required offset. The days of WYSIWYG were far in the future!

The presentation of Figure 10.53 is less cluttered than that of Figure 10.51, but the strong connection between the curves seems weaker. In the end, some decisions are just not clear-cut.

TRACING EXAMPLE 3: INVESTMENT COMPARISON

Now, suppose instead of different things, like rain rate and corona loss, the two parameters being plotted are essentially the same. Consider the two hypothetical investments shown in Figure 10.54.

Both investments have gained in value over the years, at what looks like maybe about the same rate. Suppose you don't know all the history of these investments, and your broker is trying to persuade you that the growth on Investment B makes it the wise choice for the future. What he shows you is the curves in Figure 10.55.

Here it looks as if Investment B is maybe doing a little better than Investment A, but there is not much to choose between them, really. Since

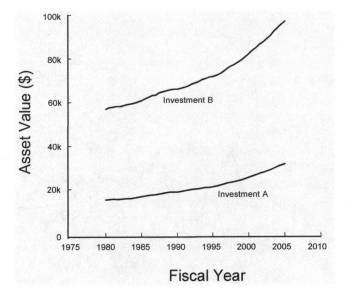

Figure 10.54

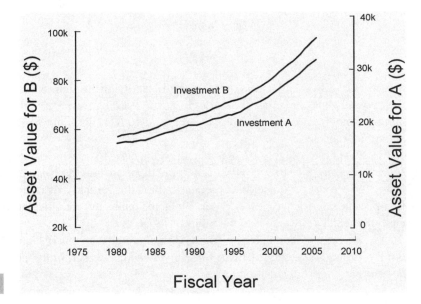

Figure 10.55

Investment B is a known polluter, but Investment A is a known green company, consideration for the planet makes you want to choose A. Is this a fair presentation?

Well, to answer that, let's look at what we did to obtain Figure 10.55. (You can check, by comparing Figure 10.55 with Figure 10.54.) We *stretched* the graph and the scale for Investment A (together, keeping the relationship fixed), and we *slid* the line for B *and its axis* down so that the two graph lines were almost together. We kept the relationship between the Investment B graph and its axis fixed, too, and we did not rescale that curve. These are all the sort of things we did with the rain/loss graph. Is it fair here? We were scrupulous about grouping each graph with its axis, so it must be OK, right?

No, it is *not* fair. We did indeed do all the things we said, but in sliding the one graph with respect to the other, we were fudging the relationship between the two sets of data. You can see this demonstrated by the fact that the zero for the right axis no longer lines up with the zero for the left axis.

Figure 10.56, produced in a graphics program, reveals what the data *really* show. Here, the zero position for the two axes has been retained, and the Investment A curve-and-scale have been stretched vertically until the curve overlaps the curve for Investment B at the beginning. *That* is fair. It represents a sort of graphical *normalization*.

It is now clear that Investment A is the better choice in terms of value growth. The curve with the suppressed zero is quite deceptive.

The difference between fairness in this case and the CO_2 situation is that here the *things* being compared were the same, whereas in the CO_2 graph the *things* were different. With temperature and CO_2 concentration, the best you could hope for would be a $y = ax + b$ kind of correlation—in other words, a

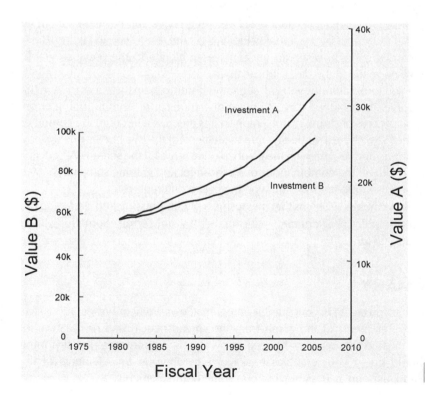

Figure 10.56

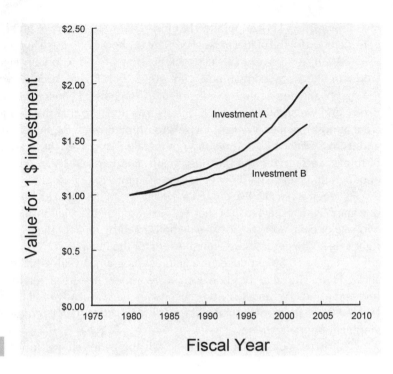

$2.50

$2.00

$1.50

$1.00

$0.5

$0.00

Value for 1 $ investment

Investment A

Investment B

1975   1980   1985   1990   1995   2000   2005   2010

Fiscal Year

Figure 10.57

scale factor and offset. And anyway, with temperature both the offset and the scale factor are arbitrary to start with.

But when the two things are the same, you expect that the offset is *zero* and the scale factor the *same*. By suppressing the zero, you can turn up the gain a little and still have the graph occupy the same space on the paper. That's the same as increasing the scale factor and changing the offset so it doesn't show.

Since we have the original data, we can go back and normalize them by dividing all the values by the initial value. We get the result seen in Figure 10.57. This, naturally enough, bears an uncanny resemblance to the previous graph, where we normalized the data "artistically."

If you look carefully, you will see some distortions on the curve for A in the artistic curve. As far as we can tell, these distortions arise from rounding errors during the process of shrinking the graph so it is much smaller than the computer screen, then stretching it vertically until the left end of the two curves line up, and then trimming and re-enlarging the graph to once more fill the screen. We got this distortion with all the combinations of spreadsheet and graphic software that we tried. If you use different software, you may get a different result.

Nevertheless, it seems that the business of compressing and stretching the graph in a graphics program may incur small but visible errors. Something to be on the lookout for.

SUMMARY

The file that you create in your graphics program (PowerPoint or Presentations) has the various elements of the graph from the spreadsheet (Excel or QuattroPro) grouped in ways that may not be ideal for your editing purposes. The first thing you should do in your graphics package is to ungroup the elements of the image and, as soon as possible, regroup some of the elements.

- In PowerPoint, you can paste as a `Microsoft Excel Chart Object`, `Picture (Windows Metafile)`, or `Picture (Enhanced Metafile)`. In Presentations, you paste as a `Metafile Picture`, *not* as a `Device Independent Bitmap` or a `QuattroPro Chart`.

- Ungroup all the elements of the picture and, as soon as possible, regroup parts that will force your operations to maintain the appropriate relationships, for example, between the data and the axes. Usually this means grouping each axis to its labels and tick marks, and grouping all the graph line parts.

- Generally, do not retain the blank white rectangle or the rectangle around the graph, as these often define a bigger space than the graph itself occupies. It is easiest to delete these parts while the material is all ungrouped.

- Separate graphs from separate spreadsheets can be combined provided you line up the data spaces accurately. But be careful: if the *things* being joined are of the same kind, be sure that the data spaces are handled properly—with the zeros lining up and so on.

- If you are scanning and redrawing an image, consider taking the opportunity to improve the quality of the picture. But be careful to be honest— use an equation to improve a curve fit, for example, or improve the quality by simply removing extraneous material to modernize the style.

EXERCISES

10.1. Use the following data to create a chart that compares the cash flow from operations for PepsiCo with that of Coca-Cola. Use either Excel or QuatroPro, then put it into a presentation software (either PowerPoint or Presentations). Make this chart fit in a single column width of 3.25 inches.

Cash Flow (in Millions) from Operations—PepsiCo (2003–2007)

2003	2004	2005	2006	2007
$4328	$5054	$5852	$6084	$6934

Cash Flow (in Millions) from Operations—Coca-Cola (2003–2007)

2003	2004	2005	2006	2007
$5456	$5968	$6423	$5957	$7150

10.2. This exercise is similar to Exercise 10.1, but with an additional five years of data. Use the following data to create a chart that compares the cash flow from operations for PepsiCo with that of Coca-Cola. Use either Excel or QuatroPro, then put it into a presentation software (either PowerPoint or Presentations). Should this chart be one column like the one in Exercise 10.1 or should it be two column?

Cash Flow (in Millions) from Operations—PepsiCo (1998–2007)

1998	1999	2000	2001	2002	2003	2004	2005	2006	2007
$3211	$3027	$3911	$4201	$4627	$4328	$5054	$5852	$6084	$6934

Cash Flow (in Millions) from Operations—Coca-Cola (1998–2007)

1998	1999	2000	2001	2002	2003	2004	2005	2006	2007
$3433	$3883	$3585	$4110	$4742	$5456	$5968	$6423	$5957	$7150

10.3. Scan and trace the image in Figure 10.58. Make improvements to font size and line weights. Should this fit in one column or two?

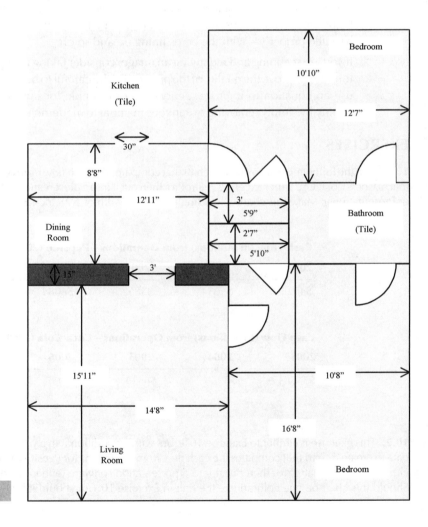

Figure 10.58

Something beginning with "P"

11

PERSPECTIVE

Look at Figure 11.1 for a moment.

Figure 11-1 See insert for color representation of this figure.

Just in case the viewer doesn't get it, this is entitled *Perspective*. It is by an Italian master, Canaletto, and was painted around 1765. You may be tempted to think that including perspective in a figure in a technical work is something that requires you to be an artist. Well, maybe. However, we think the work of Canaletto shows us that it's OK to use technology in art, so why not use art in technology?

Try to imagine a young Canaletto, somewhat struggling, making a living selling pictures to the tourists in Venice (and there were a lot of tourists in Venice back then). The paintings are selling well enough, but a guy can only do so many pictures in a given amount of time. How to crank out more?

Canaletto hits on the bright idea of using the *camera obscura* (Italian, meaning "take a photograph and I'll punch your lights out") to improve his productivity. *Camera obscura* was a wardrobe-sized device where light coming in the front was focused on the rear wall. But remember, back then the Box Brownie hadn't been invented. (And just where do we suppose the English word *camera* came from?) The thing was, not only had the camera not been invented, neither had the light-sensitive film. (The need for these went sort of hand-in-hand.)

So here's our theory of what Canaletto did. (Since it is *our* theory, you might want to consider that there may be alternative histories.) He put his canvas at the back of the *camera obscura* and sketched on it the outline of the image projected from his subjects. He needed a subject that would stay still for a long time while he did this, so his subjects tended to be buildings and so on, rather than racehorses and people walking down stairs. (These had to wait for the invention of a much faster film!) Then, with the outlines ready to be filled in back at the studio (by candlelight), he moved to another location and did it again.

So as well as using the *camera* as a way of recording an image, he also invented painting by the numbers and the picture postcard. However, since the *patent* had only recently been invented (in 1421, in Florence, but news traveled slowly in those days), the only way he could make money out of the idea was to bang out lots of pictures. Which is just what he did for several years, until he had painted all the interesting views of Venice.[1]

Figure 11.2 is another example. It is the *Arrival of the French Ambassador in Venice* and was done in the 1740s. Note the composition, the line, the detail. This is the stuff truly great jig-saw puzzles are made of!

Just in case you were wondering whether any of this was really relevant to the real world of technical writing, consider the image of a hypothetical power substation, in Figure 11.3. A picture very much like this[2] was originally drawn for a paper reviewing optical current measurements for electric power, in a

---

[1] What he did next was odd. Perhaps because many of his customers in Florence were English visitors, in 1746 he moved to England. There, the perpetual rain and cloud cover rendered the *camera obscura* rather useless, and Canaletto's abilities seemed to wane. At one point, his rivals accused him of being an imposter, so bad was the product. It was not long before he went back to Italy.

[2] Some odd colors arose in the original image through a file conversion error. The original was drawn (in 1994) in various shades of gray. When imported into more recent software, the grays all became colors. Evidently, the original file specifications of the grays were misinterpreted. More about this sort of problem in the next chapter. While it would be a simple matter to fix the image, we decided to leave it colored.

Figure 11-2 See insert for color representation of this figure.

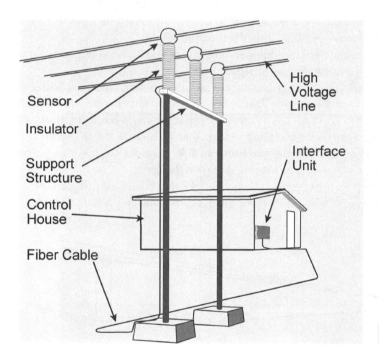

Sensor

Insulator

Support
Structure

Control
House

Fiber Cable

High
Voltage
Line

Interface
Unit

Figure 11.3

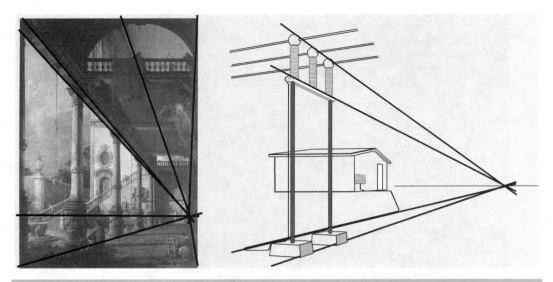

Figure 11.4

two-column format.[3] (Note the compactness achieved by the bend in the fiber cable, and the way the control house is behind the support structure.) The picture was evidently so good it was also "borrowed" for a book on power system protection, where it was reproduced unaltered in a single-column layout.[4]

Even a casual glance at these pictures will tell you 25% of what you need to know about perspective to be able to use it in your drawings. There is a "vanishing point" where parallel lines appear to meet. This vanishing point is closer than infinity, and may even be within the picture!

We can illustrate this by adding some construction lines to *Perspective* and to the picture of the substation as in Figure 11.4.

Notice a couple of things about these pictures. In each, there is a horizontal line at the level of the viewer. This is the "eye-line." It becomes the horizon when it goes to the far distance, as the standing observer is but a small object at the distance of the horizon. Next, notice that in both of these pictures, a good deal of the action is above the horizon. If it were not so, the impression would be that the observer was looking down on the scene.

This brings up the question of what you need to know that just glancing at these pictures doesn't tell you. Is there always one vanishing point, and where

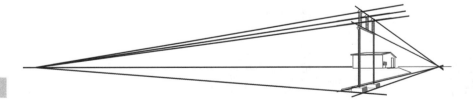

Figure 11.5

[3] *Optical Current Measurements: A Review,* Fiber Optic Sensors Working Group, Fiber Optics Subcommittee, Power Systems Communications Committee and the Emerging Technologies Working Group, Power Systems Instrumentation and Measurements Committee of IEEE. *IEEE Trans. Power Delivery* **9**(4), 1778–1788 (October 1994).
[4] Paul M. Anderson, *Power System Protection,* IEEE Press, Piscataway, NJ, 1999.

does it go? In *Perspective*, there is one, and it is within the frame of the picture. In the substation picture, it looks as if there is one, and it is outside the frame of the picture by quite some distance.

Actually, there are two vanishing points in the substation picture. We know this because the power line at the top and the foundation at the bottom are not drawn parallel to the horizon. The front of the second building in *Arrival* is parallel to the horizon, and so there is only one vanishing point. The view in Figure 11.5 shows where the second vanishing point must be in the substation image.

We are not great artists, nor is the computer a *camera obscura*. The question is therefore fair: How do we go about locating the vanishing points?

Well, there are a couple of answers. Since we are using a computer, we can add and delete construction lines, and we can zoom in and out a great deal and generally tweak things to create the desired effect. One approach is based on the use of a plan view of the object we want to draw. The other requires some measure of artistic talent. Let's deal with the technical approach first.

PERSPECTIVE: THE TECHNICAL APPROACH

For the hypothetical substation, the plan might be like the one in Figure 11.6. This is somewhat of a simplification, of course. Just the main elements are shown, so as to keep the idea clear and the page from being too cluttered.

We already know the picture we're aiming for. The viewer is standing south and east of the support structure (only the foundations are shown here) looking at the control house. We can add a line for the viewer to stand on and

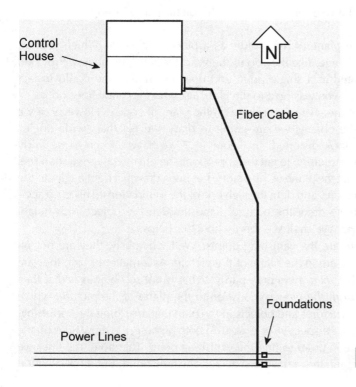

Control House

N

Fiber Cable

Foundations

Power Lines

Figure 11.6

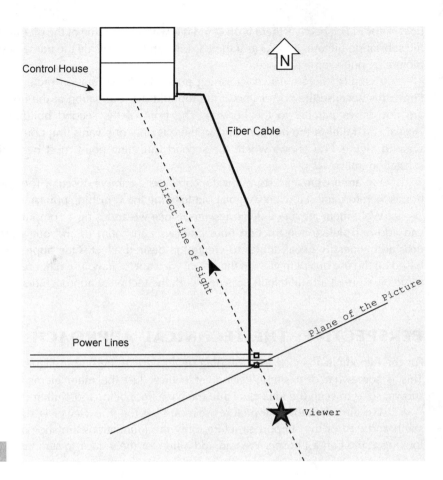

Figure 11.7

to look along, and a line for the plane of the picture at right angles to this, as in Figure 11.7.

Imagine the plane of the picture as a piece of glass on which you are going to draw the image, like they do in the War Room in the movies. Provided the viewer is located at a single point and does not move, she could in fact instruct a colleague who was next to the glass just where to draw lines on it.

We can put the viewer anywhere on the plan, of course. However, if we get too close to the objects we are going to draw, we get that "wide angle" effect, and things look distorted. In Figure 11.7 we have chosen to be fairly close to the support structure, to get the appropriate height effect, as we shall see.

Note also that the plane of the picture has been drawn at right angles to the centerline of the picture, and right through one of the foundations. This is no accident. We deliberately drew the picture plane through the object with height closest to the viewer. We shall see shortly how this helps.

Now, where are the vanishing points? Well, obviously, they are not off at infinity at all: they are in the plane of the picture. As a matter of fact, they are on the eye-line. That fixes them only partly. What is not so obvious is that they are also on lines *from the viewer's location* to the plane of the picture, which are *parallel to the structures* and objects in the plan that are doing the vanishing.

Think back to *Perspective* or *Arrival*. In both of these, we drew lines on the plane of the picture in order to find the vanishing point. Therefore, the lines we drew were not parallel; they converged at the vanishing point. But if the situation had been the real three-dimensional one, rather than the projection onto the flat plane of the picture, the lines would have been parallel to the buildings. In other words, the lines we drew were two-dimensional projections (in the picture plane) of what would have been straight-line boring graffiti on the buildings, or clothes-line strings in the air, supported by invisible telegraph poles marching off into the distance.

That might sound complicated. It isn't. On the plan, we can easily add a couple of lines to the plan to locate the vanishing points. Figure 11.8 shows two of these lines.

The new lines are parallel to the sides of the building, and one of them is parallel to the power line. Although these new lines do tell us where the vanishing points must be, they are still not completely helpful. We can add more lines (from the viewer) to show where in the plane of the picture various features would appear. We need do this only for the major features, such as the edges and corners of buildings. We added some of these lines in Figure 11.9.

Note that in Figure 11.9 we've disappeared the star marking the viewer's location. This is because we want to get the various lines accurately through this point.

Now it helps if we rotate the whole figure until the plane of the picture is horizontal. Figure 11.10 is a rotated version of Figure 11.9.

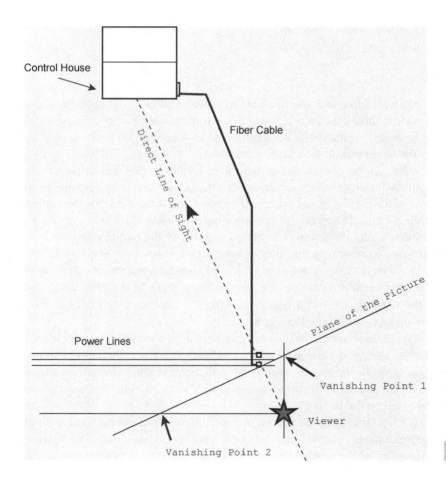

Figure 11.8

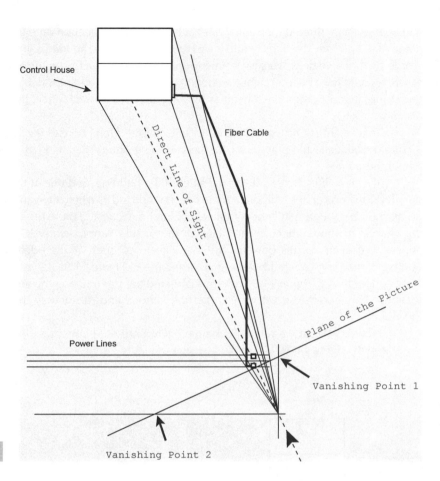

Figure 11.9

Now we can zoom in on the area where the multiple lines cross the plane of the picture, and we can add some lines to mark the crossings. These lines should be exactly vertical (use the appropriate key) and should be fairly short.

The effect should look like Figure 11.11.

Here, we have made the vertical lines a little thicker just to help them stand out. We have made the one line from the viewer that is at right angles to the plane of the picture (now horizontal) a dotted line. This should be the center-line of the picture. Things on this line are what the viewer is looking directly at.

Now, select the line representing the plane of the picture, and the new vertical lines representing the intersections. Leave everything else out of the selection. This is easily done using the lasso or rubber-band method, now that the group is horizontal on the screen. Copy them to the clipboard, and delete the diagram (or start a new one). Paste the clipboard stuff back in. You will end up with something like Figure 11.12.

Well, you won't exactly see that: we have added some convenient identifying labels. We need a name for this set of lines: let's call it the marker group. Now you have, in the marker group, the horizontal positions of all the major features of the picture. It is simple to grow them vertically into the objects you need to make up the picture.

Well, relatively simple. There is the small matter of getting the height at least approximately right! It turns out that the scale of the original plan drawing can be used to set the height of the perspective view. Remember we placed the

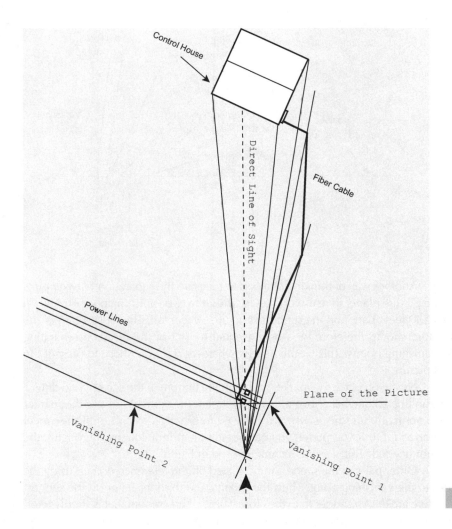

Figure 11.10

plane of the picture right through the middle of the support structure? Well, since this is the closest object, it does not diminish because of the lines to the vanishing point. In other words, if the scale of the original plan was, say, 10 feet to the inch, and the support structure is 50 feet tall, it can be drawn 5 inches high, provided the separation of all the other objects in the picture plane are drawn to the same scale.

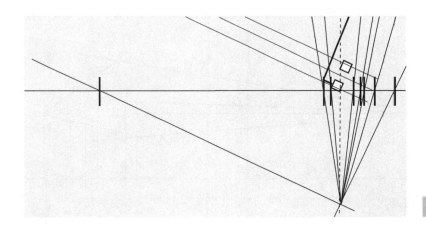

Figure 11.11

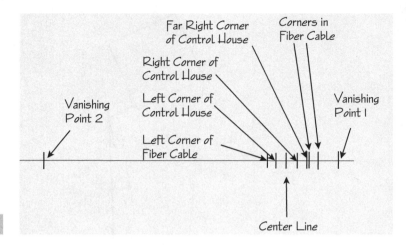

Figure 11.12

Another way of thinking of this is to measure the separation of two points on the picture plane, in real-world units. Thus, the two vanishing points might be (say) 250 feet apart, and the support structure 50 feet tall. The closest part of the structure should therefore be a height equal to 1/5 of the distance separating the vanishing points. This would be true whatever size you chose to present the final picture.

Next, you place the marker group on the bottom of the screen, and draw a horizon line somewhat above it. You can stretch the vertical lines of the marker group vertically up (this is why we drew them exactly vertical) until they meet the horizon. Or, if you prefer, you can stretch them individually by roughly the amount needed. This was the technique used in Figure 11.13.

Other parts of the scene can be scaled off the reference object, using the lines to the vanishing point. Thus, the second and third insulator on the support structure are drawn shorter than the closest one, by an amount that is simply referenced from a line to the appropriate vanishing point, as shown in Figure 11.14.

One of the objects is separate, however, and its height in the diagram cannot readily be obtained from its relationship to the others. That is the control house. However, we can readily calculate the height on the page if we know the real height, since we know (from the scale plan) the distance of both the support structure and the control house from the viewer.

It may be worth noting a couple of things before we proceed. First, the number of vanishing points is set by the artist and by the orientation of the objects being drawn. With nice, organized street layouts, such as the one in

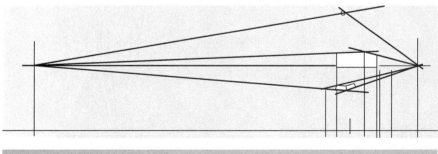

Figure 11.13

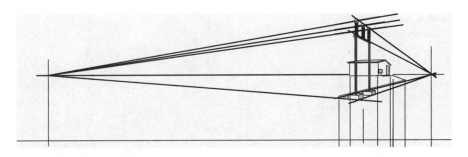

Figure 11.14

Arrival, the artist was able to use only one vanishing point. Similarly, to create the impression of being inside the arcade, he used only one in *Perspective*. If there is only one vanishing point, it is common to place it in the picture. This gives the viewer the impression of being part of the scene. However, the vanishing point can also be located outside the picture. The substation, for example, could have been drawn with only the one vanishing point at the right, as in Figure 11.15.

Suppose we put the vanishing point inside the picture, by rearranging the elements, as in Figure 11.16. This rearrangement makes for a wider picture, perhaps more suitable for a two-column layout, but it leaves the viewer wondering if something interesting will happen in the middle, if we just stare at it long enough. This happens because there is nothing at the vanishing point (except the horizon), and yet the eye is drawn there.

Evidently, we need to think a little bit like an artist if our image is to have appeal!

A further point to note. The computer allows us to use vanishing points that are a long way from one another, but it is usually important to have them on the same horizontal line. In Figure 11.17, the two vanishing points don't line up. Don't you think there is something just a little odd about the picture? There is a little of M. C. Escher in Figure 11.17, don't you think?

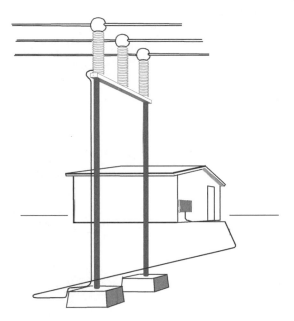

Figure 11.15

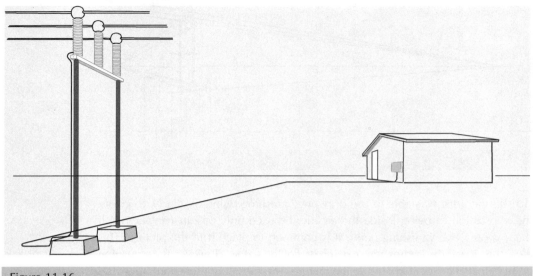

Figure 11.16

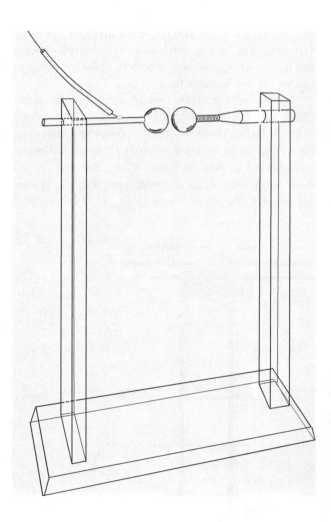

Figure 11.17

PERSPECTIVE: THE NONTECHNICAL APPROACH

Now that we've got the straightforward (if technical) way to draw with perspective out of the way, let's look at the purely artistic approach. (Not that you can avoid all need for artistry by following our technical approach. Unless you draw an enormous number of construction lines, you are going to do some interpolation and mild fudging to get your picture!)

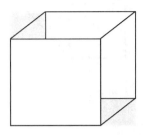

Figure 11.18

Suppose you want to draw a box. Not a block diagram box, but a normal ordinary box that any cat would be proud to sit in. You start by drawing a square. You add another square in front of that (so that it partly covers the first square), and you join the corners. What you get is Figure 11.18.

It doesn't look much like a box, does it? Well, of course, it's only a wire-frame diagram at this point. We need to make it look solid. So you draw the back couple of lines and remove the original square, as in Figure 11.19.

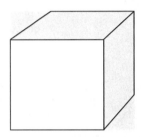

Figure 11.19

Still doesn't look like a box. No self-respecting cat (and did you ever meet a cat that wasn't self-respecting?) would sit in this box. It would certainly be curious, however. This looks like a box that gets bigger the further from you it gets!

That impression is, of course, an illusion. We are so used to judging the closeness of things by perspective, that when we draw something that does not obey the usual rules, it looks odd. What we want is something like Figure 11.20, of course.

A different angle, and solid—different enough to look right. So the question is how to draw it on the computer. Not that the world is full of boxes, mind you. But it turns out that boxes are easy to draw on the computer, and they can teach us the rules of perspective.

First thing is to remember the vanishing point. Since this is an imaginary scene, we arbitrarily choose to put the vanishing point way off at the top right of the screen, as you can see in Figure 11.21.

Figure 11.20

Then, we zoom in on the area at the bottom left of the screen, where the cube is, and add the few lines necessary to

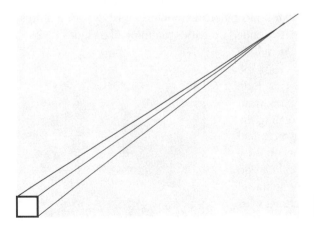

Figure 11.21

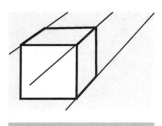

Figure 11.22

complete a cube. Figure 11.22 is a close-up of the cube as it is being drawn.

There are still a few differences between the object in Figure 11.22 and the three-dimensional cube in Figure 11.20. First, the perspective is a little warped. Second, this is not solid. Let's address the perspective question first.

We drew a square and a vanishing point, added some construction lines to the vanishing point, completed the drawing, and removed the construction lines. What we *should have done* is add another vanishing point, as in the substation picture. If our cube had been facing a different way, we would have had lines going in a different direction to a different vanishing point. Or, if we had had two cubes, facing in different directions, we would have had four different vanishing points, as in Figure 11.23. While this might look complicated, it really isn't. And the cubes end up looking real, as you can see in Figure 11.24.

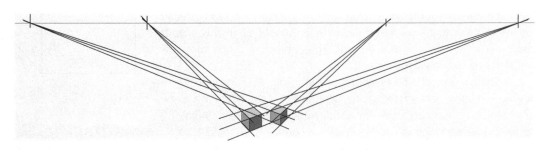

Figure 11.23

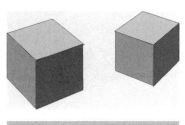

Figure 11.24

We can use this experience to generalize a little. The locus of the vanishing points is a horizontal line. In fact, it is the horizon. Strictly speaking, it is a line at the same height above the ground as the observer's eye, but since the observer is an incredibly small object at this distance, the surface of the ground and the observer's eye-line can be regarded as indistinguishable.

But hold on, here! We've sort of skipped something. How do we know how *deep* the boxes are supposed to be?

That doesn't follow obviously from the vanishing point discussion. We could have just as easily ended up with something like Figure 11.25.

Here the left "cube" is a little long in the direction away from the viewer, and the right cube is a little short. Unfortunately, unless you want to go through the effort of creating the plan drawing, and using the technical approach, there seems to be no easy answer to the problem. All you can do is give the drawing an honest appraisal after you've drawn it. Sometimes it is

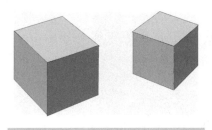

Figure 11.25

hard to see small errors. If so, don't worry. If you don't see the errors, it is unlikely that your readers will either!

In fact, the original of the substation picture was drawn this way. When we used it to demonstrate the technical method of drawing with perspective, it became apparent that there was a small error in the perspective. The plan was drawn and the "correct" version of the drawing was produced, and is used here. The original, in two published documents (the first of which is quite frequently cited), has never been criticized for its errors of perspective.

PERSPECTIVE: THE *Z* DIMENSION

We can create the illusion of height or depth by separating the usual horizon from the eye-line. Imagine there is some kind of view, and you are a camera looking at it. If you tilt down, the horizon line will appear to move up. For example, Figure 11.26 is a view down some stairs. In the view on the left, we get the sensation that we are standing at the top of the stairs, looking down. The eye is drawn to the center of the picture, which is down the stairs. A handrail is on our left along a landing, as well as one on our right that goes down the stairs. The picture on the right adds some lines to indicate where the vanishing points are. One is at our level and straight in front, but it is up and out of the picture. The other is down the stairs and in the picture. It is directly below the "usual" one on the horizon. If we draw a vertical line joining the two vanishing points, we can see that we are standing slightly to the left of the center of the stairway.

Note that a rectangle has been added from the end of the rail on the left to the corner of the rail on the right. This rectangle ensures that the height and location of the two rails match up realistically.

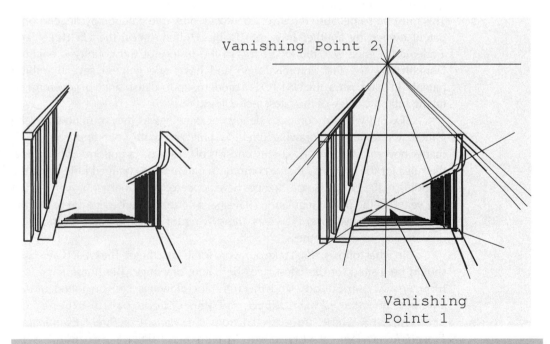

Figure 11.26

One can make a general rule. There is no limit to the number of vanishing points of a picture, and they can be inside the picture, outside it, to the left or right, and up or down.

It is, of course, the *objects* in the picture that tell where vanishing points can be drawn: in any picture there is one straight in front of the viewer, at the location of vanishing point 1 in Figure 11.26, but unless there are stairs or some other object pointing to it, it will go unnoticed.

Strictly speaking, there should be a vanishing point at the top of the picture and another at the bottom, where lines extending the verticals of the banister would meet. For simplicity, we have omitted these vanishing points—they are important only if the vertical extent of the picture is large enough to demand better representation.

That is about all we are going to say about perspective, except this. An outstanding example of vertical perspective, with no perceptible vanishing points, can be found in the opening seconds of the cartoon movie *Lion King*. In one scene, the view is straight down, and the perspective is given purely by the relative sizes of the objects seen (birds) and their relative speeds (the closest ones seem to move faster, just as the telephone poles closest to the railway track seem to be hurtling by, while the distant mountain is barely moving).

Compared to that, our task of doing stationary perspective seems simple.

PATENT DRAWINGS

The societal value of creativity is recognized by governments everywhere in the granting to the inventor of exclusive rights to manufacture his invention. Strictly, the patent grants to the inventor the right to exclude others from making or using his invention. (Thus, a patent does not actually bestow any guarantee that the owner will be allowed to make or use the patented invention. This must be so because the invention covered by one patent may depend on a patent owned by another inventor.) In the United States, the US Patent and Trademark Office was founded early in the history of the country (Benjamin Franklin was the first commissioner) and has since granted several million patents. In many ways, the US PTO is a modern, up-to-date branch of government, taking full advantage of the latest technology.

Except when it comes to figures, or drawings as they are known in the Office. Or rather, *the* drawing (there is only one with your application, no matter how many figures it actually consists of). To make a uniform appearance, the rules for the drawings that accompany a patent have changed little since the founding of the Office. Some changes have appeared, most particularly in the last few years, as the patent regulations have been "harmonized" with the rest of the world. Nevertheless, over the years, the striving for uniformity has given patent drawings an air of quaintness.

In what follows, don't take our word for it, take theirs. The US PTO website should be visited for the latest rules on patent drawings. The home page is at `http://www.uspto.gov` and the rules for drawings can be found (today, anyway) at `http://www.uspto.gov/web/offices/pac/doc/general/drawing.htm`. These are extracted from the *Manual of Patent Examination Procedure*, available at `http://www.uspto.gov/web/offices/pac/mpep/mpep.htm`.

In the version of the rules that is on the web today, the last line is "Applicants are advised to employ competent draftsmen to make their drawings." Fat chance! We all know that failure to do this is why this book even exists. Our hope is that if you read this, at least the competent draftsman employed by your patent attorney will not have to start from scratch.

Basic Principles

Although the drawings produced for patents often look old-fashioned, the principles behind them are sound: they have to be clear, understandable, and reproducible. These are the same tenets that have guided us in this book, and indeed much of what we have offered in the way of advice can be applied to patent drawings.

You more or less *must* have a drawing in your application. While the Office rules say that you must have a drawing (i.e., a set of figures) if one is required to understand the invention, the actual interpretation is that if a drawing is *possible*, whether or not needed for understanding, you will be required to produce one.

What Must Be in Your Drawing

Not only do the rules tell you that you must have a drawing, they tell you what you must put in it. You must include, as a minimum, every feature that is *mentioned in the claims*. Actually, this is a help, as the requirement could have been to include everything mentioned in the *description* of the invention. The claims specify what is new, and the requirement is to have all this included in your drawing. This could be a lot less than all the material in your description, which could, after all, include a good deal of what is called the prior art.

The requirement to draw what is in the claims is counterintuitive, however. It means that you cannot be sure you have an adequate set of figures until the claims are drafted. Since the claims are what define the protection afforded by the patent, they are often not finalized until they have been iterated several times between the inventor, the patent attorney, and the Patent Office. You may be wasting your time trying to prepare drawings until the claims are at least part-way finished.

You might guess at what is going to be in the claims, but unless you have them in front of you, it may be hard to remember that you are not producing drawings to "teach" (as the lawyers say) how your invention works. The "how it works" is in your description, not in your claims. So as you move forward, check your drawings against your claims.

Peculiarities of Patent Drawings

Your drawing(s) can include block diagrams, timing diagrams, electrical circuits, flow charts, and, of course, mechanical drawings. The drawings can be plan, elevation, section, perspective, or even exploded views. Symbols that are ordinarily used in such drawings may be used, so long as such use does not cause confusion. (For example, the symbol for a hydraulic pump may seem similar to the symbol for an electronic mixer. Usually, context will make clear what is intended.)

Given this wide range of possibilities, there is need for a few peculiarities to make patent drawings uniform. This is somewhat justified by the fact that a figure from your drawing will appear in the *Official Gazette*, and on the first page of your patent when it issues. (So it might be a good idea for you to deliberately make one figure that you feel is representative of the invention. Thus, if your invention is, say, a new mechanism for a robot arm, you might want to guarantee that an image of the mechanism is used as your representative drawing, rather than a flow chart that shows some part of its operation.)

One of the unifying ideas in patent drawings is the way parts are identified in the figures, that is, the callouts. Parts that are identified in the drawing are labeled with *numbers*. So, one thing you can do to make your diagram consistent with a patent drawing is take off all the handy labels that clarify and explain and replace them with numbers. Clarification and explanation will be in the text. For example, there is an image of a hypothetical electric field mill in Figure 11.27.

The diagram of Figure 11.27 would become compatible with the requirements of the Office if we replaced all the labels and made a few changes to the artwork, with a result as shown in Figure 11.28.

So what have we done in Figure 11.28 to achieve this antique look? First, we've removed all the shading. Solid black is forbidden (it does not copy well) and solid gray is not much better. So the old rules of shading using a pen and ink have been applied. Vertical surfaces get vertical shading lines, horizontal surfaces get horizontal shading lines, and curved surfaces get lines parallel to an edge. In most views, the light is supposed to be coming from the top left at an angle of 45°. (Actually, we're allowed more latitude in perspective views like this one, but the same rule does not hurt.) In perspective views, an object's edges closest to the eye are drawn slightly heavier than the edges furthest away.

We have, of course, replaced the callouts by numbers and used a suitably rococo font (this one is one of a series called, for some reason, English111). There is actually no restriction on what font you use. It just seemed to us that the classic-looking patent drawings did not use modern sans serif fonts.

In fact, the drawing in Figure 11.28 would still not be acceptable to the US PTO. The rules call for short leaders, so the diagram of Figure 11.29 is more in line with their requirements.

We think it's kinda quaint, don't you?

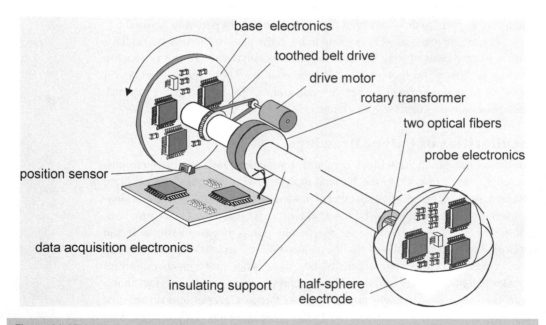

Figure 11.27

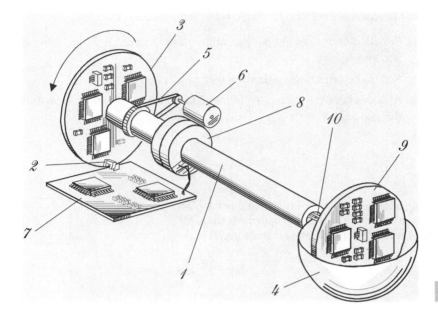

Figure 11.28

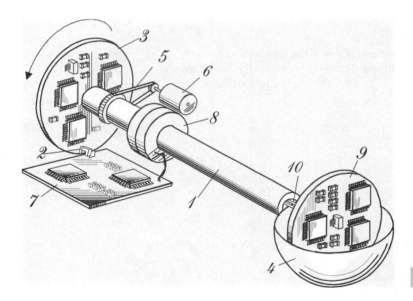

Figure 11.29

SUMMARY

Perspective is rarely needed in a technical drawing, but when it is, don't be put off by the difficulty.

■ If you have a plan drawing, and you need moderate accuracy, you can produce a perspective drawing using the guidelines presented early in this chapter.

■ If you don't need particular accuracy, you can sketch something out on the computer and add some lines to vanishing points just to refine the drawing.

For patent drawings, the PTO rule is the same as ours: clarity.

- ■ Provided your drawing is clear and will copy well, it is likely to be acceptable.
- ■ Recall the principles of drawing using India ink.
- ■ Always check the Web page of the PTO for the latest information and the latest rules.

EXERCISES

11.1. Draw the view that the observer would see from the star in the plan drawing of Figure 11.30. Place only one vanishing point in the picture. All the buildings are the same. They have flat roofs and are three stories high.

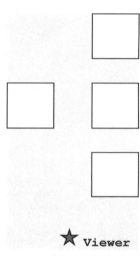

★ **Viewer**

Figure 11.30

11.2. Imagine you are high up in one of the tall buildings that create the "canyons" of New York City. Draw a picture of how it might look if you stared down one of the canyons, and left the skyline just visible at the top of the picture.

11.3. Take one of your drawings of an object, and modify it for use as a patent drawing. Show the "before" and "after" versions.

File formats and conversions

12

"*The German language speak I not good, but I have numerous connoisseurs me assured that I her write like an angel.*" Actually, this is not a machine translation, but a "double translation" of a speech before the Vienna Press Club in 1897. The author was Mark Twain. *He* gave the speech in German, and then translated it back into English. At that point it has lost something, although from the humor point of view it very much has gained!

CONVERSION PROBLEMS

This is about the quality that you get when you try to open a PowerPoint file in anything except PowerPoint, or a CorelDRAW file in anything except CorelDRAW. Some of the basics seem to be intact, but the subtlety may have gone. For example, consider the graph in Figure 12.1, extracted from some recent IEEE transactions.

The subject is the variance of Gaussian noise, and the estimator of IEEE 1241 is being compared with a maximum likelihood estimator (MLE) and the Cramer–Rao lower bound (CRLB). Sensing that the curves were hard to distinguish even with the callouts, the authors added a caption that explains how the different curves can be distinguished: "Variance of (solid curve) IEEE 1241 estimator and (dotted curve) MLE. The continuous curve reports the corresponding CRLB."

We cannot be sure, but it seems highly likely that in the original graph one of the lines was indeed dotted, and became a solid line during translation to the format required for the printer. This is exactly the sort of thing that bedevils file format conversion.

Depending on which programs you are translating from and to, it may be worse: one of the most annoying messages your computer can produce goes something like "File format not supported."

Sometimes this can mean that a significant amount of effort has just been wasted in producing a graphic file. If you get that kind of message more than a couple of times, you *are* going to look for a solution. Since they don't like you

---

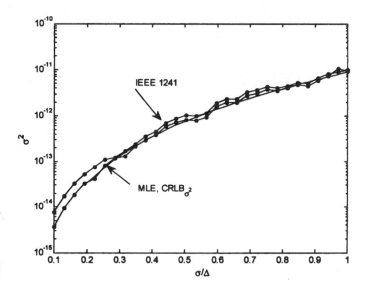

Figure 12.1

taking guns on airplanes, going hunting for the geniuses that produced your software is really not an option. Our plan is to help you without your having to wear
one of those attractive electronic bracelets around your ankle!

TWO QUESTIONS

So what's going on here? There are many (dozens) different file formats for
graphics, and they are not all compatible. Two questions must be answered:

- Do we really need *all* these formats?
- Why aren't they interchangeable?

Actually, it gets a little more complicated. While you are dealing with your
graph in your graphics program, the file has one format (sometimes called the
native format). When you stick the picture into the word processing program
that you use, it may change. And as if that wasn't enough, there is a growing
trend for what you see on the screen to depend on what *printer* you have selected.
While it makes sense to show you, in a WYSIWYG program, where the page
breaks will be, and so on, there seem to be some unfortunate side effects. Since
the screen is now driven by information translated from the printer driver
output, something is lost in translation, and the screen image can be quite discon-
certing. This seems to be even more true if you have one of those "port replicators"
that let your little laptop use a nice big screen. Seemingly, the extra translation
takes something away from the image quality.

Let's go back and look at the question of interchangeability first. There are
really two reasons why files are not always translatable. First, graphics file formats
are not all interchangeable because they are designed to do different things, and
they are designed to handle different kinds of information. Second, while every
file format is defined (somewhere), the people who write the code that handles
the information sometimes take shortcuts, and sometimes just make mistakes.
For example, the shortcut may take advantage of some particular feature of the
operating system. But then, the operating system gets updated in order to fix a
bug somewhere, and suddenly the shortcut won't work any longer. As for the

mistakes, well, we're all human. The shame of it is that the *managers* of the people writing the code don't test the software enough to discover the mistakes, or just don't care.

So, forgetting the human aspects of the problems, why do the differences exist? Some file formats are designed (say) to control plotters to generate a specific output. These files will contain instructions such as *pen up* or *pen down* that only have meaning when the plotter is moving from one part of the image to another without leaving a line. Other formats are designed for the exchange of information that will allow some other software to generate the appropriate image, putting a photograph on paper, for example. In this case, since the final output may be via a laser printer, pen instructions would have no meaning.

Evidently, then, graphics files cannot really be considered in isolation. They are created by a *writer* module and they are used as input to a *reader*. The writer may be in a graphics program, a computer-aided design package, a scanner, or a data acquisition system. The reader may be embedded in an output device such as a printer, or it may be part of your word processing system, or it may be a *filter* designed specifically to convert one format into another.

Today, people try to make the graphics file itself independent of the reader. The idea is that the same file can be used with any number of readers, simply by switching to the appropriate driver software. And, as we said, sometimes there are unwanted side effects. Nevertheless, device independence is here to stay.[1]

FILE VARIETIES

At the highest taxonomic level, graphics files come in two basically different kinds. We could think of them as low level and high level. A low-level file consists of a description of the image on a pixel-by-pixel basis. This is known as a *bitmap*, or a *raster* image. Such files readily generated as the normal output of digital cameras and scanners. In contrast, a high-level file describes the image in some shorthand way, using lines and curves and areas. It is generally called a *vector* file. Such files are the normal output of design programs that have a human input.

In a bitmapped file, the image is considered to be simply a flat surface to be described. It is divided into pixels arranged in horizontal rows and vertical columns (i.e., a raster), and the file contains a description of every pixel. If the image is black and white, the pixel description can be as simple as a one or a zero. If the image contains a gray scale, then each pixel must be described by an intensity. For example, 8 bits may be used to describe the pixel grayness. If the image is color, there are several ways to describe each pixel. One way is to give the intensity of each of three primary colors, with a depth of 8 bits each. This means that each pixel requires 24 bits to characterize it. If the image size is 300 pixels by 300 pixels (say, from a scanner with 300 dpi resolution, and an image area of 1 inch square), the file size must be at least 270 kB $(300 \times 300 \times 24 \div 8 = 270,000)$. This is without counting the additional data

[1] It is, of course, impossible to make a reader or a writer that is completely independent of the operating system. These programs begin by making calls to the OS to open the file in question! It is possible, however, to create a graphics file that makes no use of the OS. Therefore, most often, the data stored in a graphics file are independent of the computer platform to be used. A file created on one kind of machine can be ported to another, and used there without difficulty. Sometimes, however, the graphics file itself is designed to use some of the computer resources, and cross-platform operation is difficult. Mac® file writers, in particular, used to do this, making life difficult for the PC user whose OS did not have matching resources.

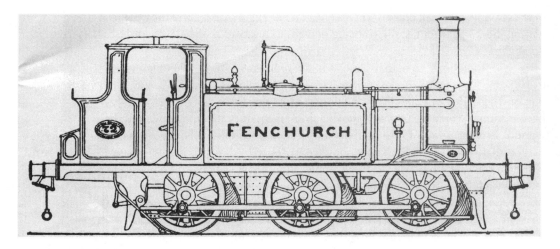

Figure 12-2 See insert for color representation of this figure.

required to define the rows and columns. This is a lot of information just for 1 square inch, and it makes no difference whether the square inch is all the same color!

Figure 12.2 is a sample picture.[2] This picture was obtained by scanning at 600 dpi in 24-bit color.

If we zoom in on an area above the "N" in the name, we can see the individual pixels, and the fact that there is color and shading. (In Figure 12.3 we are looking at an area 45 pixels wide, from an image scanned at 600 dpi. That means the image is a close-up of an area 0.075 in.—2 mm—wide.)

We could, of course, have scanned the original in 8-bit gray scale, as in Figure 12.4, or even in true 1-bit black and white, as in Figure 12.5.

Figure 12-3 See insert for color representation of this figure.

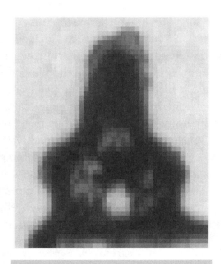

Figure 12.4

[2] Some readers might find the locomotive in the picture familiar. In fact, *Fenchurch* entered service with the London Brighton and South Coast Railway in England in 1872: younger brother *Stepney*, who later achieved fame in the *Thomas the Tank Engine* stories, started operations 3 years later. Both locomotives are in semi-retirement at the Bluebell Railway Preservation Society, south of London.

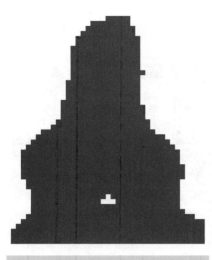

Figure 12.5

As you would expect, the file size decreases as you move from color to gray to black and white. (Note that we must maintain a distinction, in this business, between the true meaning of *black-and-white* and its more common meaning in photography.) The original file size (24-bit color) was 18.1 MB. The gray version was (as you would expect) about one-third of that at 6.3 MB, and the black-and-white version was a mere 758 kB.

So much for bitmaps. Files in the other major class of files, the vector files, create an image made up of lines (straight or curved) and bounded areas that are described mathematically. Just what description is used differentiates between the various formats. It is possible, for example, to describe a (mathematical) circle by specifying the coordinates of the center, and the radius. A physical circle, on the other hand, has thickness and (possibly) color. Thickness can be added as a parameter to the definition of the circle, or two radii (inner and outer) can be specified. Other shapes can be constructed as line segments.

As we saw in Chapter 9, it is possible to create a vector file by *tracing* a bitmap image (with software designed to do this). The vector version of the little locomotive, based on curves chosen by the software, occupies only 81 kB. As you can see from the blow-up in Figure 12.6 of the mechanism next to the dome, all evidence of the original pixels has disappeared. And while the individual shapes and curves may not be what an artist would have chosen, the final product, in Figure 12.7, is surely as good as the original for most purposes.

The reason for preferring a vector file (if it is possible to produce one) is often the size of the files involved. While there is some loss of fidelity in going from the full color version to the final vector monochrome version, the file size has reduced more than 2 orders of magnitude.

A smaller file has several advantages: it will take up less of your hard disk (or whatever storage medium you are using), it will load faster, and it will transmit over a network faster. So let's hear it for vector files!

Actually, in spite of the apparent superiority of vector files, the ubiquitous Web has made very effective use of bitmapped graphics. The problem with vector files is that you need some very particular software to interpret the curves and lines, and everyone has their own idea what works best. It is the competitive nature of this variety of formats, and the failure to adhere to standards, that lie behind the problems we will be discussing in this chapter.

The variety of formats is not an issue limited to vector files. If you can install the right software in the users' computers, you can *compress* bitmap files

Figure 12.6

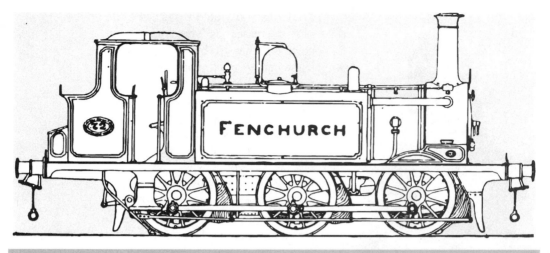

Figure 12.7

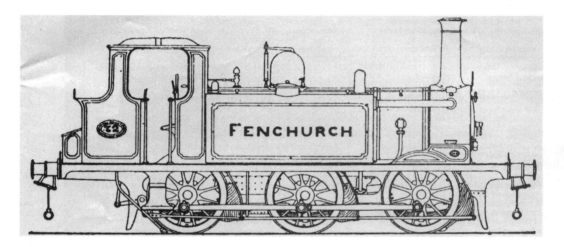

Figure 12-8   See insert for color representation of this figure.

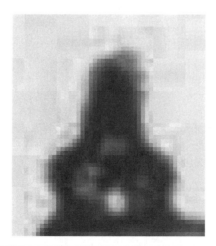

Figure 12.9

and send them over a network without requiring long transmission times. Figure 12.8 is a color version of *Fenchurch* compressed so that the file size is only about 312 kB.

A closeup of the valve, in Figure 12.9, shows some effects due to the compression, but the whole image is surely acceptable.

So it seems that in the "competition" between vector files and bitmaps, file size may not always be a determining factor. In fact, users of the Web have taken the notion of compression and have come up with several new and compact formats for bitmap files. And for whatever reason, while there are many bitmap formats, the process of converting one format to another is generally more successful than attempts to convert between vector formats.

BACK TO THE TWO QUESTIONS

Clearly, the two kinds of files contain different kinds of information. A bitmap is a pretty good way of capturing a photograph, but it's a *really* bad way to describe lines or text. A bitmap is like a fax: the letters and lines are not letters and lines but dots. Lots and lots of dots. Terrible for editing, since the lines and letters have no real existence in this kind of file.

So the answer to the question about files not being interchangeable is that, at least for the two major categories, the files contain very different kinds of information. There is no obvious and completely general way of converting one to the other.[3]

It is no easier to give a straightforward answer to the first question, which was—if you can remember that far back: Do we really need all these formats? Actually, there are really two questions. Given that an image is stored in a file as a bitmap, do we really need all the bitmap formats? Given the existence of one vector file format, do we need all the others?

While bitmaps are always bitmaps, in fact there are many different ways to represent the data that can improve the efficiency, and make the files smaller. As for vector files, there are two reasons why there are many formats: capabilities and ownership.

The different formats sometimes *do* offer different capabilities. For example, the DXF format used by AutoCAD (and others) contains scaling/dimensioning information that a simple drawing program might not need.

As for ownership issues, it seems that during the evolution of the personal computer, every company that had a software product went through a phase in which it imagined it could hold its customers by creating noninterchangeable file formats. At one point, the people at WordPerfect, for example, went so far as to change the specifications of the WPG files apparently so as to make more difficult the life of the companies that offered products competing with their own DrawPerfect program. DrawPerfect seems to have vanished, but we are left with the legacy of two unequal WPG formats.

---

[3] Actually, this is a bit of an overstatement. A fax machine converts an image (however generated) into a bitmap, and there is plenty of fax software to do the same to images stored only in files. And the reverse process can be done, usually with less success. The process of optical character reading (OCR) converts an image into the ASCII letters, more or less accurately. A tracing program (CorelTrace is an example) can create a vector file from a bitmapped image. A scanned drawing can be converted into lines, though the process may do some surprising things, and you may end up with a lot of lines where one would do.

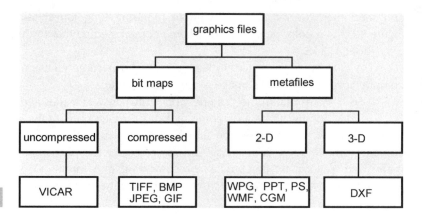

Figure 12.10

A FILE TAXONOMY

The situation is summarized by the taxonomy shown in Figure 12.10.

Mind you, even this is not a complete picture.[4] There are many file formats not specifically included here that can handle both bitmaps and vector files (e.g., CorelDRAW or Adobe Illustrator), and there are capabilities not shown in this arrangement. Some vector file formats will not support color, for example. Some have poor support for embedded bitmaps. But the ones that are included here are representative of what you will likely encounter. Note, by the way, that all of the formats shown are proprietary (though widely used) except for CGM, which is an ANSI and ISO standard.

A SAMPLE FILE FORMAT

It will be instructive to spend a while looking at the way the formats in common use are specified. Let us begin by making up a generic file format, to see what must be included.

A graphic file is created by the writer software in (at least) two parts. First, there are some *header* data, then there is the *body* of the file. The header tells the reader (or the OS) about the remainder of the file: at least what kind of file it is, what program and what version number created it. Some file formats permit variations, such as whether data compression is used and, if so, what kind. Some formats store data differently if the image is monochrome or color; this information is often included in the header. In other words, the header includes all the information needed to set up the reader software appropriately.

The body of the file contains the graphic information itself. There may be several sections within the body, for example, if the image is built up of several planes or layers. Each layer could be treated as a separate body (respecifying everything) or there could be some shared information in common between layers.

---

[4] Two formats that are becoming popular are PNG (Portable Network Graphics) and DejaVu. PNG files follow an open standard, seem to be quite compact, and are of good quality. More on PNG later. DejaVu is a proprietary format from LizardTech™. It is excellent at capturing and compressing (extraordinarily!) circuit diagrams, text, and photographs. Like the well-known Acrobat® of Adobe® (pdf), there is a free viewer that allows zooming and printing. This one makes use of your Web browser and offers no way to edit the images.

In addition to the purely graphical information in the body, there may be text. If there is, the body of the file must contain not only the text, but also a specification of the font used, and any special formatting that has been applied (e.g., italics or bold).

There are lots of details we have glossed over. If there is text, how is its location on the image to be stored? Are regular shapes (such as rectangles) to be stored as a specification of only the (x, y) coordinates of diagonally opposite corners? (This is a very compact way of defining a rectangle, but it means that rectangles must have sides that are parallel to the sides of the screen.)

Figure 12.11 shows what our generic file format might look like inside.

Some programs, written with spiffy presentations in mind rather than technical graphics, allow neat nongraphical features. For example, slides may fade or slide in and out, "buttons" may be placed on the image so that when you click on them they perform some action, such as going to the last slide, or the first, or launching another application. Not many serious graphics packages seem to care about this sort of capability, but as we saw in Chapter 6, it can be an attention grabber with the right audience. Information about what function is to be performed under what circumstances must be stored somewhere in the body.

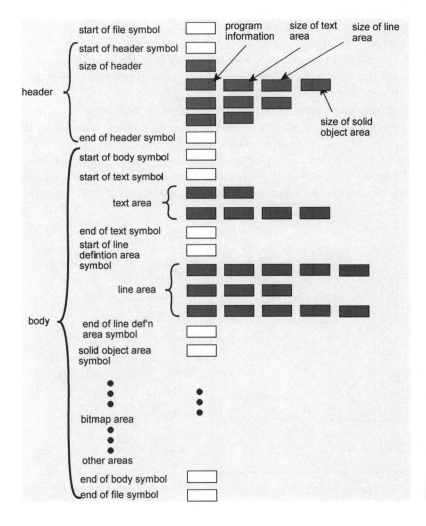

Figure 12.11

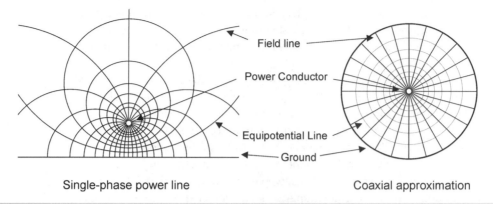

Field line

Power Conductor

Equipotential Line

Ground

Single-phase power line   Coaxial approximation

Figure 12.12

Thus, it seems that even a simple graphic file, using our generic format, is inevitably a complex entity.

It is probably fair to say that there are not, in fact, any good technical reasons for *all* the file formats, only for some of them. The good news is that many of them can be converted from one format to another. The bad news is that the conversion may not be entirely successful.

Figure 12.12 is an example of a graphic that was made for a glossy magazine. It was produced in CorelDRAW, where the left part of the figure was part of a larger image, the remainder of which was cropped. All the lines in the left part are circles except the vertical straight line.

A file was sent to the publisher, but there must have been something wrong somewhere, because when it appeared in print, it looked like it does in Figure 12.13, which was scanned from the pages of the magazine.

The printed version had a pale gray-green background, which we have converted to gray here. Note that the circles on the left are no longer circular, nor are they of uniform thickness. Some have partly disappeared. The vertical

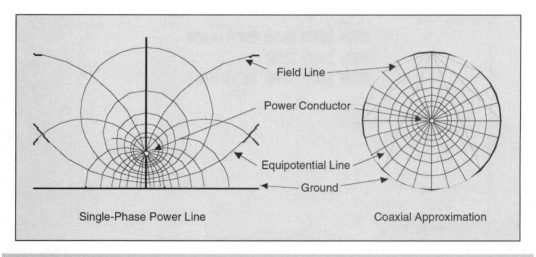

Field Line

Power Conductor

Equipotential Line

Ground

Single-Phase Power Line   Coaxial Approximation

Figure 12.13

line down the center of this part of the diagram has suddenly assumed special importance—at least, it got thicker!

The outside edge of the outer circle on the right has a distinctly moth-eaten look, and a set of straight lines has appeared from nowhere joining some of the intersection points on this circle. (Look at the top of the outer circle of the right figure.)

We have no idea what software committed these crimes. Perhaps it is as well.

SOME REAL FORMATS: A QUICK LOOK

Most graphics for use on the Web itself (i.e., for screen display) are based on the bitmap Graphics Interface Format (GIF, originally of CompuServe®). By reducing the number of available colors to 256 (i.e., a total of 8 bits for color instead of 24), and by using a compression algorithm, the format achieves about a 4:1 overall compression compared to an uncompressed bitmap. Since most of us are not on the Web just so we can gaze at somebody's cute logo, or be dazzled by their advertizing, this reduction in file size is important.

Another popular bitmap format is JPEG, named after the committee that approved it, the Joint Photographic Expert's Group (though who decided these were experts, we do not know). JPEG files allow 24-bit color (which implies larger files than GIF), but because they use a different compression algorithm, they achieve a 20:1 file size reduction. You may notice a sequential improvement in image quality during download and decompression of such files. The JPEG definition is revised from time to time, as new algorithms are devised and more processing power becomes available to compress the images. (The *Fenchurch* image in Figure 12.8 was JPEG compressed.) The latest, a considerable departure from its origins, is JPEG-2000, with the extent jp2 (or jpx if certain extensions are used). This format uses wavelet compression, recognizes and preserves text, and is so good it really is worthy of study all on its own. (But not here. Do a Web search.)

The problem that JPEG and other compression methods solve is the reduction of file size, so as to reduce the download time on a system that is bandwidth limited. There is, however, a trade-off. Most compression algorithms are "lossy," meaning that information is lost with compression. With some (including JPEG), you can choose the degree of compression and judge whether the loss is acceptable. (JPEG-2000 offers a lossless compression option.)

TIFF (tagged image file format) file structures are only slightly like our "generic" format. They allow for multiple images, some with other encoding (such as JPEG), but they do not store vector data. TIFF was designed (in the early 1990s at a company called Aldus, that later merged with Adobe) to be a holder for data from scanners. At first, only black-and-white images could be represented. Later, the format expanded to include gray scale and then color. The platform was originally intended to be the Mac, but more recent versions include information that copes with the different byte-order convention adopted by Intel (of PC fame) and Motorola (of Mac fame). TIFF files are now quite interchangeable between operating systems.

In fact, TIFF files are portable between applications as well as platforms, and seem capable of keeping up with a changing environment. As a result, the file structure is complex, and TIFF files are not always as small as one would hope. The file size and the processing needed to access the data result in slow file operations, usually considered a small price to pay for more or less universal usefulness. Most graphics programs can produce TIFF files (i.e., they include a TIFF writer). Not all TIFF writers are created equal: for example, some will allow you to control the resolution, others will not.

However, since TIFF files store their information in raster format, any text in the file is not editable by a graphics program. As output to a printer, or as an embedded graphic in a word processing program, the raster format is fine, but a graphic that is being worked on should not be stored in TIFF. Indeed, most graphics programs do not ordinarily store the data in any of the formats we have mentioned so far. They store the data in their native format. To produce a JPEG or TIFF file, you have to "export" or "save as" the appropriate format.

CGM (computer graphics metafile) format is a US standard (ANSI X3.122-1986) and an international standard (ISO 8632:1992) that can handle both vector and raster images. It is supposedly supported by all the major graphics packages, but in fact its use is increasingly uncommon. In our view, the major graphics program companies (such as Adobe and Corel) put little effort into supporting the format, preferring instead to concentrate on other file types.

A metafile (as in CGM) is a "file of files" or a "file of commands." The term is usually taken to be equivalent to "vector" file, but strictly speaking, a metafile (such as CGM) can handle either image type. One cannot help feeling that CGM was created to be the one format that could handle the output of any graphics package, and retain all the features that go with it—and the competitive world of computer software could not abide the thought! Shame, really, as this would have given a much needed capability to exchange mixed-format files between applications.

A relatively new format is being developed to meet the need, and more. SVG, for Scalable Vector Graphics, describes itself as "a language for describing two-dimensional graphics and graphical applications in XML." It is an open standard, which means the details of the format are available to all. It also means that at least some of what is important about the concept lacks the clout (and resources—think LINUX) of a single big supporter. Theoretically, the SVG format allows proprietary information (metadata) to be embedded in a file, so that the application that created the image could use the SVG file as its intermediate storage. Frankly, it all sounds very good. Perhaps it will be one day—right now it seems to be a bit underwhelming.

The specification includes the ability to do animation, and seemingly SVG is finding application in cell phones. Sadly, it looks as if a perfectly good version could be developed for cell phone use without ever generating a version that will help those of us who want to produce good scientific graphics. When we tried to export a graphic (one that was created from scratch) from CorelDRAW into SVG format, we got told in a little dialog box that there were two "issues." One of them was "Invalid IDs found" with the following useful clues:

```
Details: One or more IDs used in the file are either duplicates or contain
invalid syntax.
```

```
Suggestions: Check all group, object and symbol instance names to
ensure they are not duplicated within file and do not start with a
number. Duplicate and invalid IDs will automatically be appended on
export.
```

It may have occurred to you, dear reader, that programmers (or "developers" as they prefer to be called) write for the pleasure of other programmers.[5] It has always been so, and probably always will. We have been guilty of this ourselves, back when we were doing a lot of programming. But to tell the *user* of a program like CorelDRAW to "check . . . all symbol instance names" is going a bit far.

Things weren't much better when we exported a file from Adobe Illustrator®. We got asked whether it was OK, as far as the fonts were concerned, to use only "glyphs." Not being familiar with the word, we looked it up in the *Oxford English Dictionary*, which told us the word came from the Greek word for "carving" and meant a "sculptured mark or symbol." Big help, that dialog box.

One day, probably long after all of us have shuffled off this mortal coil, software will not be regarded as something written by the high priests to bamboozle the rest of us, but will be viewed as something that should work quietly in the background, and never require the understanding of its user. Does the checkout person or the shopper need to know how a point-of-sale device works? Why do these idiot programmers think we should know (or care) about whether the graphics application is creating invalid syntax. Not our problem, guys, *it's yours*. Unless you get your act together, those of us using graphics programs to generate publication quality scientific and engineering graphics will have to manage without SVG.

Enough ranting, back to the file formats.

PNG (portable network graphics) format is another "open source" format, designed to replace the GIF after the owners of that format announced that royalties would be required for using it. This one has been around long enough to become a standard: ISO/IEC standard 15948:2004. Like GIF and TIFF, PNG is only capable of storing information as a bitmap. But if you find an implementation that actually works (they don't all), the features (e.g., you can control the resolution of the image) and the quality are good.

BMP (Windows bitmap format, or OS/2 bitmap format) is a basic device-independent format that can handle up to 24 bits per pixel (i.e., good quality color) and may allow compression. This format is old and well supported.

DXF (data exchange format), originating in AutoCAD as the native format, has expanded over the years and been adopted or supported by several companies. The format includes the capability to add dimensioning information, as well as bitmap and vector graphics and text. However, as AutoCAD develops the format as a native format, they have added features that they have not documented well for the external world.

FILTERS

The bad news in all of this is that there really isn't a generally useful, completely interchangeable file format that will support all of the features of all of the graphics packages. The best you can hope to do is save your work in its native format, and hope that you can translate it when you need to.

Most often, the conversion process can provide an interesting challenge. The tendency these days is for the various graphics programs to call on various *filters* provided by the operating system or by a separate vendor. These filters may be more or less well integrated into the graphics package: often the first

---

[5] There are many examples. One of the more egregious ones was the hiding of Flight Simulator software in the 1997 version of Excel. It says something for the quality control at the company that it went undiscovered until after the program was released, doesn't it?

you know about them is when you see the cheerful greeting "The filter does not support this version of the file," or something equally optimistic.

If this happens, do not give up! Sometimes a path directly from format A to format B is impossible, but a circuitous route with intermediate stops is allowed. For example, the WPG files produced by·CorelDRAW could not be imported into AccuDraw, a program (alas, now extinct) specifically tailored to WordPerfect (DOS version 5x). But a program called Presentations (part of the current Corel family) could handle the import from CorelDRAW, and the export to AccuDraw.

If you have a program that converts one format to another, *save it!* You never know when it might come in handy. We would go further. It may serve you well at some time to go on the Web and find a conversion program. There are many out there, and some are even free!

Why bother with all this effort? First, because you may have an old file you want to reuse rather than redraw; and second, because some programs are just better at some things. And third, everyone has a favorite application. Maybe you cut your teeth on *this* program, or just like the way *that* one does certain operations. Chances are good that a conversion path can be found between any two formats, though the chances are also good that something will get done wrong in the process. We can just hope the defects are few and easily fixed.

So far, we have talked about file conversion. We will return to this vexed topic later in the chapter. Meanwhile, there is another process in the sequence of getting things from the file to the paper, and that is file *translation*. When a word processing program embeds a file into itself, it translates it.[6] It may be that what we described earlier as conversion errors are actually translation errors. (On the other hand, wrong is wrong. The exact location of the error is only important to know if you need to fix a bug in a long conversion path.)

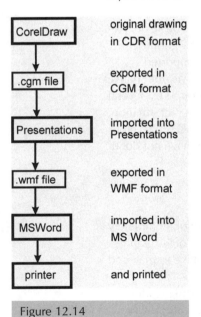

CorelDraw	original drawing in CDR format
.cgm file	exported in CGM format
Presentations	imported into Presentations
.wmf file	exported in WMF format
MSWord	imported into MS Word
printer	and printed

Figure 12.14

The diagram illustrates the point. Assume your collaborator on a paper has given you some graphic image she drew using CorelDRAW. Because she knew your word processor wouldn't read the CDR format, she exported[7] the file in the CGM format. If you are using Microsoft Word, you are out of luck. While MSWord will handle Windows metafiles, it won't work with the standard computer graphics metafile format. Nor will PowerPoint, the usual graphics companion to Word, open CGM files.

All is not lost! Launch Presentations, import the CGM file, and save it as a WMF file. And do not be surprised if some of the text is reformatted along the way as a bonus. We said earlier that the conversion process is not always perfect. The sequence is shown in Figure 12.14.

There is a simpler way, but it does not always leave you with a file you can edit. If you have *any* software that runs under Windows and can handle the input format in question, open the file. Select everything (usually `Ctrl-A`). Copy everything to the Windows clipboard (usually `Ctrl-C`).

---

[6]You can test this idea yourself. Delete everything except an embedded graphic from a word processing file, and try to open the remaining file segment with your graphics package. It probably won't work. On the other hand, you might succeed by double-clicking the image. Often the translation process leaves header information intact, so the computer operating system can backtrack to the creating software.

[7]The term "export" is not used by all software. Some perform the same function under the "save as" option.

Now you can `Paste` or `Paste/Special` the image into your target application (such as your graphics package or Word) and you should be able to edit it.

We have not discussed the Windows clipboard so far. The capability of the clipboard to copy information has been increasing over the years, and these days it is likely worth trying the clipboard first when you need to move some graphic information from one format to another. Sometimes, the process is flawless, but sometimes you will need to fix something. Still, it is definitely worth trying.

The clipboard is a way of (automatically) creating a metafile matched to the objects selected. The file extent that corresponds to the clipboard is WMF (or EMF in the 32-bit world), and files with this extent can be used on other Windows machines, as a way to exchange clipboard contents. The clipboard is indeed a useful and valuable capability.

We will look at the process of conversion again later. In the meantime, we have found it so useful to use intermediate programs that we will give you a summary of the input and output formats offered by four of the popular packages. Note that we have chosen to show what the programs *do not do* for reading and writing.

Some of the capabilities that exist are "unadvertised." For example, the screens of Presentations, which ostensibly has no "import" capability, do not advertise the fact that it can read a PowerPoint file. *But it can.* All you do is set the file type to `All Files` (*.*) and select the PowerPoint file you want.

In the following table, the four columns on the right are headed by the names of the more commonly used programs. The table shows which of the file formats in the leftmost column those programs *cannot* deal with for input or output.

File	Meaning	Adobe Illustrator®		CorelDRAW®		PowerPoint®		Presentations™	
		Input	Output	Input	Output	Input	Output	Input	Output
`.ai`	Adobe Illustrator®					No	No	No	No
`.bmp`	Windows® bitmap					No	No		
`.cdr`	CorelDRAW®		No			No	No		No
`.cgm`	Computer graphics metafile					No	No	No	No
`.dxf`	Data exchange format					No	No		No
`.plt`	Plot (HPGL)	No	No			No	No	No	No
`.ppt`	PowerPoint®	No	No		No				
`.svg`	Scalable Vector Graphics					No	No		No
`.vsd`	Visio®	No	No			No	No	No	No
`.wmf`	Windows® metafile								
`.wpg`	WordPerfect® graphic	No	No			*	No		

The asterisk * shown in the input column for PowerPoint means you need to install a filter from the Microsoft Office Premium disk, as the default installation does not include it. When you have done this, the conversion feature still may not work, however.

As a matter of fact, this list is far from complete. We have tried to cover the most commonly encountered formats and the more popular programs. We have included only file formats that might be useful to exchange data between programs, which means that the TIFF is not included, as it supports only bitmap data. (And anyway, all the programs shown can import and export TIFF files, though not necessarily with equal capability.)

As the table shows, the "big" graphics programs (Illustrator and CorelDRAW) have more capability to import and export file formats than the "little" graphics programs (PowerPoint and Presentations). That makes them more useful for importing graphics as part of a conversion sequence.

While PowerPoint is probably the most used of the graphics packages, it is not capable of directly opening files from many other formats, or of saving to many others.[8] Perhaps when you are the biggest, you don't have to acknowledge the competition. However, other programs can import directly from PPT files.

Sometimes, merely going through the sequence—dropping some information from the clipboard into a program such as PowerPoint, perhaps ungrouping the information, and then reselecting everything and dropping it into a different application—can have the effect of avoiding a bug that you get with a direct transfer. What this probably means is that an import translation somewhere has the bug, and another somewhere else does not. This kind of trick, using the clipboard to get information in and out of an intermediate application, is probably worth a try if you are having trouble getting a graph from one set of software to work properly in another.

Another thing to consider is the use of an application from another field. Some formats (not so far discussed) are not meant for the exchange of data between programs, but are sometimes used for this anyway. An example is the PLT format, normally used to output information to a plotter (or a printer) that uses HPGL, or Hewlett–Packard Graphics Language. Some data acquisition software, LabWindows®, for example, will print graphs on HPGL plotters or will save the same data as a file.

If you have such a file and are interested in modifying it before embedding the information in a report or paper, you must first convert it to a format that your graphics program can use. In the case of HPGL files, it is fortunate that there are self-contained programs to do this. (GRAPHCNV that was bundled with WordPerfect 5.1+ is an example.) Figure 12.15 is an example.

Fortunately, after the HPGL-to-WPG conversion process, the text is available as text, and the rectangular boxes as rectangular boxes, so editing the image for clear printing is straightforward.

In fact, you could fake the GUI part of this figure in your graphics program, but the data from the vibration would not look genuine without very considerable effort.

---

[8]For example, although the program will indeed save files as TIFF format, it will not (as far as we can tell) allow you to control the resolution of the image, which seemingly defaults to something very coarse.

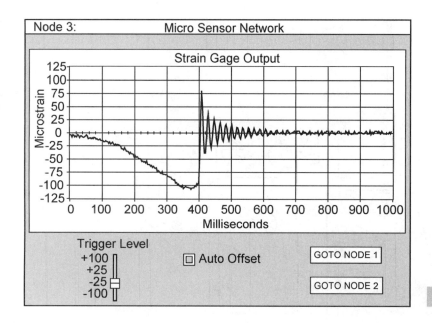

Figure 12.15

We mentioned earlier that after the conversion of the file from HPGL to WPG format, the text was still available as text. This is not always so when files are converted or exported. Sometimes you will be asked whether you want the text exported as text or as curves. Sometimes you will not be offered a choice—you will get curves.

If you need to edit an image, and the text is in the form of curves, you might as well not bother with the text. This is what happened a few years ago when one of us drew the plans of a house, for bureaucratic approval of a proposed remodel. The original drawing was done using one of those inexpensive house-plan programs, because they can save an enormous amount of work when it comes to detail. (They know that a wall is two lines, for example, and they know how to draw a door. They can also produce elevation drawings from data entered as the plan is drawn.) The program saved its results in a DXF file, but the text was stored as curves. We needed to end up with a file for a plotter (for a large-scale drawing), and the program only knew about printers. The plotter, and the program to use it, belonged to a friend.

In the end, the process was as follows. First, the existing house was drawn in the house-plan program, with no text. Next, the DXF file was imported into CorelDRAW, where text was added (sometimes at odd angles—this was an odd-shaped house), and the remodel changes were made. The plumbing and wiring diagrams were made at this point, starting with portions of the overall house plan. Finally, the drawings were exported (again in DXF format) to a floppy disk and taken to our friend. He imported them into AutoCAD, where they were scaled for the plotter. The final output was to quarter-inch-to-a-foot scale, and in nonfading ink (unlike the original house blueprints). A section of the product is shown in Figure 12.16.

This sequence might sound like a lot of work, but actually, once we had figured out what the best process was, it became quite simple. When the city asked us to modify one of the six or so pages in order to show how a particular construction problem was to be solved, the change was made in CorelDRAW, a

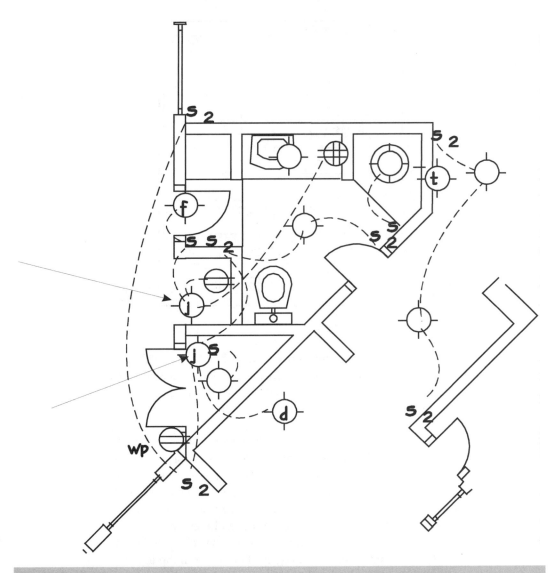

Figure 12.16

new floppy disk produced, and the new page printed by AutoCAD without diffi-culty. The remodel was approved.

SUMMARY

File conversion is needed when collaborators don't use the same software. It may be needed to make graphics in a form that can be integrated into your word processing software. While the capability to do file conversions is gradually getting better, there are still plenty of bugs.

■ When you need to convert a file format, first try using the clipboard as a mechanism.

- If the amount of fix-up work that you are faced with after that looks daunting, try reading the file directly into the target application.

- If it still looks like too much work to fix up, try exploring a route that involves an intermediate format, or an intermediate application. There will be many possible combinations, and one of them may be just what you want.

- If a conversion is forced on you because your publisher demands a particular file format for graphics, be sure to read the converted file back into another program or two. Your publisher's software may be quite capable of adding bugs, so it's good to remove all the ones you know about.[9]

EXERCISES

12.1. What are the two principal kinds of graphic files, and what are the principal differences in how the information is stored?

12.2. Broadly speaking, which kind of file should you use to store your graphics in, if you can't use your software's native format because you want to exchange it with a collaborator? Why?

12.3. Draw (or find) a single-variable graph in PowerPoint. Convert it into the format of another graphics program that you own. Open the file. Count the differences from the original. Try some other way to access the information in the file.

12.4. Repeat Exercise 12.3 for another pair of programs.

[9]This book was produced with graphics from many applications. All were converted to TIFF at 600 dpi for the publisher.

Style matters

13

Writers of all kinds have their own style; so do composers and players of music, and so do architects. And, of course, so do painters and sculptors, the people who most of us think of when we hear the word "artist." By producing our own technical graphics we are doing the work that used to be done by the artists in the company graphics department. We are surely free to develop a style of our own, too.

The main purpose of the kind of graphics we have examined in this book is to inform and persuade, and so it is paramount to present the information accurately and clearly. But our success at influencing and persuading can be affected by our style. We aim to tell the truth, of course, but in a way that will get the reader's attention and be convincing. Ultimately, whether or not we get our message across is something that happens in the eye and the mind of the reader. Since our style can interest the reader, or confuse her, or simply bore her, it is up to us to arrive at a style that will improve our chances.

When it comes to the written word, style is considered so important that there are many books on the subject. Some, like the famous manual by Kate L. Turabian (*A Manual for Writers of Term Papers, Theses, and Dissertations*), go into the overall organization of a technical document, and then get into detail on such seeming trivia as footnotes. Others, like the *Chicago Manual of Style*, go into a great deal of detail and will get into spelling and punctuation as well. The US Government has its own *Government Printing Office Style Manual*.

Such guides are general. Some technical or scientific organizations take it upon themselves to define a style that their particular authors are required to adopt. The American Psychological Association, for example, has a very detailed Publication Manual. (It is so good that writers in other fields often use it for guidance with citations.)

These style manuals are minutely prescriptive[1] on matters of detail. They will tell you that there should be a period after the abbreviation for "editors" and another between the names and the date in a journal citation, so that "Blow, T., & Mann, D. (Eds.). (1999). Wind and Weather . . ." is the required punctuation for a journal issue citation.

---

[1] Meaning that they tell you what you *should* do, not what people *actually* do. We point this out because most dictionaries, unlike style books, are actually *descriptive*.

Then there are the style guides that tell you the right way to write certain words or phrases that might cause difficulty. The *New York Times Manual of Style and Usage* and the *Pocket Style Book* of *The Economist* are in this category.

At a yet higher level, there are books that will help you with style in the more common meaning of the word: the clarity and effectiveness, even the beauty of the written expression. Among the more well known of these are the updated works of the Fowler brothers, the excellent *Modern American Usage* of Wilson Follet, and the many books and essays of Jacques Barzun.

The list is, in fact, endless, as new books are coming out all the time. Some of these style books are very readable and entertaining (try Kingsley Amis' *The King's English*); others are simply authoritarian statements of formula.[2] Style books cover a spectrum of guidance. Some even deal with graphics. Usually there is not much more than a passing reference. Sometimes as much as a chapter.

Even if a style manual prescribes (or constrains) many of the options for a graphic, you will probably find that you still have some latitude. This is where your own individual style is allowed to show. If you have been following the advice of this book, you will be producing clear graphics whose meaning is unambiguous.

DEVELOPING YOUR OWN STYLE

To save time, you may once in a while decide that rather than starting from scratch, your next graph starts with editing a previous graphic, or at least borrows from one. Once you do this, your graphics will start to take on a certain sameness. You are establishing your own style.

It is good practice, and we strongly urge you to do this, to save copies of some of your best-looking graphics in a special file on your computer—or on your memory stick, if you work in more than one place. This file of graphics provides examples for you to edit (and remember, it is usually quicker to edit a graph than to start one from scratch) and will play a role in getting consistency in your work.

Within reason, your individuality can be allowed to manifest itself. There is a tension between, on the one hand, the desire for uniformity on the part of the organization publishing your work, and on the other hand, your desire to establish a style of your own. But if the published work will bear your name, you surely have the right to be satisfied with the result!

Be choosy. Study the pages of, for example, *The Economist*. You will see many good technical graphics and a very consistent style. The same statement can be made with respect to *Scientific American*, *New Scientist*, and others. But the styles are not the same. What are the things that make them consistent? What makes them different?

Study the Transactions of your professional institute; see if you can discern a consistent graphics style. You may find it difficult.

In terms of the overall "look," let us begin by using *The Economist* as representative of good quality. The newspaper tends to size its graphics very small: typically to fit within a single column of the three-column format used for body text.

---

[2] And you'd better read carefully. A citation valid in one journal may not suit another. While most journals expect you to cite the authors' names, the publication dates, the journal name, the page numbers and so on, some journals (particularly in the world of physics) may be offended if you include the *title* of the article! What does this imply about physicists?

The graphs are usually separated from the text by having a pale blue background color. Against this color, the text stands out in black, and the lines of the graph in color, usually darker blue with a black outline. The magnitude of the vertical scale is indicated by white lines that cross the graph area behind the graph lines, and the line showing the vertical axis itself is generally omitted. The caption is usually minimal (see Fig. 13.1).

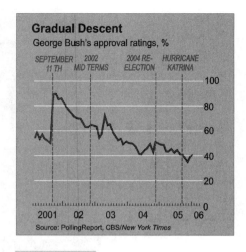

Figure 13-1 See insert for color representation of this figure.

Now open the pages of those Transactions. The chances are good that within a few pages you can find graphs with spidery lines that would disappear if copied once, freeway-width lines, unlabeled axes, text too small to read, and text set in so many fonts it looks like a ransom note. You can find graphs with dotted lines across the graph area, with no lines across the graph area, and a host of other inconsistencies. Sometimes you can find all these varieties in the same paper!

Just think, by developing your very own style, you can add to the cacophony!

But if you are careful, you can instead add a voice of harmony and unity. We have already hinted at some of the ways. Your figures will be consistently similar to one another, they will be in concert with the layout of the text, and they will be as small as possible. Therefore, if the paper is $8\frac{1}{2}$ inches wide, or A4 width (210 mm), and the format is a single column, you can produce a rather large figure without it seeming oversized. While the normal guideline for the shape of a graphic is that the height should be about 3/4 of the width, you could relax this relationship somewhat with the aim of saving space and avoiding a large amount of white space at the sides of the figure.

If the figures you are using really do not need so much space as to occupy the full width, you could try the effect of placing two figures side by side, or (if the style rules of the journal or association allowed it), you could place the caption to the side of the figure, instead of below it.

In fact, the optimum sizing of figures for a single-column layout is a little challenging. This book has some examples that result in largish amounts of white space. This occurred because, for the most part, the figure size had to be fixed independently of the paper size in order to make a point. Look in your textbooks: old and new books alike have white space at the sides of the figures.

We found, in a recent IEEE journal, an example of what *not* to do to solve the problem. Figure 13.2 was actually used in a two-column layout, but it was spread across the center of the page and occupied more than the width of a column.

As you can see, the figures have been "squished" vertically or stretched horizontally. In the original, the rectangles are 4 inches wide and 1.1 inches high. It may be that the graphics were produced by a program that sized its output for a single column, but the author wanted to expand the presentation to make some detail visible. Or, since the original graphic seems to have come from a dot-matrix printer, but the body text is from a laser printer, perhaps there

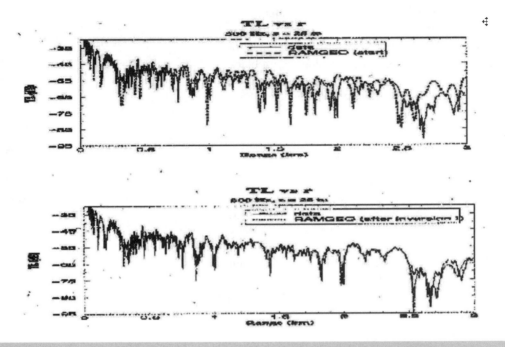

Figure 13.2

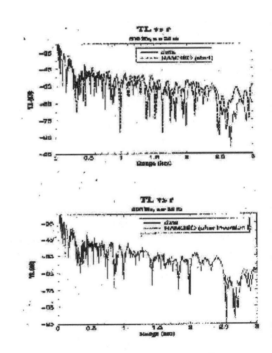

Figure 13.3

was a printer compatibility problem. At any rate, a doubling of the height : width ratio makes the text look more normal, as shown in Figure 13.3.

Some graphics packages will not allow text to be stretched or squished like this, and perhaps that is appropriate. The result of distorting the text is remarkably unappealing. Distort the graph, by all means, but leave the text proportions alone!

SOME (GENERAL) ALTERNATIVES

Let's get down to brass tacks. Let's look at how you might develop your own style. Figure 13.4 is a graph produced by a spreadsheet. As it happens, the data are for the resistance of a thermistor as a function of temperature, but that makes no difference for our present purposes.

This is what appears in the spreadsheet. You could put the same thing into a document simply by copying and pasting. And everybody reading your paper would know that you used a spreadsheet to generate the graphs, *and that you gave it no further thought.* Perhaps without realizing why, your reader would recognize the *style*: the combination of a gray backdrop to the graph area, horizontal lines across the graph, but no vertical ones, a meaningless identifying key for the single graph presented, and a color for the graph line and its data. The rectangle around the outside, which also adds no information, may not register.

Suppose you spent a minute or two changing the "look" in your spreadsheet program or your graphics package.[3] You could make a few improvements, and get Figure 13.5. Here, the spreadsheet style still comes through, and though the unnecessary identification of the curve ("res" over on the right in Fig. 13.4) has been removed, the general appearance is much the same.

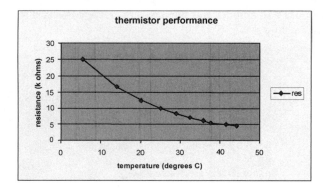

Figure 13.4

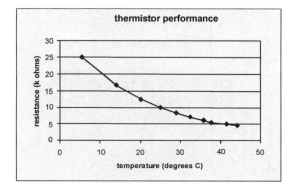

Figure 13.5

[3]You could do some of the things shown here inside your spreadsheet program, and you *should,* saving the style for future use. But when it comes to things such as separating the axes, you could not get the required result without using your graphics package.

Let's do a little more, as shown in Figure 13.6. Here, we've thickened up the graph line and the axis lines, and removed the horizontal lines across the face of the graph. The thing is losing its original identity. Let's go further, as shown in Figure 13.7.

This graph no longer looks like a spreadsheet output. All we did was separate the two axes, remove the caption from across the top, and make the "fill" on the little diamonds white instead of black. For most purposes, this graph would be considered publication quality. Try doing some graphs this way.

The idea of separating the axes may grow on you. The joined axes we are all so familiar with came from the days of graph paper, and the desire not to waste any of that valuable resource, so they have been with us a long time. Maybe it's time for a change.

Of course, you could go the other way, and *add* lines instead of removing them. Most people would consider extreme a graph like the one shown in Figure 13.8. Here, we have added a paler background than the spreadsheet already provides for the graph area, and made two sets of dotted lines across the graph.

The added dotted line feature is something we could also have done to the version with the separated axes. We could then have removed the solid lines that indicate the location of the axes, and the little tick marks, and arrived Figure 13.9.

Maybe, maybe not. Without the axis lines, it seems to be missing something. Mind you, it isn't really missing anything. All the *information* of the original is still there. Perhaps we should start a trend, and see if we can get accustomed to looking at graphs with no axis lines.

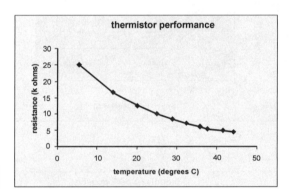

Figure 13.6

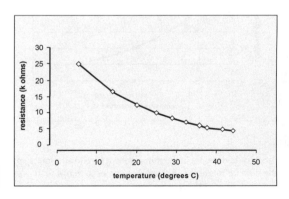

Figure 13.7

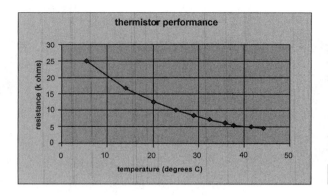

Figure 13.8

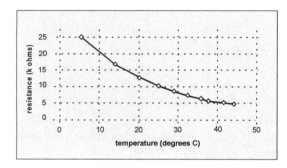

Figure 13.9

That isn't so outrageous, really. Compare the graph in Figure 13.9 to the graph in Figure 13.1 from *The Economist*. There was no vertical line for an axis on that graph, but there was a line for the time axis. Since *The Economist* had a colored background, they could use white for the lines. But it is clear that they have abandoned the notion that a graph must have a vertical line for an axis.

Another option for indicating some of the detail across the graph area without creating the impression of a Scottish kilt is to add a small dot at the locations in the graph area where axis lines would intersect, if they were there. To our eye, it looks rather like the graph (Fig. 13.10) has a communicable disease. You can also do this with lots more dots, only done fainter. Then it just looks like the ink smudged.

Of course, your choices are many, and the amount of work may depend on where your data came from and what you choose to do with the software at your disposal. If you invest in a graphics package that is intended for science and engineering use (and some of these are now available at very reasonable prices), you can go from style to style very simply. Consider Figure 13.11.

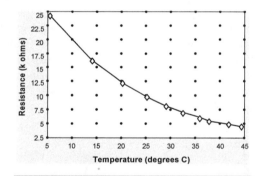

Figure 13.10

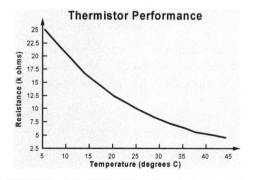

Figure 13.11

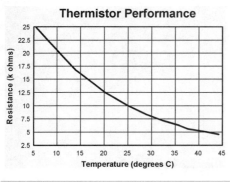

Figure 13.12

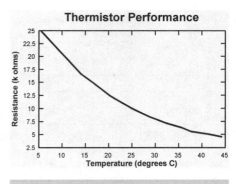

Figure 13.13

With that image on the screen of one of the low-price graphics packages, the style shown in Figure 13.12 was obtained with a single mouse click.

And the style in Figure 13.13 was obtained with but one mouse click, too.

The choice is up to you. Provided the graph is clear and copyable, and unambiguous in meaning, it would be acceptable to us. You may need to adapt to suit a particular publisher, but for the most part you can make your own decisions about the details, as shown in Figures 13.14 and 13.15.

You might think that the conventional graph with just two lines for the axes looks a little "exposed" on the top and the right, especially when it is inserted into a two-column paper:

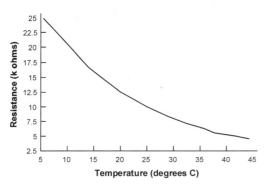

because somehow it seems undefined and vulnerable on those two sides. One solution might be to adopt the solid color background of *The Economist.*

Figure 13.14

An alternative might be to reverse the default shading from your spreadsheet so as to provide a sort of border:

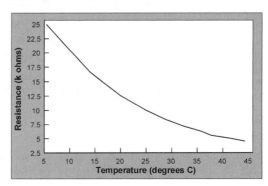

This is not bad-looking, and it has the advantage that you can save the style in your spreadsheet program, so it will not be necessary to do any editing in your graphics software.

Figure 13.15

YOUR OPTIONS

Once you have decided what you like, it might be a good idea to write down your various choices, so you can maintain consistency. Here are the things you will be noting for graphs:

Graphs		
Surrounding box	Yes/no	Line weight
Axes	None/one/two	Line weight
Axes joined	Yes/no	
Lines across graph area	Yes/no	Style, line weight
Unlabeled tick marks	Yes/no	Line weight
Ticks	Through/outside/inside axes	Line weight
Graph line		Line weight
Callouts	Yes/no	
Background color	Yes/no	

Text		
Font	Axis labels	Size
	Numbers	Size
Key	Yes/no	

The question of color has been mentioned before. If you can achieve your objective without the use of color, then do so. Color (at least at the time of writing this) is not readily copied and adds to the difficulty of producing high-quality and uniform documents. Shades of gray might be less visually attractive but may well archive better.

Not mentioned in this set of things to consider is the use of "3-D" in graphics. We have already dismissed this as a nonsensical waste of ink in earlier chapters, but it will bear repeating. Do not add shading to create a gratuitous impression of depth. Do not add a dimension that has no meaning and no scale! That is not a style you should want to adopt. Your graphs do not need vertical stabilizers any more than a 1959 *Coupe de Ville* did.

ADAPTING SOMEBODY ELSE'S STYLE

Consistency partly makes up for lack of color. As you will see in the next chapter (Case Studies), a set of graphs drawn with a uniform style can give a very pleasing appearance even without color. In contrast, graphs that were not unified in style after they were collected by the lead author from the co-authors give the impression of disarray and disorganization.

This comment applies to graphics other than graphs. For example, if you need to use circuit diagrams, you could insert something like Figure 13.16 into your document. This image says everything that has to be said about the circuit: what the parts are, what their designations are, what the pin connections are, and even what the voltages are. It also very clearly comes from a program that, if it isn't PSpice, is very much like it.

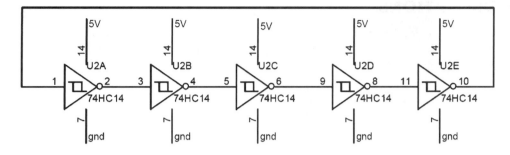

Figure 13.16

So the style is the one decided on to make PSpice work well. For the purposes of a publication, it is workable, but not really as good as it should be. You could try fiddling around with the image in a graphics program, as shown in Figure 13.17.

This is better, but it still looks machine generated.

And since the image is a bitmap, you can't really reuse bits of it in other diagrams. If you put a diagram together using parts from your ready-made catalog of parts, you can assemble something like Figure 13.18 in a couple of minutes. Now, it's not greatly different from the previous diagram, but the fact that the parts and the interconnections use different line weights somehow makes it more attractive.

Generally, with circuit diagrams, you don't have the freedom to invent your own style completely.

A colleague of ours decided that the symbol commonly used for resistor, ⎯⎯\/\/\/\⎯⎯ had too many wiggles, and he invented his own version, ⎯⎯\/⎯⎯ with less than half the usual number. This symbol even appeared

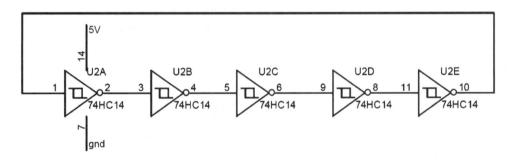

Figure 13.17

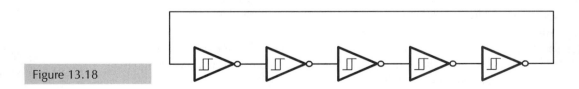

Figure 13.18

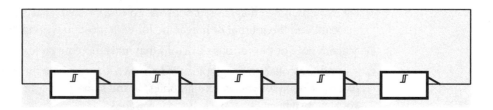

Figure 13.19

throughout a book that he wrote on circuit analysis. However, anyone familiar with the ANSI/IEEE standard symbols will recognize that the "new" symbol is actually the symbol for a relay coil! If we are to avoid confusion, we must remember that there is a limit to the number of ways we can put squiggles on paper, and that the meaning of many of the squiggles has been defined by our ancestors.

You are constrained by the rules of where you are going to publish, and by the need to convey the information you want to convey. Thus, if you are going to publish in Europe, you should check whether there are any differences in the way things are represented there as compared to the United States. For example, the IEC version of the previous figure would be as shown in Figure 13.19.

But within these various sets of rules there is some flexibility, and you can develop you own "look" to a surprising extent.

SOME ADVICE

If you feel unsure about what looks good, study your journals and (better yet) magazines and books. The peer-reviewed journals are developing some unfortunate tendencies, in the hope of saving a few bucks. But the monthly magazines have the need to maintain circulation, and the editors know that appearance is an important part of creating an impression of value. Whatever your field of endeavor, it's a fair bet that the glossy magazines have some good graphics, with a consistent style.

Now that you know what to look for, "borrow" as much of the style as you feel you want to. You shouldn't think of this as plagiarism; after all, you're not stealing someone's technical ideas, you are flattering them by imitating their style. And you're not passing your stuff off as having been done by Picasso, even if there is a slight cubist look to it. Naturally, you shouldn't borrow the parts of a style that don't make sense—for example, the 3-D chart or the shadow behind a histogram. Extract what works for you.

SUMMARY

As always, the basic rule is to strive for clarity and avoid ambiguity. Within the constraints of where you publish, you can still establish your own consistent style.

- For your graphs, consistently use (or don't) a surrounding box, put the tick marks inside the box (or outside), and so on.
- For your text, make your choice of font consistent, and its scale relationship to the rest of the graph.

- For various other diagrams, choose a consistent style that suits the material and the journal or magazine for which you're writing.
- Make a note of the various decisions that make up your style.
- Assemble a collection of graphics of your own that can form the starting place for many of your future graphics, at the same time establishing your own style.

EXERCISES

13.1 What are the principal differences between the styles of the two graphs in Figures 13.20 and 13.21?

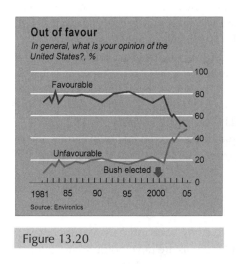

Figure 13.20

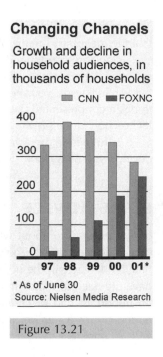

Figure 13.21

13.2 How many examples of wasted ink can you find in Figure 13.22? What is missing?

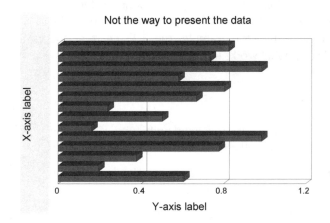

Figure 13-22   See insert for color representation of this figure.

13.3 What is wrong with the graph in Figure 13.23?

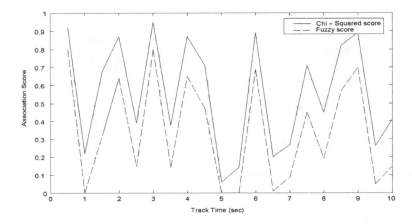

Figure 13.23

13.4 The two graphs in Figures 13.24 and 13.25 appeared in the same paper. What would you change?

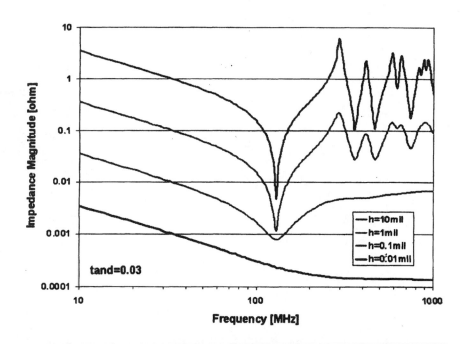

Figure 13.24

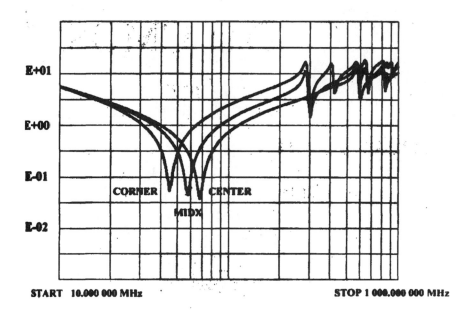

Figure 13.25

Case studies

14

$\mathbf{B}$efore we start this chapter, we need to make a disclaimer. We are going to present some examples of graphics that would never have made it. Some of the material really needs a lot of work, even though it came from engineers and scientists who thought the stuff was OK. Since some of our colleagues (unwittingly) produced some of the examples, and since we do not want to find that strange things have happened to our offices, or that there is suddenly a large dog in one of our cars, as a result of publishing this book, let us just assume that all the material you are about to see is anonymous, and in any case was never meant for publication. Even if we all know that isn't true.

CASE STUDY NUMBER 1: VOLTAGE REGULATOR

Figure 14.1 shows an extract from an IEEE paper.[1] The topic is a dynamic voltage regulator, and the figure shows the action of such a device in controlling a distorted voltage waveform on a power system. Note that the figure is small enough to appear in a two-column format, as it did in the paper. This is a very compact presentation of a lot of information.

We might have done things a little differently, of course. It would have been more complete to make the caption big enough to tell the reader what the four parts of the graph represent. As it is, the descriptions are all in the text. But this is a minor matter, and surely one that is for the authors to decide as they wish.

Look at what *is* shown in the graphs. In two of them, there are three curves, each showing the distorted waveform on each of three phases of the power supply. Both of the first two graphs have the same vertical scale, from −2 to +2 per unit. Nevertheless, the authors chose to show a scale for each graph: their choice.

[1]A. Ghosh and G. Ledwich, Structures and control of a dynamic voltage regulator, In *Proceedings of IEEE PES 2001 Winter Meeting*, January 28 to February 1, 2001, Volume 3, pages 1027–1032.

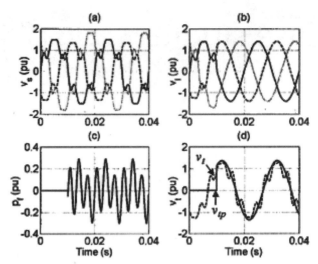

Figure 14.1

Fig. 4 Ideal DVR performance with distorted source

All the graphs have the same horizontal axis, from 0 to 40 ms. The top two graphs show only the numbers 0, 0.02, and 0.04, while the bottom graphs show also the fact that the dimension is time, and the units are seconds. This approach is a little inconsistent, but it seems to work here because the scale labels are all the same. (If any of the graphs had shown a different set of numbers, the simplifications here would not have worked, and the result would have been confusion in the mind of the reader.

All in all, this is a very effective use of graphics.

CASE STUDY NUMBER 2: BASEBALL PERFORMANCES

In his book *Full House: The Spread of Excellence From Plato to Darwin* (Harmony Books, New York, 1996) Stephen Jay Gould, a biologist, discussed baseball, one of his passions. He examined the reasons why a season batting average of 0.400 has disappeared from the scene (the last was Ted Williams of the Boston Red Sox in 1941).

He is making the case, in the chapter and in the book, that competition can give the effect of decreasing variance. He shows that the season batting average has stayed constant over the years at about 0.260, but the variance has decreased, so that exceptional numbers such as Williams's are now rarer.

There is a limit to performance, he reasoned, that was imposed by the strength of our muscles and bones, and our vision and coordination. But how to make this clear to the reader? He presented the graph in Figure 14.2.

You may or may not agree with Gould about the reasons for the demise of the 0.400 season. Some baseball fans point to changes in the rules, and changes in the strategies adopted by batters, pitchers, and coaches over the years as explanations. They may be right: after all, the object is to win the game, not to get a good BA, so there is no particular reason why BA numbers should not decline.

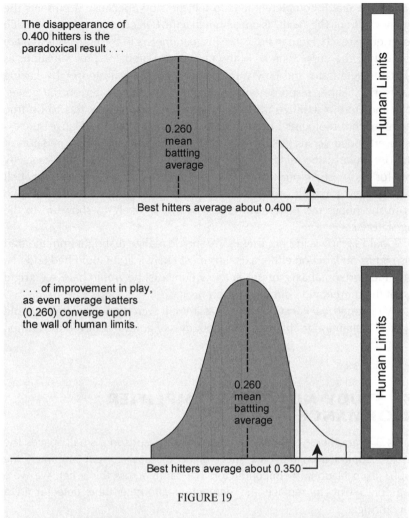

The disappearance of 0.400 hitters is the paradoxical result . . .

0.260
mean
battting
average

Human Limits

Best hitters average about 0.400 ⎤

. . . of improvement in play, as even average batters (0.260) converge upon the wall of human limits.

0.260
mean
battting
average

Human Limits

Best hitters average about 0.350 ⎤

FIGURE 19

Four hundred hitting disappears as play improves and the entire bell curve moves closer to the right wall of human limits while variation declines. Upper chart: early twentieth-century baseball. Lower chart: current baseball.

Figure 14.2

But you would certainly be right in thinking there was something not quite right with the graph. The problem starts with the fact that the horizontal axis is not labeled. It seems to be a scale of batting average, with perhaps 0.170 on the left, and 0.400 near the right of the bell curve. But in this case, the gap between the curve and the "wall" should be bigger on the lower chart, not smaller, on the assumption that the wall has not moved.

One might make an analogy with evolution: after millennia of natural selection by evolution, prey and predator are closely matched in performance, and their populations become stable. After decades of careful selection by coaches and managers (a process that operates much more rapidly than natural selection), pitchers and batters are closely matched. That may indeed account for the disappearance of the 0.400 season.

But the graph completely fails to tell the story, because it presents the wrong information. The "wall" is not a limit that the bell curve can only approach but never quite reach. Human limits are just *parameters* in the equations that are represented by the curve. Surely, in the case of the predator/prey situation, as the predator gets faster the prey must respond by being faster, or by having more offspring. Either response will lead to success in passing on the genes, and will therefore restabilize the populations. We don't know this to be true (since we are not biologists), but it seems to us likely that the relative success of improved speed versus greater fecundity will be evident in the predator-to-prey ratio. Similarly, the relative success of pitchers changing strategy (e.g., by striving for more strike-outs rather than getting the batter to put the ball in play, and by pitching around strong hitters after they have thrown two strikes in the hopes the batter will take a swing) is likely to show up in the batting statistics.

Gould's graph is just not useful. We should realize that right from the start because it has no label on either axis. It is most likely that if Gould had added a label for the horizontal axis, or put in a few numbers, he would have seen right away that the presentation did not support his case.

We learned back in Chapter 1 that if there were two axes, both should be labeled. Inattention to this rule does nobody any good—not even the author.

CASE STUDY NUMBER 3: AMPLIFIER PERFORMANCE

Figure 14.3 is another excellent example of many graphs in a small space. The graphs are copied from the data sheet of an ultralow-current amplifier, the LMC6001 from National Semiconductor. We came across this when we were looking for a way to replace the defunct electrometer-tube detector in a megohm bridge.

The page is typical of the data sheets from that company (and representative of others) in that there are many graphs on a page, and the variation of many parameters can be seen. Before it became so convenient to make such information available on the Web, National used to publish books, about $6\frac{1}{2}$ inches wide by $8\frac{1}{2}$ high, for the data sheets and application notes. The graphs of Figure 14.3 are reproduced as if they were printed on $8\frac{1}{2}$ by 11 paper. The figure is cropped to fit the smaller print area here.

Let's look at the graphs. The first looks linear, at first glance. Closer examination shows that the vertical axis is logarithmic, however. By arranging the scale to be marked only at the factors of 10 (a reasonable thing, when there are so many decades to cover), the tick marks are evenly spaced. The 6th, 7th, 8th, and 9th graph use the same approach.

Evidently the graphics folk at National did not think there would be enough information to make it obvious that the scales were logarithmic if they did not add more lines on graphs 5 and 9, however.

Note the use of callouts within the graph area on graphs 3, 4, and 6. On graph 4 it was necessary to add leaders.

While there are some differences between the graphs in the way they are drawn, altogether they do retain a unity of style and a clarity of expression.

Typical Performance Characteristics $V_S = \pm 7.5V$, $T_A = 25°C$, unless otherwis

Input Current vs Temperature

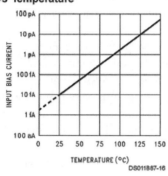

DS011887-16

Input Current vs V_{CM} $V_S = \pm 5V$

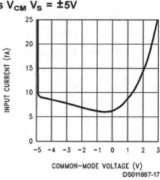

DS011887-17

Supply Curren vs Supply Vol

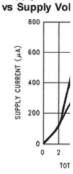

Input Voltage vs Output Voltage

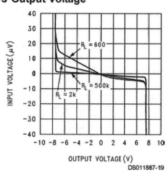

DS011887-19

Common Mode Rejection Ratio vs Frequency

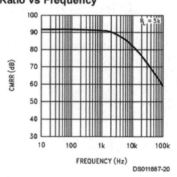

DS011887-20

Power Supply Ratio vs Freq

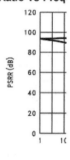

Input Voltage Noise vs Frequency

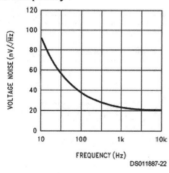

DS011887-22

Noise Figure vs Source Resistance

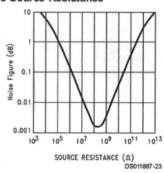

DS011887-23

Output Chara Sourcing Cur

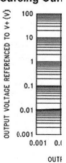

Figure 14.3

CASE STUDY NUMBER 4: RADIO COMMUNICATION IN ICE

Consider the graphs presented in Figures 14.4 and 14.5. These are the output from simulation software, rather than experiments. The figures (at this size) were intended for one column of a two-column format. The idea is to examine the possible use of radio in ice.

Because the presentation is rather small at this scale, Figure 14.6 presents the graph in Figure 14.4 again, blown up a bunch in Figure 14.6.

Study Figure 14.6 and Figures 14.4 and 14.5 for a while.

We already fixed one thing. When the graphs were first embedded in the file for this chapter, the horizontal size of the figures was the same, but the vertical size of Figure 14.5 was about 10% taller than Figure 14.4, even though the numbers on the axes were identical.

How many other things can you see that have to be fixed before submitting for publication?

Here is our list. It may not be complete.

1. There is *too much clutter* because of the lines all across the graph area, the top, and the right side.

2. It is *hard to decipher* which line belongs to which frequency/antenna combination.

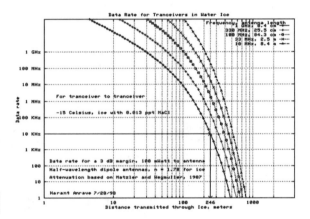

Figure 14.4

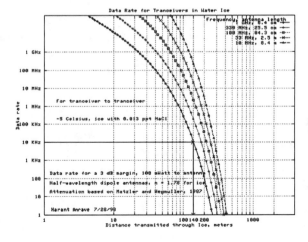

Figure 14.5

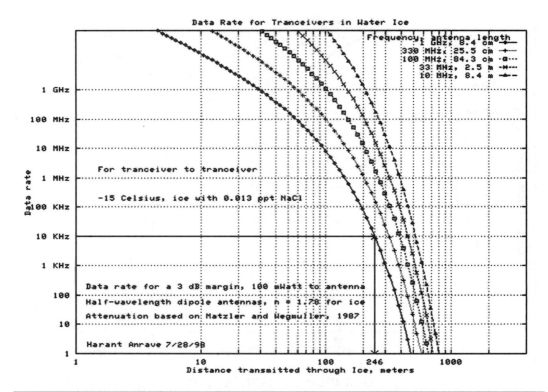

Figure 14.6

3. The *writing is not legible* at the two-column scale, yet the information in the curves does not warrant taking up any more of the page.

4. The key, which should not be in the graph area anyway, contains *redundant information*. For each frequency, there is only one antenna size, so one piece of information is enough to specify a line on the graph.

5. The *units* (Hz) for the vertical axis *should not be repeated*. As it is, the units appear in most (but not all) labels.

6. There is a *citation in the graph area*. It should be in the text.

7. There is a *caption as part of the top of the graph.*

8. Some kind of *description* of the graph appears *in several places.*

9. It is *hard to see a difference* between the first and second graph.

10. The *author's name* and a date appear *in the graph area.*

11. There is a *repeated typo*—tranceiver.

And finally,

12. *More than half the data are meaningless.* The figures purport to show the data rate that can be used over various distances in frozen salt water, for various conditions. Presumably the details, such as bit error rate or received signal level are in the text. However, the data rate seems to exceed the carrier frequency for a good deal of the space covered by these graphs, and that is nonsense! In fact, discussion with the author revealed that above a data rate of 10 kHz the graph can be discarded.

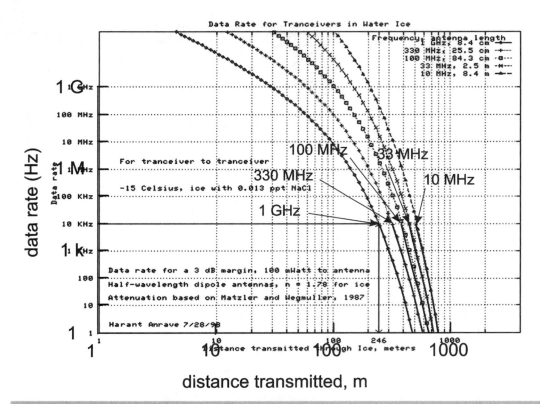

Figure 14.7

If we fix some of these problems, by tracing over the graph (Figure 14.7), and get rid of the original, we end up with Figure 14.8, which is all very well, as far as it goes, but it highlights another problem. Most of the graph area actually contains no information. All the data are for distances above 100 m, so the decades from 1 to 10 m and from 10 to 100 m might as well not be there. Similarly, and we knew this earlier, none of the data above 10 kHz mean anything, so there might as well be no axes corresponding to this region.

Another edit yields Figure 14.9.

Here, we have put some more tick marks on the horizontal axis to reinforce the idea that it is a logarithmic scale.

The labels for the lines look pretty bad stuck out like some kind of floral arrangement, so let's do what we can to make them better (Figure 14.10).

This is rather overwhelmed by the line identification information on the right. If we enlarge the graph as shown in Figure 14.11, but not the text, it may look better.

Good enough. Suppose we produce a version of Figure 14.5 that has been through the same process. It would look something like Figure 14.12.

Now, the difference between this graph and the previous one is obvious, right? The curves have moved to the left. All of them, by the same amount. There is not much more information in the five lines than in any one of them, since the relationship between the five curves has not changed appreciably.

In fact, discussion with the author revealed that this whole second set of curves was included in the paper to show the effect of temperature. We suggested

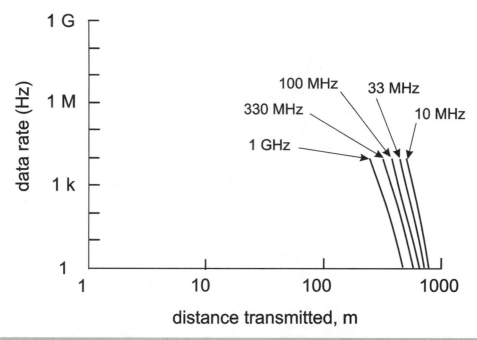

Figure 14.8

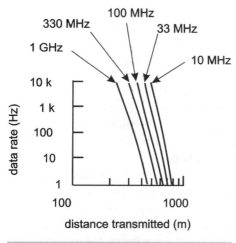

Figure 14.9

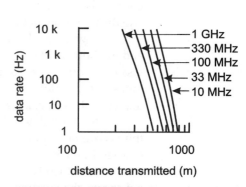

Figure 14.10

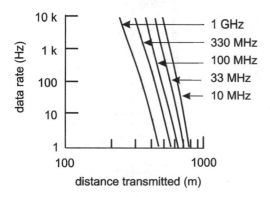

Figure 14.11

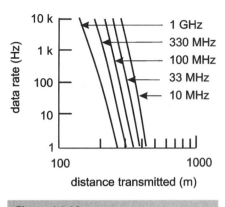

Figure 14.12

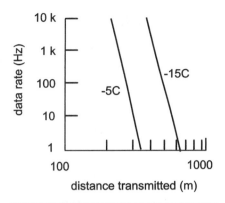

Figure 14.13

an alternative: show only two curves, corresponding to the same frequency but at two different temperatures. Why not select the 100-MHz line, since this is central.

The result looks like Figure 14.13.

We might be tempted to try to derive a curve of distance against temperature, for a given data rate, but that would be a risky undertaking with only two data points. Ice is known to do odd things at around $-4\,°C$, so better stick with the original simulation.

So now, instead of putting the two graphs in Figures 14.4 and 14.5 into the paper, we're inserting the graphs in Figures 14.12 and 14.13. Much clearer. Not only are the graphs more legible, their functions are more obvious. The first graph shows the effect of frequency, the second the effect of temperature. Of course, the caption will be required to state the temperature of the first graph, and the frequency of the second.

Now, think about what these graphs show. The vertical axis is data rate, and the horizontal one is distance. If we are designing a radio communication system for use in ice, we are probably free to specify both of these parameters. Indeed, if we intend to use a series of repeaters to get information from under the ice to the surface, we will have to specify both.

But the reality is that the designer will probably select a data rate first, as this is governed by other considerations, such as the total amount of information expected, and the time allowed to recover it. (This is why the original graphs were marked so as to indicate the distance corresponding to a 10-kHz data rate.) Given a data rate, the question would then be: What is the maximum distance that I can

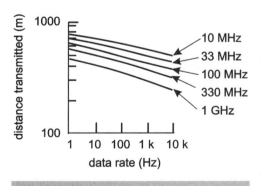

Figure 14.14

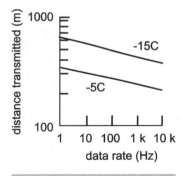

Figure 14.15

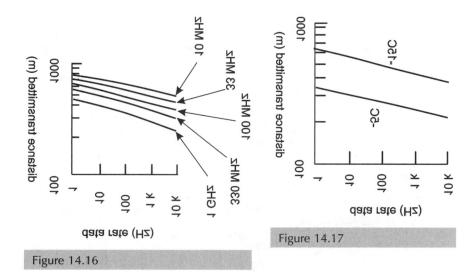

Figure 14.16

Figure 14.17

allow between repeaters? Graphs like Figures 14.14 and 14.15, with the dependent and independent axes switched from what we have seen so far, answer that question better than the previous two, although the information in them is identical.

To derive these graphs, from the earlier ones, just rotate counterclockwise and flip left to right. Be careful to get all the text out of the way, however, or you may end up with Figures 14.16 and 14.17.

These are not so convincing if you put them in a paper. On the other hand, these graphs at least confirm that the rotation and flipping of the original graphs did not change the nature of the curves.

CASE STUDY NUMBER 5: NUCLEOTIDES IN DNA

Now let's turn to some graphs of a different kind.

Figure 14.18 is from a paper, "Gene Regulation for Higher Cells: A Theory," by Roy J. Britten and Eric H. Davidson, that appeared in *Science*, Volume 165, pages 349–357. (The material was redrawn and retyped for clarity of reproduction and is reproduced here the same size as it appeared in *Science*.)

There are, as always, a few things that could have been done better. For example, the horizontal axis at the top of the graph does not *quite* match the scale of the one at the bottom. There is an unlabeled data point between Bacteria and Fungus. Let's face it, it's hard to write a whole bunch of stuff and not make any mistakes, or take a couple of shortcuts. Mere detail.

The authors themselves recognize the main problem with this graph: the vertical "axis" has no axis! *The ordinate is not a numerical scale, and the exact shape of the curve has little significance*, they say in the caption. Why then have an ordinate at all? To show an increasing level of "organization." The authors are proposing to explain gene regulation. It is their thesis that some of the increased amount of DNA in organisms to the right of the diagram has to do with regulating gene activity. It is therefore necessary to show that there is indeed a correlation between the amount of DNA and the degree of "organization" of a species.

The addition of the dotted line does much to reinforce this view. The eye is drawn away from the data points, several of which do not support the thesis

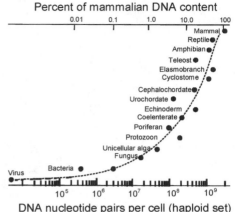

Percent of mammalian DNA content

DNA nucleotide pairs per cell (haploid set)

Fig. 3 The minimum amount of DNA that has been observed for species at various grades of organization. Each point represent the measured DNA content per cell for a haploid set of chromosomes. In the cases of mammals, amphibians, teleosts, bacteria, and viruses enough measurements exist to give the minimum value meaning. However for the intermediate grades few measurements are available, and the values shown may not be truly minimal. No measurements were unearthed for aceola, pseudocoela and mesozoa. The ordinate is not a numerical scale, and the exact shape of the curve has little significance. The figure shows that a great increase in DNA content is a necessary concomitant to increased complexity of organization.

Figure 14.18

at all. (And, of course, this work was done before complete genomes had been sequenced, and the Human Genome Project revealed that there are several "lower" forms with genomes up to 3 times *bigger* than the human genome.)

But we are not here to argue science, only graphics. The interesting but meaningless sweep of the curve evidently caught the eye of Carl Sagan, who was writing a book with a related theme, *The Dragons of Eden*. Sagan thought he would add some meaning to the correlation by using a genuine vertical axis. So rather than arrange the data by some vague notion of "organization," he used *time* instead: in this case, the time of origin of a species, on the assumption that more recent species are more organized, genetically speaking.

Suppose Sagan had taken the data and plotted it on a logarithmic scale, covering perhaps the period since the Earth was formed, about 4 or 5 billion years ago. The result would be that all the data at the end of the curve would be compressed. This would be unfortunate, as this is where most of the modern taxa would be shown.

For example, to cover the period in which life existed on Earth on a logarithmic scale, the decade from 1 billion years after planet formation to 10 billion years after could be used. There would be some blank space over on the right, but not much. However, the last 100 million years (roughly, since the Age of the Dinosaurs and the origin of mammals) would be a barely visible dot occupying the last 1% of the timeline. All multicellular life is compressed into about the last 10% of the line. In other words, almost all of the taxa in the Britten–Davidson graph would be crunched into the last little bit of graph space. It would be hard to make any sense of it.

So Sagan reversed the measurement of time and presented the data as "years ago." He retained the same kind of horizontal axis that Britten and Davidson had used, but his trick of reversing time expanded the more interesting recent years, while compressing the earlier ones.

Figure 14.19 appeared in *Dragons of Eden*. (As before, the material was redrawn and retyped for clarity of reproduction and is reproduced here the same size as it appears in the paperback edition.)

The careful observer will note a few problems. The top two vertical intervals correspond to time before the universe existed. Small wonder there are no

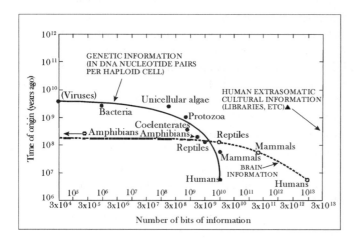

The evolution of information content in genes and brains during the history of life on Earth. The solid curve, which goes with the filled circles, represents the number of bits of information contained in the genes of various taxa, whose rough time of origin in the geological record is also shown. Because of variations in the amount of DNA per cell for certain taxa, only the minimum information content for a given taxon is shown, the data being taken from the work of Britten and Davidson (1969). The dashed curve, which goes with the open circles, is an approximate estimate of the evolution in the amount of information in the brains and nervous systems of these same organisms. The information in the brains of amphibians and still lower animals are off the left edge of the figure. The number of bits of information in the genetic material of viruses is shown, but it is not clear that viruses originated several billions of years ago. It is possible that viruses have evolved more recently, by loss of function, from bacteria or other more elaborate organisms. If the extrasomatic information of human beings were included (libraries, etc.), that point would be far off the lower right edge of the chart.

Figure 14.19

data here, but why even have an axis? One might also note that most of the points have moved to the right, perhaps as a result of transcription errors. These are details. More importantly, one is left wondering how to read the graph.

After his extensive caption, Sagan includes these words on the page facing the diagram: "*We see from the chart that there was a striking improvement in the information content of organisms on Earth some 3 billion years ago, and a slow increase in the amount of genetic material thereafter.*" A page or so later, there is mentioned the possibility that there is a practical upper limit to the amount of genetic information that DNA can accommodate. You might wonder just what is being referred to. We did.

It is amazing how the familiar process of interpreting a graph can lead to a wrong conclusion. Time, with the very unusual units of "years ago," is the vertical axis on Sagan's graph. One might be forgiven for thinking that the "striking improvement" must be due to the sudden appearance of viruses at about 4 or 5 billion years, with a ready-made complement of 3×10^4 bits of information. But the graph in this region is very flat and smooth, and we cannot deduce anything about a "striking improvement" here, as this is the edge of the graph! Nor do we see evidence of a practical upper limit.

It must be that Sagan had in mind something else. We will probably never know what. We suspected, however, that the bizarre nature of the time axis being vertical and the scale being "years ago" was misleading in the extreme. We were curious, too, about the shape of the curve that is evidently intended to fit the data. No explanation is offered for what it is, or why it was chosen. Is it an artistic curve intended to show a trend? Is it a least-squares fit of some equation?

To answer these questions, we chose the simplest reasonable model for the growth of genetic information: ordinary compound interest. Since the Earth got started about 5 billion years ago, the same as on Sagan's graph, we assumed a cell with 3×10^4 "bits of information" appeared 4 billion years ago, the same as on Sagan's graph. We then "grew" the information content at an annual rate of $(1 + 3.18 \times 10^{-9})$. That is, each year's content is 1.00000000318 times the year before. (We had to allow fractional "bits" to make this work when the numbers are small. This might be the same as whole bits in a fraction of the population.) The result is shown in the graph in Figure 14.20. The solid line is Sagan's curve, the dashed one is ours.

The "interest rate" controls the left–right extent of the graph. The date of origin controls the vertical extent. Our 30,000-bit viruses got onto the planet at

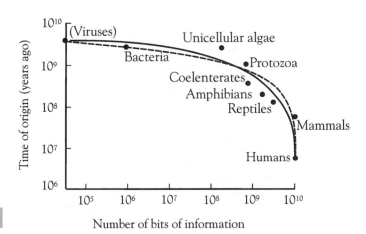

Figure 14.20

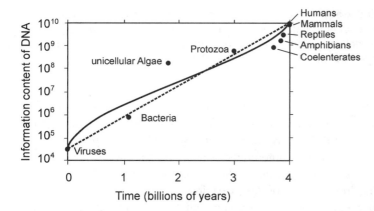

Figure 14.21

the same time as Sagan's viruses, and exponentially got more DNA. Figure 14.21 shows how our exponentially growing information looks with a linear and horizontal time axis and a logarithmic ordinate. Once again, the solid curve is Sagan's, the dashed one is ours.

Admittedly, we adjusted things so that our curve fit the original curve. But this is not really an example of fitting an arbitrary equation to measured data. The various assumptions constrained what we did. The *shape* of the curve comes out of the assumption of compound interest. The *location* of the upper left comes out of the assumed origin time of the virus, and the assumed number of bits of information it contained. After that, there is only *one* parameter we are free to vary, and that is the interest rate.

The curve fits the data nearly as well as the original curve. This means there is nothing here to imply a change in the rate of growth. If anything, the assumption must be that things are continuing as they have always done, and the number of bits of information in genes will continue to increase exponentially. In fact, the number of bits has about doubled in the last 200 million years.

The turn-down of the curve, interpreted by Sagan as a slowing down of the growth, is purely an artifact of his presentation. The presentation, using vertical years-ago time, is so unfamiliar that any conclusion based on the *appearance* of the curve is almost bound to be misleading. If the author cannot read the graph correctly, what chance has the poor reader?

There are two lessons here. One is that graphs that show human beings as some kind of pinnacle of evolution tend to represent wishful thinking, rather than any deep truth.

The other lesson is that you should choose your axes to show what you want, and be very careful if you choose to use something unusual!

CASE STUDY NUMBER 6: THE GRAPH BEHIND THE MMR–AUTISM CONTROVERSY

In an article under the heading "False Alarm: Autism Isn't Really Running Riot. It's All in How You Interpret the Figures," the magazine *New Scientist* printed the following:

> In particular, Fombonne says, many people have misinterpreted a graph that appears to show a sudden rise in the cases of autism during the 1980s. The graph in fact shows the number of people known to have autism in a single

year, 1991, plotted against the year of their birth. The rise in cases for birth dates nearer the present could reflect the rising population and improved diagnosis among young children. Portraying the data in this way is very misleading, Fombonne says. "Trying to link this with MMR is complete nonsense."

Apparently, a link was suspected by some between the combined measles, mumps, rubella vaccine (MMR) and autism. Critics of the idea were blaming misinterpretation of a graph. Naturally, we were interested in how a graph could be so bad that it would be widely misinterpreted. That's what this case study is about.

The quotation above appeared in *New Scientist,* February 17, 2001 (page 17). Backtracking in the pages of the magazine, we found the name Andrew Wakefield, a London gastroenterologist, and a citation in *The Lancet.* There we found the figure[2] and caption shown in Figure 14.22.

We tracked down the link to the California Department of Developmental Services. Here we found Wakefield's source (Figure 14.23), in a report to the California legislature.[3] (The original was in color.)

Now, the authors of this diagram are at pains to point out (right above the diagram) that "data points in Figure 1 do not show how many persons entered the system in a given year, but how many already in the system were born in a given year."

We redrew *The Lancet* graph to make it more legible and obtained Figure 14.24.

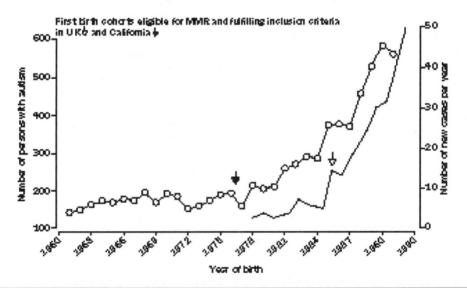

Figure 14.22

[2] *The Lancet* seems to have trouble with graphics, at least in the version available over the Web. The text is very crisp and sharp. The graphics, for the most part, appear to be low-resolution (75 dpi?) files. Our extract is from Volume 354, September 11, 1999, page 950.
[3] "Changes in the Population of Persons with Autism and Pervasive Developmental Disorders in California's Developmental Services System: 1987 through 1998." A Report to the Legislature, March 1, 1999.

Figure 1 - Distribution of Birth Dates of Regional Center Eligible Persons with Autism

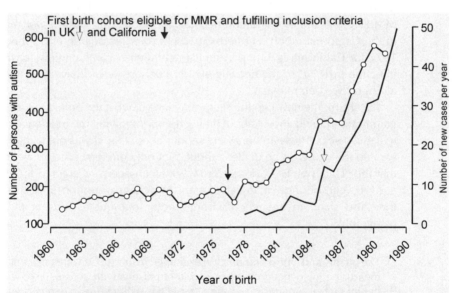

Temporal trends for autism in the USA (California*) and the UK (north-west London)
In 1998 the expected numbers of newly diagnosed autistic children in California should have been 105–263 cases, according to DSM-IV; the actual figure was 1685 new cases. The temporal trend in north-west London is almost identical, although the rise is delayed by about 10 years. The two countries use the same diagnostic criteria. The sequential trends are consistent with the timing of introduction of MMR to both regions.

*Data from Department of Developmental Services, Sacramento,1987–98 (www.dds.ca.gov) .

We noticed right away that the ordinate label for the California data has changed. But there's more. Altogether, the following things have changed from the California version:

1. The label for the California data has been changed. The word "enrolled" was deleted, so it reads "Number of persons with autism."

2. New words have appeared at the top of the graph, introducing MMR and something about "inclusion criteria."

3. There are now two vertical scales. The new one is on the right, to be used with the London data, labeled "Number of new cases per year."

4. The new scale on the right starts at zero, and the suppressed zero of the original left scale has been retained. Even so, the bottoms of the two axes have been aligned.

5. The old figure caption above the graph, with its words about "Distribution of Birth Dates" is gone, replaced with a new caption at the bottom of the graph. The new title for the graph, in boldface, is "Temporal trends for autism in the USA (California) and the UK (north-west London)."

6. The abscissa is still labeled every three years, which was an odd choice to start with, but the last two labels (1960 and 1990) are clearly wrong.

Let us make a few comments on some of these changes.

First, the omission of the word "enrolled" from the ordinate label is significant: it broadens the meaning from the scope of the original.

Second, the new words at the top of the graph, describing the arrows on both curves as indicating the "First birth cohorts [that were] eligible for MMR . . ." are a major addition because the subject of MMR is not mentioned in the California report. Furthermore, the arrows include an error: the arrow at 1977 for California is wrong. American children became eligible for the MMR injection in 1971, so the first eligible birth cohorts would have been those born a year or two before that.

Third, the new caption flies in the face of what the authors of the original graph said (quoted above) about their graph *not* showing the incidence of autism in a given year. Those authors were serious about that: it was not something they wanted the reader to be unclear about. Not only did they state above the graph that the "Data points . . . do not show how many persons entered the system in a given year, but how many already in the system were born in a given year," they also state, in the report's Conclusions (our quote starts at the second paragraph):

> The quality and type of information examined in this report were not suitable for measuring incidence in the population of persons with autism.[4] Ascertaining the incidence for autism and the other PDDs will require carefully controlled research. Furthermore, it is far beyond the capability of this Department to undertake such studies . . . .

[4]The *incidence* of a disease is the number of new cases of that disease per unit population over a given time (typically a year). It is thus a ratio and a rate, for example, 3.5 cases per 1000 per year. The extract from the California report shows that the authors stress that theirs was not an incidence study. Nor is it what is called a *prevalence* study. Prevalence is the number of cases per unit population at a given time. It is thus a ratio (such as 3.5 cases per 1000), but not a rate. (It is sometimes called a *cross section*.)

The cause(s) of the increase in the population of persons with autism served by the regional center over the past 10 years is unknown. The sheer complexity of this phenomenon prevents any clear conclusions about the exact determinants of the increase. Speculation about the rise in numbers is abundant, but such speculation is not based on scientific research and typically leads to debate and controversy when offered as a cause . . . .

What we do know is that the number of young children coming into the system each year is significantly greater than in the past, and that the demand for services to meet the needs of this special population will continue to grow . . . .

The authors recognize an increase in demand for the Department's services (whether due to increased incidence or increased population is not addressed), and they are trying to help the State of California figure its future needs. They are concerned with their system and the numbers it must serve, not with causes. This is a report to the legislature, not a scientific paper. That is why the vertical axis of the California graph is labeled "Number of *Enrolled* Persons with Autism." (Our italics.)

Up to this point, we had assumed that the London data were Wakefield's, since Wakefield was based in London. We are encouraged to think this because nowhere in his *Lancet* letter does he describe the origin of the data. He introduces them by saying: "The figure juxtaposes the data from California with those from north-west London." In fact, it turns out that the data are "borrowed" from a paper by Taylor et al.,[5] the last reference in the letter, and represent cases identified in a search of the records of "eight North Thames health districts." The Taylor paper does not mention London anywhere, and does not state that the districts were (or were not) in London. Since the Thames originates in Gloucestershire and flows through Oxfordshire on its way to London, perhaps you just have to *know*. Figure 14.25 is the graph from which the data are evidently obtained.

(Here, we have adjusted the contrast to make it more readable.) It seems that the curve used is the top one, the one labeled: Core + atypical.

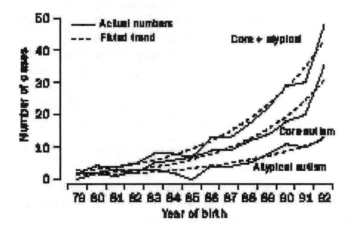

Figure 14.25

[5]B. Taylor, E. Miller, C. P. Farringdon, M.-C. Petropoulos, I. Favot-Mayaud, J. Li, and P. A. Waight, MMR vaccine and autism: no epidemiological evidence for a causal association, *The Lancet* **353**, 2026–2029 (1999).

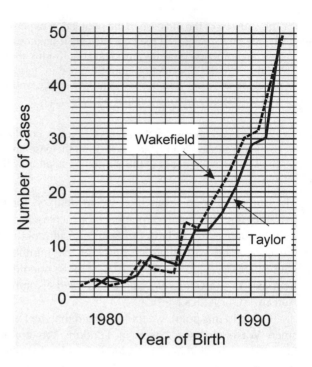

Figure 14.26

By tracing over the curve, and superposing the curve on the graph in Wakefield's letter, we were able to verify that this was *very likely* the source of the data. However, the accuracy with which Wakefield and his colleagues traced the data is not impressive, and we cannot be certain. The curves in Figure 14.26 allow point-by-point comparison of the data as presented by Wakefield and the data as presented by Taylor and co-authors.

Even allowing for the poor quality of the graph in the Web version of *The Lancet*, the copy is not very accurate. But there does not seem to be any particular purpose or bias to the errors: the Wakefield starting point is off by a year, but captures the correct number of cases (2). His ending point is evidently off by one case (50 instead of 49) but the year is correct (1992). The overall impression is certainly similar, and a reasonable assumption is just that the draftsman was not very careful. This impression is supported by the error in the labels of the abscissa, where (evidently using the copy-and-paste approach to generating the text, the draftsman forgot to edit the labels at the end of the curve; see Figure 14.24 and numbered comment 6 presented earlier).

However, there are some other, more serious, inconsistencies between the data as presented by Taylor and as reproduced by Wakefield:

1. The vertical axis label of this graph was changed from "Number of Cases." It was given as "Number of new cases per year."

2. The caption for this graph, "Core and atypical autism cases under 60 months of age and fitted trends by year of birth 1979–92" has been removed and replaced by the "Temporal trends" caption.

Some further comments are in order.

First, the change in ordinate label for the California data from "Number of Enrolled Persons with Autism" to "Number of Persons with Autism" does indeed broaden the meaning. But the change from "Number of Cases" to "Number of new cases per year" is more serious. It implies a *rate* is being measured. The Taylor study would have to have been an *incidence* study for this to be the case. Like the California study, it was not. Nor was it a cross-sectional *prevalence* study, since the data are not expressed as a fraction of the population.

When you think about it, a label that says "new cases per year" is actually inconsistent with an abscissa label that reads "Year of Birth," because it overspecifies the data. The plane defined by the two axis labels would imply that in London, in (say) 1987, 13 cases were born and entered into the system. This could, of course, be true if the diagnosis were made in the first days of life, but typically the diagnosis is not made until at least the second year: in fact, with one exception, the earliest diagnosis in the London data was made when the patient was 18 months old, and most were made much later. Not one of the cases was a "new case" in the year of birth.

Second, the label "Core and atypical autism cases under 60 months of age ... 1979–92" is an odd one. (The label means, in fact, that the patient was 60 months old or less at the time of diagnosis.) There are 204 cases represented in the graph, out of the 494 cases that were "identified" by Taylor and co-workers in their search of the records. Presumably the restriction to cases under 60 months of age at time of diagnosis accounts for the reduced number. However, it seems fair to say that a presentation restricted to the youngest 40% of the cases of autism found in "eight North Thames health districts" is not directly comparable with a presentation of all the enrolled cases in a large and populous state. In one case, the data are persons enrolled in a support system; in the other, they are persons born since 1979 and diagnosed by age 5. These two sets of restrictions are not comparable, so the data they represent are not comparable.

We can therefore see that Wakefield made a large number of alterations to the presentations of the data. Some of these changes give the appearance of mere carelessness, others suggest motive. Whether deliberate or not, most are graphically inappropriate:

- Comparing data on a scale with a suppressed zero with data on a scale with a zero
- Changing the labels on various axes
- Using data as if from an incidence study: using them in a way explicitly said to be inappropriate by the original author
- Removing captions from graphs and replacing them with inappropriate ones about temporal trends
- Copying graphical data inaccurately
- Labeling axes by copy and paste, and failing to edit the pasted labels
- Comparing data sets that are not comparable

Some of the nongraphical errors are suggestive of carelessness, too. For example, the California data are attributed in the text to the "Office of Developmental Services." In fact, the organization is the *Department* of Developmental Services. The "London" data are not attributed at all; and the apparent congruence between "eight North Thames health districts" and "N.E. London" is not established.

Nevertheless, whatever Wakefield and his co-authors did to change the presentations, the *first* mistakes in this whole graphics business were made in the *original papers*. Each set of data was represented by its original authors as a time-series graph. This way of showing the data was likely to lead to a misinterpretation of their meaning. The authors of the California report and the London *Lancet* article should have used scatter plots or a bar chart or (best of all) a *population pyramid*.

The use of a bar chart, instead of a graph, is justified by the fact that the distribution of birth dates of people enrolled in a program is not a time-series trend. It is a distribution. None of the original authors are plotting the time evolution of a function, or even the sampled values of a variable: they are showing recorded independent statistics.

This topic was discussed in Chapter 3, where a test for whether a bar chart or a scatter plot should be used is given. The points that represent the data in the graph area should not have been joined, *because the values of the ordinate change if the sampling frequency is changed*. For example, had the data been presented on a basis of month of birth rather than year, the numbers would have been about 12 times smaller. The interpolation implied by "joining the dots" is simply not valid.

But a population pyramid is an even better way to present the data. Remember the population pyramid for Japan that we saw in Chapter 2 (see Figure 14.27)? This is a very rich presentation, showing data from 2004 and 1935, separating male and female citizens, and exposing odd features in the size of the population growth each year. This is the way that the autism data *should* have been presented. Each of the original papers could have presented

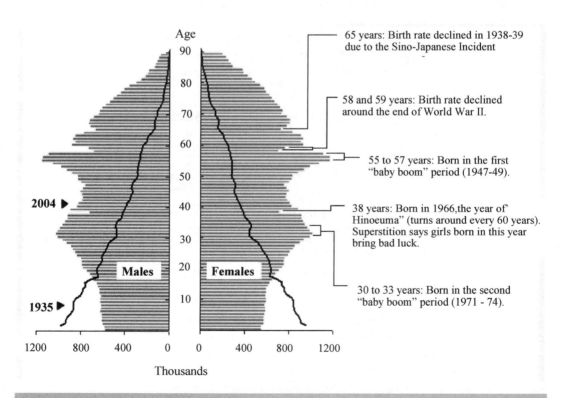

Figure 14.27

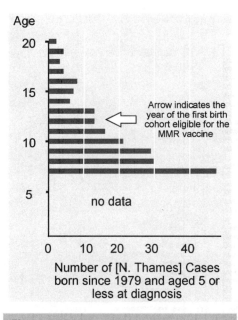

Figure 14.28

a pyramid showing male and female persons. Lacking those details, we shall give the examples without the breakdown by sex (Figs. 14.28 and 14.29).

The caption for each of these pyramids could indicate that the data were analyzed in 1998 for patient records up to 1992. They could have been combined as shown in Figure 14.30.

Likely this kind of presentation would not have lent itself to quite so much misinterpretation. Even if you try to read the graph as a vertical trend, it is evident that the growth in the California

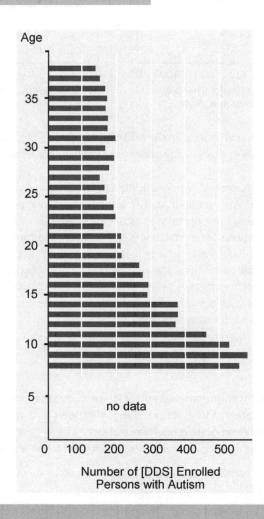

Figure 14.29

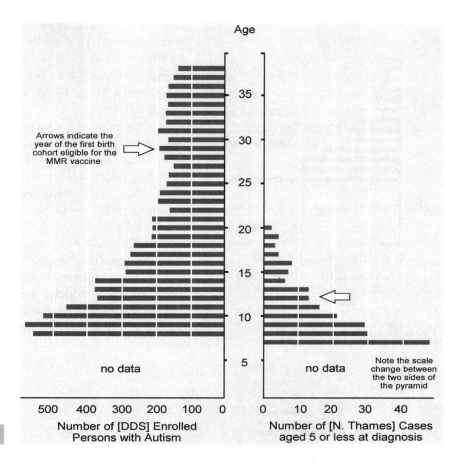

Figure 14.30

Age

Arrows indicate the year of the first birth cohort eligible for the MMR vaccine ⇨

⇦

no data

no data

Note the scale change between the two sides of the pyramid

500 400 300 200 100 0   0 10 20 30 40

Number of [DDS] Enrolled Persons with Autism

Number of [N. Thames] Cases aged 5 or less at diagnosis

numbers started several years after the MMR injection was begun, and the growth in the Thames numbers was already under way when the MMR policy was adopted there.

Back in Chapter 2 we encountered Florence Nightingale, and her coxcomb diagram. An excellent statistician, she is also credited with saying: "To understand God's thoughts we must study statistics, for these are the measure of His purpose." It seems that is still good advice for doctors.

The debate over MMR and autism has exercised a number of people in the last few years. In all probability, the debate will last as long as funding can be obtained for researchers to pursue the supposed connection. One cannot help feel, however, that a better job of presenting and understanding the graphics would have avoided a lot of wasted effort.

SUMMARY

We have seen that well-prepared graphs can be quite small and contain a large amount of information. However, we have also seen that failure to attend to detail can mislead even quite renowned authors.

For the graphic user, the following lessons are to be learned.

■ Choose carefully the kind of graph you use to show your data.

■ Even if you select an apparently normal kind of graph, be sure your selection is appropriate. Had the original authors of the autism report selected the population pyramid instead of the time-series graph, it would have been much harder to misinterpret their birth year distribution data to bolster the case for similarity between the two disparate data sets.

■ Once you have decided on the appropriate presentation, remember all the basic rules about labeling axes and so on.

■ Strive for a consistency of style in the graphics you produce for any one work. This will not only make the work more readable, it will also create the impression that you have paid attention to the details.

■ If you are going to quote someone, do it accurately. That applies both to verbal and graphical statements. If you are going to the trouble of tracing and redrawing a graph because you do not have the original file, be sure you make no changes.

■ Even a small change can give the impression you are trying to mislead. Instead of supporting it, that can undermine your case.

EXERCISES

14.1. Study the graph in Figure 14.31 for a moment. It shows the decreasing cost of memory and the increasing speed of processors over time. Something like it appeared in the *IEEE Spectrum* in an issue that contained an article about Gordon Moore of Intel. Don't you think that the graph gives the impression that nothing much was happening in the world of semiconductors between about 1980 and 1995? Is this the case? What mistakes did the graph's originator make: that is, what changes would you make to fix the graph?

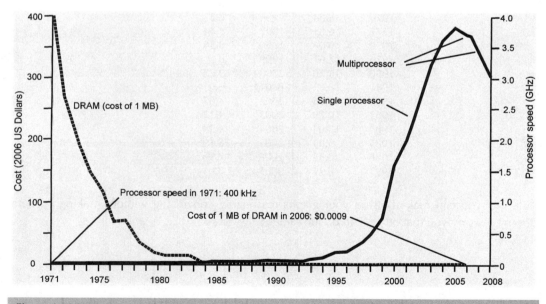

Figure 14.31

14.2. Study your local newspaper or news magazine for a week. Can you find any graphs with mistakes? Do you think the mistakes were accidental?

14.3. Study your local newspaper or news magazine for a week. Are there any graphs that really impress you with their ability to make a case clearly and unambiguously? What features are particularly effective?

14.4. The data below are annualized data representing the Fed prime rate. Make up a graph that presents the case that there has been a long-term downward trend in the rate. Next, make a graph showing that the rate has generally been flat. Both graphs should end with 2007. (*Hint*: You don't have to use the complete data set.)

Year	Prime Rate	Year	Prime Rate
1956	3.77	1961	4.5
1957	4.2	1962	4.5
1958	3.83	1963	4.5
1959	4.48	1964	4.5
1960	4.82	1965	4.54
1967	5.63	1966	5.63
1968	6.31	1988	9.32
1969	7.69	1989	10.87
1970	7.91	1990	10.01
1971	5.73	1991	8.46
1972	5.25	1992	6.25
1973	8.03	1993	6
1974	10.81	1994	7.15
1975	7.86	1995	8.83
1976	6.84	1996	8.27
1977	6.83	1997	8.44
1978	9.06	1998	8.35
1979	12.67	1999	8
1980	15.26	2000	9.23
1981	18.87	2001	6.91
1982	14.85	2002	4.67
1983	10.79	2003	4.12
1984	12.04	2004	4.34
1985	9.93	2005	6.19
1986	8.33	2006	7.96
1987	8.21	2007	8.05

Note that you can make both of your graphs reasonably convincing without making up data. Doesn't it worry you that others can do the same?

Summaries

15

In this chapter, we bring together the summaries at the ends of the preceding chapters, for handy reference.

CHAPTER 1—BASICS

We looked at several kinds of graph, and examined some of their peculiarities. For all graphs, the basic rule for the author is to strive for clarity and avoid ambiguity.

- Think about what it is you want your graphic to say, and how best to arrange things to that end.
- Let the data make the case. Don't distort the graph to make your point. Don't use areas or volumes in simple numerical comparisons.
- Aim for a neat, clean appearance, with balance and good proportions. Varying line widths, making the graph line heavier than the axes, is a good way to improve appearance.
- Minimize the size of the graph, resizing the words if necessary.
- Control the amount of clutter in the graph. For example, use tick marks sparingly, and don't feel obliged to label all of them.
- Don't feel obliged to have lines around the graph on all sides, or all the way across the area.
- Clutter control is a matter of judgment and style, so the details are up to you.
- Think about what happens when the graph is copied or converted to microfiche.
- If there are two axes, label them both.
- Keep the graph within an imaginary rectangle, but don't feel you have to draw the rectangle.
- More or less fill the rectangle range with data, unless you have several similar data sets to compare.
- It is not essential to join axes—but avoid mid scale breaks.
- Look out for step functions.
- If you can avoid using a key, do so. If you must use a key, make sure it is clear.

The Right Graph. By Harold Kirkham and Robin C. Dumas
Copyright © 2009 John Wiley & Sons, Inc.

CHAPTER 2—WHICH KIND OF GRAPH?

There are three common chart families: histograms and bar charts, *x-y* plots (also called scatter graphs), and area charts (pie charts or donut diagrams). While each of these has several species within it, some generalization is possible.

■ A *histogram* is used to present statistical data for a single variable.

■ A *bar chart* can be used for numbers that change with time, or to compare two (or more) sets of data. Bar charts are the one kind of graph that can be rotated to be vertical or horizontal.

■ An *x-y plot*, or *scatter plot*, is used to show how one parameter varies as a function of another. It is usual to have the independent variable horizontal, though it is conventional in some fields (e.g., oceanography) to have an obviously *vertical* parameter (such as depth) shown vertically.

■ An *area chart* can be used to show how some entity is divided among its constituent parts.

■ Consider using a graph any time you are considering a table of data.

■ In general, it is not good practice to add a third dimension to graphs unless there is a third parameter involved. The gratuitous use of "3-D" creates the impression of a less-than-serious piece of work.

After these general guidelines, there are some specifics.

■ To show linearity, do not use an *x-y* plot with the output as a function of the input. Show instead the *error* as a function of the input.

■ To show a parameter with a large dynamic range, use a logarithmic scale.

CHAPTER 3—CONNECTING THE DOTS

The question of connecting the dots is far from simple to answer.

There are some situations where connecting the dots is definitely *wrong*. You need to know how to avoid this mistake. *In general, do not join the dots.* Find another way to show the data.

In particular, do not connect the dots with a line if the data (a) represent different *kinds* of things or (b) *interact* with the horizontal axis. This last is an example of the *Nobel prize* problem, and a bar-like histogram is appropriate.

There are some situations where connecting the dots is definitely *needed*. If you are trying to show a trend, but the data are rather sparse, or irregular, show and join the dots, provided the dots (a) represent the same *kinds* of thing (e.g., samples) or (b) are at least a little predictive or capable of extrapolation. These are examples of the *support for the death penalty* situation.

There are also some situations where you have a free hand and may do as you please. If the data are a time series, where the same parameter is sampled regularly, the connecting lines may highlight the data. However, joining the dots by straight-line segments is almost never correct, and certainly not if you can *see* the line segments, except that in the case where the results contain significant noise, the connecting lines can serve to make the noise (rather than a trend) more evident.

However, in both of these situations, if you can do a curve fit or at least a linear regression, you will not need to connect the dots.

Altogether, the answer to the question "Should the dots be joined?" evidently depends both on the data and on what you are trying to show.

We recommend you try to develop a bias *against* the practice.

CHAPTER 4—THE NONDATA PARTS OF THE GRAPH

The basic rule is strive for clarity and avoid ambiguity.

- Choose a font for your graphics that complements the one used in the body text—but not necessarily the same font.
- Make sure the type is the right size in the final graphic, about one point size less than the body copy, if you are using a sans serif font.
- Don't write everything in the graphic in uppercase.
- Label graph lines in the simplest way possible.
- Label axes with conventional scales, and don't use multipliers.
- Rotate the label for the vertical axis, but not the axis numbers.
- Keep captions appropriate and relevant.
- Don't explain abbreviations or acronyms in the graphic.
- Don't put leader lines parallel to lines in the graphic.
- As always with graphics, once you have a set of words formatted correctly, copy and edit them, rather than typing from scratch and reformatting.

CHAPTER 5—GETTING THE MOST OUT OF YOUR SOFTWARE

Getting your software to do what you want, and do it efficiently, is essential if producing graphics is not to become burdensome. It is worth learning your software well—you might even benefit from cracking the binding on the manual.

- Learn as many two-hand shortcuts as you can. Using the mouse to select options from a menu is slooooooooow!
- Whenever possible, copy and edit graphic elements, instead of drawing them from scratch.
- PowerPoint and Presentations will cycle through all objects with the Tab key, once you have selected an object.
- Lassoing only works when an entire object or group is lassoed.
- Use the Control key shortcuts when possible.

    Ctrl-c copies the current selection into the clipboard.

    Ctrl-x copies the current selection into the clipboard and deletes the original.

    Ctrl-v pastes the contents of the clipboard.

    Ctrl-z undoes the last editing operation.

    Ctrl-s saves the file.

    Ctrl-d works to make duplicates of your selection and move. This does not work in all applications.

- Draw a box, and while it is still selected, start typing. What you will get will be the default font, centered on the box. This works in PowerPoint and Visio.
- Holding the Shift or Ctrl key while you draw a line constrains the angle in most programs.
- Holding the Shift key while you stretch a line or drag an object or group keeps the direction constant (some programs).
- Suggestion: *Learn to avoid precision control and adjustment in your work*—let the computer do the work for you.
- Remember the rules of layout:

    Graphics and tables *must* be called out.

    They should be inserted *in order as soon as possible* after the callout.

    The arrangement should not impede the *flow* of the paper.

    The *font size* should match the environment.

- Insert graphics into MS Word by using a text box.
- Insert graphics into WordPerfect by drawing (or pasting) the graphic into Presentations and using the Insert/Graphics/From File sequence.

CHAPTER 6—PRESENTATIONS OR HOW TO SUCCEED IN BUSINESS

Getting your presentation in good shape is a matter of remembering the fundamentals, and being organized. Start by writing an objective statement for the presentation. It should be based on what you expect of your audience.

- Remember the basic rules for viewgraphs:

    Use bulleted lists.

    Avoid sentences.

- Build up logically:

    Start with an outline.

    Use thumbnails.

- Adjust your presentation to your audience

- Make your style decisions:

    Use a repeating background.

    Use sans serif font.

- Use color in a presentation even if you don't in the paper:

    Make the slide "work" first in B/W.

    Use color for appeal or for clarity.

    Beware readability issues with colored lines and text.

    Some colors have low contrast against other colors.

    Fidelity of color is not maintained by all projectors.

- Use the power of the software, such as auto-timing:

    But don't overdo the fades and slide-ins.

    Use branching, action buttons.

    Use graphs and diagrams as part of presentation, but minimize equations.

    Use the things you can do in a presentation that you can't do in print—such as animation or turning gridlines on and off.

- Consider presentation integration and your final thoughts.

- Check for rhythm and pace; count the slides on each subtopic.

CHAPTER 7—AN INTRODUCTION TO SPREADSHEETS

Spreadsheets allow ready analysis of data and allow graphs to be created quickly and easily. For creating graphs with a spreadsheet, there are three steps:

- Get the numbers.
- Create the graph in the spreadsheet.
- Fix the graph to be publication quality.

Most of the fixing of the graph will take place in the spreadsheet program. Some work may have to be done in a graphics program.

CHAPTER 8—USING SPREADSHEETS: EXCEL®

Excel has many convenient features as far as creating graphs is concerned. Editing graphics in Excel is much more efficient than doing it in PowerPoint because Excel can remember the changes you made and reproduce them later for another graph.

- Add the `Drawing` toolbar to your custom version of Excel.

- Select the zoom level you want before you edit a chart. Changing zoom is not an option once you have a chart element selected.

- Excel does not do a good job with pie charts. Anticipate editing them in PowerPoint.

- To have Excel automatically choose your first column of data as x-axis labels in a bar chart, put an empty cell on the top of the column, or make all the cells into labels by starting them with a single quote mark.

- Get accustomed to using the `Chart` pull-down menu, and to looking at the window in the top left of the screen that indicates what you have selected.

- You can select elements in turn on your graph by using the up–down arrows (for broad selections, such as an entire series of data) or the left–right arrows (for more detail, such as a single data point in a series).

- Right-clicking on selected items brings new capabilities to change the appearance of your graphs.

- If you override the automatically selected scale on an axis, the change will be remembered. Usually, it is worth deferring this step until you have saved your chart `Type`.

- To create a box around your graph, with tick marks on all sides, you need to add a `Secondary axis`. Often, it is best to add a dummy set of data for the axis. Start with the vertical axis. You can select a data series and use the `Format` pull-down, or you can right-click on a data series, and go through the `Format Data Series` sequence to add another vertical axis.

- Once you have added a second vertical axis, you have two ways to add tick marks to a top horizontal axis: click in the `Chart area` (outside the `Plot area`), and select `Chart Options` from the pull-down menu, or right-click in the same area, and select `Chart Options` from the dialog box.

- Format the color and border of the dummy data so they do not show on the graph.

- You can add a data table under a bar chart.

- When you have finished making a graph the way you want it, consider adding it to the selection of user-defined `Types` under the `Chart` pull-down. The time saving next time around will be considerable.

The following changes to a graph will not be remembered:

- To add callouts and comments, use the text and arrows available in the Drawing toolbar.
- To make white (or any other color) lines on top of your graph, use the Drawing toolbar.

CHAPTER 9—USING SPREADSHEETS: QUATTROPRO®

QuattroPro has many convenient features as far as creating graphs is concerned. Editing graphics in QuattroPro is much more efficient than doing it in Presentations because QuattroPro can remember the changes you made and reproduce them later for another graph.

- It may not be worth adding the Drawing toolbar to your custom version of QuattroPro, because the changes you make cannot be copied out of the spreadsheet environment into your word processor.

- Select the zoom level you want before you edit a chart. Changing zoom is not an option once you have a chart element selected.

- QuattroPro does not do a good job with pie charts. Anticipate editing them in Presentations or PowerPoint.

- To have QuattroPro automatically choose your first column of data as x-axis labels in a bar chart, make all the cells into labels by starting them with a space or make the first one into a label by starting with a single quote mark.

- Get accustomed to using the Chart pull-down menu, and to using the Properties Selector on the Property toolbar. This is by far the easiest and most consistent way to move through the graph.

- Right-clicking on selected items is often an alternative way to get to the Properties Selector.

- If you override the automatically selected scale on an axis, the change will be remembered. Usually, it is worth deferring this step until you have saved your chart Template.

- To create a box around your graph, with tick marks on all sides, you can choose to have axis labels on both sides. This works for the x-axis and y-axis. However, you will get what you asked for: *labels* on both sides. Removing them to leave only the tick marks is tedious.

- When you have finished making a graph the way you want it, consider adding it to the selection of user-defined Templates under the Chart pull-down. The time saving next time around will be considerable.

The following changes to a graph will not be remembered:

- To add callouts and comments, use the text and arrows available in the Drawing toolbar.

- To make white (or any other color) lines on top of your graph, use the Drawing toolbar.

CHAPTER 10—FIXES USING GRAPHICS PROGRAMS

The file that you create in your graphics program (PowerPoint or Presentations) has the various elements of the graph from the spreadsheet (Excel or QuattroPro) grouped in ways that may not be ideal for your editing purposes. The first thing you should do in your graphics package is to ungroup the elements of the image and, as soon as possible, regroup some of the elements.

- In PowerPoint, you can paste as a Microsoft Excel Chart Object, Picture (Windows Metafile), or Picture (Enhanced Metafile). In Presentations, you paste as a Metafile Picture, *not* as a Device Independent Bitmap or a QuattroPro Chart.

- Ungroup all the elements of the picture and, as soon as possible, regroup parts that will force your operations to maintain the appropriate relationships, for example, between the data and the axes. Usually this means grouping each axis to its labels and tick marks, and grouping all the graph line parts.

- Generally, do not retain the blank white rectangle or the rectangle around the graph, as these often define a bigger space than the graph itself occupies. It is easiest to delete these parts while the material is all ungrouped.

- Separate graphs from separate spreadsheets can be combined provided you line up the data spaces accurately. But be careful: if the *things* being joined are of the same kind, be sure that the data spaces are handled properly—with the zeros lining up and so on.

- If you are scanning and redrawing an image, consider taking the opportunity to improve the quality of the picture. But be careful to be honest—use an equation to improve a curve fit, for example, or improve the quality by simply removing extraneous material to modernize the style.

CHAPTER 11—SOMETHING BEGINNING WITH "P"

Perspective is rarely needed in a technical drawing, but when it is, don't be put off by the difficulty.

- If you have a plan drawing, and you need moderate accuracy, you can produce a perspective drawing using the guidelines presented early in this chapter.

- If you don't need particular accuracy, you can sketch something out on the computer, and add some lines to vanishing points just to refine the drawing.

For patent drawings, the PTO rule is the same as ours: clarity.

- Provided your drawing is clear and will copy well, it is likely to be acceptable.

- Recall the principles of drawing using India ink.

- Always check the Web page of the PTO for the latest information and the latest rules.

CHAPTER 12—FILE FORMATS AND CONVERSIONS

File conversion is needed when collaborators don't use the same software. It may be needed to make graphics in a form that can be integrated into your word processing software. While the capability to do file conversions is gradually getting better, there are still plenty of bugs.

- When you need to convert a file format, first try using the clipboard as a mechanism.

- If the amount of fix-up work that you are faced with after that looks daunting, try reading the file directly into the target application.

- If it still looks like too much work to fix up, try exploring a route that involves an intermediate format, or an intermediate application. There will be many possible combinations, and one of them may be just what you want.

- If a conversion is forced on you because your publisher demands a particular file format for graphics, be sure to read the converted file back into another program or two. Your publisher's software may be quite capable of adding bugs, so it's good to remove all the ones you know about.

CHAPTER 13—STYLE MATTERS

As always, the basic rule is to strive for clarity and avoid ambiguity. Within the constraints of where you publish, establish your own consistent style.

- For your graphs, consistently use (or don't) a surrounding box, put the tick marks inside the box (or outside), and so on.
- For your text, make your choice of font consistent, and its scale relationship to the rest of the graph.
- For various other diagrams, choose a consistent style that suits the material and the journal or magazine for which you're writing.
- Make a note of the various decisions that make up your style.
- Assemble a collection of graphics of your own that can form the starting place for many of your future graphics, at the same time establishing your own style.

CHAPTER 14—CASE STUDIES

For the graphic user, the following lessons are to be learned.

- Choose carefully the kind of graph you use to show your data.

- Even if you select an apparently normal kind of graph, be sure your selection is appropriate. Had the original authors of the autism report selected the population pyramid instead of the time-series graph, it would have been much harder to misinterpret their birth year distribution data to bolster the case for similarity between the two disparate data sets.

- Once you have decided on the appropriate presentation, remember all the basic rules about labeling axes and so on.

- Strive for a consistency of style in the graphics you produce for any one work. This will not only make the work more readable, it will also create the impression that you have paid attention to the details.

- If you are going to quote someone, do it accurately. That applies both to verbal and graphical statements. If you are going to the trouble of tracing and redrawing a graph because you do not have the original file, be sure you make no changes.

- Even a small change can give the impression you are trying to mislead. Instead of supporting it, that can undermine your case.

Index

The Right Graph. By Harold Kirkham and Robin C. Dumas
Copyright © 2009 John Wiley & Sons, Inc.